Kidd Copper Property Case
In Defense of a Younger Earth
Conventional Wisdom Challenged

Front cover: *Borehole 103, photograph by Andy Weller showing colored layers.*
This photograph shows features of rapid stratification.
Rapid Stratification supports a 'Younger Age' for Planet Earth.

Loreen Michele Sherman, MBA

Copyright 2010-2011 Star-Ting Incorporated. All rights reserved. Worldwide.
ISBN-13: 978-0-9810999-4-1 (Star-Ting Incorporated)
ISBN-10: 0981099947

Kidd Copper Property Case
In Defense of a Younger Earth
Conventional Wisdom Challenged

Library and Archives Canada Cataloguing in Publication

Sherman, Loreen Michele
    In defense of a younger Earth : Kidd Copper property case conventional wisdom challenged / Loreen Michele Sherman.

Includes bibliographical references and index.
Issued also in electronic format.
ISBN 978-0-9810999-4-1

    1. Sediments (Geology)--Ontario--Sudbury Region-- Observations--Analysis. 2. Sequence stratigraphy.
3. Earth--Age. I. Title.

QE571.S54 2011   551.3'0409713133   C2011-904151-0
Printed in Canada
10 9 8 7 6 5 4 3 2 1

Preface

"The site you will be visiting this year is owned by Crowflight Mineral Inc and is known as the Kidd Copper property … You have been supplied with four air photographs and one hyperspectral image that show the site's evolution from 1970–2002" (McMaster University, August 2004, Assignment #9 Mine Tailings).

This scientific report on Kidd Copper property and its case was written in response to a Geo3Fe3 Assignment #9 defined by McMaster University professors in 2004 where certain boundary conditions were set to conduct primary research on the site. The Kidd Copper property was owned by Crowflight Minerals Inc. at that time. Five essential steps were followed:

1) The professors defined the population by giving the University students the opportunity to walk over the entire site.

2) The students were directed to create a map with details of the different processes that [they] observed in the different areas.

3) They were to use a spade at several locations to observe what is occurring in the subsurface.

4) Then to use the air photos to describe the changes that have occurred at the site over time.

5) To submit a field report of their findings from the primary research that each student collected.

Afterwards, Loreen (Rudd) Sherman assimilated the various log entries on her own, independently of the students with their written consent but without their knowledge of her suggested hypothesis to summarize the findings in this report. This report takes the primary research collected by several geology students and integrates the findings to record in a collective the summary from the field entries, which is used to support the theory that rapid stratification can support a 'younger age' for planet Earth.

Editor's Note: the use of personal names or brand names in a peer-reviewed paper is for the identification purposes only and does not constitute endorsement by the named subjects or McMaster University on the subject matter presented by the author's views on the age of the earth represented by this report.

Executive Summary

Background

"We are still a considerable distance from the goal where geologic timescale calibration is achieved by precise direct astronomical tuning or radiometric age dating of all successive stage, zonal, or magnetic polarity chron boundaries. In fact, it is doubtful if the rock record on Earth harbors all the precise age information. This sparse skeleton of age control, especially prior to ~30 Ma (as of 2004), leaves considerable room for interpolation in construction of a geologic time scale" (Gradstein, *et al.*, 2004, p. 455).

"Classically, the concept that the present is the key to the past; the principle that conventional geologic processes have occurred in the same regular manner and with essentially the same intensity throughout geologic time, and that events of the geologic past can be explained by phenomena observable today" (Licker, 2003, p. 384; Garrels & Mackenzie, 1971, p.65).

The aim of this report is to interpret the geologic time scale of the past rock records by applying the knowledge of the geologic sedimentary processes on the chosen study area of Kidd Copper Property. The parameters are optimal and constitute an outside natural observatory because the composition of parent rock and the initial time of deposition had been documented historically.

Analysis

This report tested 120 soil samples taken in an area encompassing 4 km^2 about 85-90% of the surface ground area covered. Field observations are compared with conventional geologic principles and laws in detail throughout this report. Conventional textbook interpretation is that uniform strata should occur when there is original deposition of sediment deposits. Lithographic logs of the vertical profiles of associations of two separate transects was examined. The sequential layering of Kidd Copper sediment composition was studied with regards to the matrices of sediment composition, mineral composition, color descriptors, size of grains, and orientation of the grains. Both rock stratigraphy and time stratigraphy were used to classify the layers of sediment.

This report uses 32 eye-witness observations of *in situ* excavations eliminating guesswork on time parameters and depositional processes. This report will consider principles of (1) original horizontality; (2) superposition; (3) lateral continutity; and many (4) other geologic parameters of conventional geologic wisdom to determine the new hypothesis that is formed by the new information found at Kidd Copper. This report tests the claim that the many layers found in sediment are a rapid, one-time process and a primary structure, so that, the new hypothesis is

H_1 = rapid and the null hypothesis is $H_0 \neq$ rapid.

Conclusion

The evidence collected at Kidd Copper was contrasted with several geologic laws and principles and found to falsify many. Field observations support rapid stratification for initial deposition of sediment. Multiple layers are recognized in other geologic formations and support that rapid processes like catastrophic events formed them, rather than a slower process over millions of years. The resulting layers of sediment deposition at Kidd Copper property did not form in thousands of years but instead rapidly. This process of multiple layers in the rock record was observed through a very short time in Earth's history. The evidence supports multi-layered deposits are one event and through rapid stratification.

Radically, the Kidd Copper property case is research with established time parameters and physical boundaries that determine precisely the geological depositional process of three different steps on the impoundment of tailings area. The 24 findings in this report are significant and support the hypothesis that stratification in the rock record supports a fast rate of deposition; thus opening the doorway to the theory that a 'younger date' of the earth to be re-examined. The geologic time scale can be re-calibrated with this new view of the geological history of the Earth. This report emphasizes the formation of multiple layers in a short period of time and challenges established concepts in landscape evolution.

Acknowledgements

Much thanks and appreciation is extended to Andy Weller for his contribution to the collection of the photographs taken on August 30, 2004 (without his knowledge to the direction or conclusion of this study). He blindly allowed the author the use of the photographs and the author is very appreciative of his gift. Derrick Li is another photographer who captured the shot of the first shovel without realizing the implication of this report. Thanks Derrick. The shovel exposes stratified layers contrary to the expected outcome of the Law of superposition. Topographic maps were provided for this study by Calgary's Map Town with complete cooperation by Paul Hopkins on March 4, 2009, who is a credit to cartographers. A special acknowledgement is extended to my husband, Paul Whittington Sherman, for the graphics, encouragement and hours of formatting and proofreading which helped immensely. If, despite the best efforts of those that have offered help, errors, omissions, or other problems remain, they are the author's sole responsibility.

DEDICATION

To the many professors, teachers and educators who taught me how to write academically, reason and query so that the culmination of their efforts is made manifest in this scientific report; a long process that underscored many life lessons.

To the educational institutions that were a part of my life:

St. Pius School, Holy Cross, Harold W. Riley, Ian Bazalgette Jr. High, Forest Lawn High School, Red Deer College, University of Calgary, Mohawk College, McMaster University, International College of Applied Linguistics, Phoenix University and Meritus University. Thank you for the gift of education where the mind can explore and learn.

Key Words: Kidd Copper property, bore holes, geologic time scale, deposition, geologic laws, sedimentary processes, layering of beds, mine tailings, principles of geology, stratification, younger earth.

Abstract

The mining on Kidd Copper property was well-known from 1936-1970. This report is based on the collection and analysis of field observations of the Kidd Copper property from August 30, 2004 of an abandoned, well-explored mine tailings waste site south of Perch Lake, Sudbury, Ontario. In this study, 32 Geology students collected data using traditional surface methods for mapping 120 boreholes over the nickel-copper mine-waste site at Kidd Copper, Ontario. A synoptic overview of the waste site area was achieved previously using "remote-sensing data" (Shang-Morris-Howarth, 2004, p. 2). Other supporting evidence was combined with the ground observations to examine the diagenesis of the site.

The author endeavors to develop the history of the site and relate it to the original deposition of sediments. Three depositional processes are observed in the site area with definitive ages: (1) The original deposition of known parental material and subsequent deposits; (2) The upswelling like a turbulent flow caused by a dam breach; and (3) The redeposition of the sediment onto the major mine tailings area and a new extension.

The Kidd Copper mine tailings waste site is an important field observatory as it allows examination of the re-settling of sediments over an obvious short period of time. The author compares field trip observations against sedimentologic laws and principles as found in current geology textbooks. These laws include: (1) original horizontality; (2) superposition; (3) nature; (4) original continuity; (5) equal declivities; and (6) correlation of facies or Walther's Facies Law. The principles include: uniformitarianism >actualism; and relative dating.

The author strives to develop this report into seven main facts: (1) The historic authenticity of the Kidd Copper property is well established and documented; (2) The dam breach resulted in redeposition of a 'primary sediment structure' on the impoundment area; (3) Examination of redeposited sediment in its diagenesis context reveals heterogeneous sediment layers as the primary structure, which is unexpected based on many current Geologic laws and principles; (4) Rapid stratification is an outcome for a primary structure; (5) The Law of original horizontality and the principle of relative dating are problematic; (6) Time correlation based on the geological time scale is problematic; and (7) The accepted old age of the planet Earth can be questioned.

Résumé

L'exploitation sur la propriété Kidd Copper était bien connue durant les années 1936 à 1970. Ce rapport est basé sur la collection et l'analyse de les observations faites sur la propriété Kidd Copper, durant le jour du 30 Août, 2004, sur un site de rebut bien exploré et *abandonné, où les dépôts de sol des résidus industriels d'un bassin de décantation se sont écoulé sur la surface. Le site est situ*é au sud du Lac Perch, *près* de la ville de Sudbury dans la provine de l'Ontario. Trente-deux étudiants de géologie ont pris part dans une étude et ont rassemblé des données en utilisant des méthodes de surface traditionnelles pour cartographier 120 trous de forages au-dessus du site appartenant a une mine de nickel et de cuivre situé sur la propriété Kidd Copper. *Un Aperçu synoptique* du site de rebut a précédemment été réalisé avec l'assistance de "données de détection à distance" [remote-sensing data] (Shang-Morris-Howarth, 2004, p. 2). Ceci et d'autres observations au sol ont été combines pour examiner la diagenèse de ce site.

L'auteur essaye de développer l'histoire du site et de faire la liaison entre *ceci* et le dépôt original des sédiments. L'auteur observe trois processus dépositionnels dans le secteur du site avec des âges définitifs: (1) Le dépôt original du matériel parental connu et des dépôts subséquent; (2) Le upswelling causé par un écoulement turbulent provoqué par un bris de barrage; et (3) La redéposition du sédiment sur le secteur principal de les produits de décantation de la mine et une nouvelle extension.

L'emplacement du rebut des produits de décantation de mine de Kidd Copper fournit un lieu d'observatoin très important car il permet l'examination du reclassement des sédiments pendant une courte durée évidente. L'auteur compare les observations d'excursion sur le terrain contre les lois et les principes de la sédimentologie comme est couramment *perçu* dans les manuels de géologie. Ces lois incluent: (1) l'horizontalité originale; (2) la superposition; (3) la nature; (4) la continuité originale; (5) les déclivités egales; et (6) la corrélation des *faciès* ou de la Loi De Walther du *Facies*. les principes incluent: l'uniformitarisme > l'actualisme; et la datation relative.

La tâche de l'auteur a été de développer ce rapport en sept faits principaux :
(1) L'authenticité historique de la Propriété Kidd Copper soit bien établie et documentée;
(2) Le bris du barrage a eu comme conséquence la redéposition d'une "structure sédimentaire primaire" sur l'impoundment du secteur; (3) L'examen du sédiment redéposé dans son contexte de diagenèse indique que la structure primaire sont des couches composées de sédiment hétérogènes, ce qui est inattendue basée sur plusieurs lois et principes géologiques courrament *utilis*és; (4) La stratification rapide est un résultat d'une structure primaire; (5) La Loi de l'horizontalité originale et le principe de la datation relative sont problématiques; (6) Le temps de corrélation basée sur l'échelle de temps géologique est problématique; et (7) L'*âge* de la planète Terre peut donc être remise en cause.

Editor's Note: If Any Discrepancies exist between French & English translations – the English holds.

Table of Contents

PREFACE 3

EXECUTIVE SUMMARY 4

ACKNOWLEDGEMENTS 6

ABSTRACT 7

RÉSUMÉ 8

TABLE OF CONTENTS 9

LIST OF FIGURES 22

LIST OF TABLES 23

KIDD COPPER PROPERTY REPORT BASED ON 2004-AUGUST-30 25

1.0 HISTORICAL AND GEOLOGICAL OVERVIEW OF SITE 28

    1.1 THE KIDD COPPER PROPERTY SITE AND ITS HISTORY ARE WELL DOCUMENTED ... 28

    1.2 KIDD COPPER IMPOUNDMENT LOCATION AND COORDINATES 30

    1.3 ELEVATION OF 850 FEET 31

    1.4 PARENT MATERIAL IS GOETHITE 33

    1.5 THREE STEPS IN THE DEPOSITIONAL PROCESS ON KIDD COPPER OFFERS ANSWERS ... 34

2.0 GEOMORPHOLOGICAL BACKGROUND – STEP 1 INITIAL – WASTE ... 35

    2.1 STEP 1: INITIAL DEPOSITION OF THE MINE TAILINGS ON THE WASTE SITE ... 35

    2.2 TAILINGS ANALYSIS: MINERAL ORE AND FINE SEDIMENT 35

    2.3 SEDIMENT DEPOSITS ON KIDD COPPER ARE NOT VARVES 35

    2.4 PRE-CAMBRIAN CRUST BANDED ORE FORMATIONS OCCUR 36

    2.5 MINE TAILINGS SMELTING AND FLOTATION PROCESS 36

    2.6 MECHANICAL-HUMAN-TRANSPORT TO DUMP MOLTEN SLAG 37

    2.7 TAILINGS WASTE DEPOSIT AT KIDD COPPER PROPERTY 38

    2.8 DURATION OF 'SILTY' TAILINGS DEPOSITS 1936-1970 39

    2.9 CLOSED SYSTEM 40

3.0 GEOMORPHOLOGICAL BACKGROUND – 2 MIDDLE – UPSWELLING ... 41

    3.1 SURFACE MINERAL MAPPING OF THE KIDD COPPER MINE TAILINGS ... 41

    3.2 1970 A SMALL LAND BRIDGE 41

    3.3 HISTORICAL CONTEXT OF KIDD COPPER DAM 42

    3.4 RETAINING DAM AT THE SOUTHEAST CORNER BROKE 44

    3.5 STEP 2: UPSURGE OF MINE TAILINGS WASTE SEDIMENTS 45

    3.6 1983 TOPOGRAPHY HAS SUBSTANTIALLY CHANGED 45

    3.7 STREAM EROSION WAS A MINOR CONTRIBUTOR 46

    3.8 1992 DATA NO MAJOR DISTRIBUTION CHANGE AFTER 1983 47

    3.9 LANDSCAPE CHANGES AS A RESULT OF THE FLOOD 48

3.10 Evolution of Mine tailings established During a Short Period in Time 49
3.11 Rock Cycle and Word Definitions 49
3.12 Kidd Copper Depositional Setting Answers Sediment Bed Question 51

4.0 MECHANISM FOR UPLIFT **53**
4.1 Velocity of Water 53
4.2 Fluid Lift Force 54
4.3 The Rate of Water Movement through Soils 56
4.4 Strength of Turbidity Currents 56
4.5 Turbidity Currents Features 58
4.6 Flow Slope 60
4.7 The Load of Streams 60
4.8 Hydraulic Gradients at the Copper Cliff Impoundment 63
4.9 Solubility 63
4.10 Fluids are Thoroughly Mixed 65

5.0 EXOGENETIC PROCESSES **67**
5.1 Biochemical Activity 67
5.2 Precipitation 68
5.3 Air Erosion 70
5.4 Physical Weathering 72
5.5 Chemical Weathering 73
5.6 Mechanical Weathering 75

6.0 ORE MINERALOGY **77**
6.1 Mineralogy Defined 77
6.2 Fine-Grain Sized Waste Products on Kidd Copper Property 79
6.3 Satellite Site – Impoundment at Copper Cliff, Ontario 80
6.4 Clastic Sedimentation 81
6.5 Spherical and Evenly Distributed 82
6.6 Colloidal Homogeneous Appearance 82

7.0 GEOMORPHOLOGICAL BACKGROUND – 3 FINAL – TAILINGS REDEPOSITED **84**
7.1 Step 3 Observable Sediment Redeposited on Kidd Copper Property 84
7.2 Sediment Deposits 84
7.3 The Dam Drains 86
7.4 Sediment Deposition like That of a Turbidity Current 86
7.5 Settling Velocity of the Particle 89
7.6 Field Capacity 90

8.0 SIGNIFICANCE OF KIDD COPPER **91**
8.1 Law Defined 91
8.2 Theory Defined 91
8.3 Inductive Reasoning 91
8.4 Scientific Approach 92
8.5 Hypothesis Defined 92
8.6 Rapid Stratification Hypothesis 93

8.7 The Multiple Layers are Not Individual Deposits 93
8.8 Possible Sources of Error 94

9.0 GEO3FE3 GEOLOGY ASSIGNMENT **95**

9.1 Methodology 96
9.2 Collection of Data 96
9.3 Sampling Frequency 96
9.4 Sampling Station Interval 97
9.5 Benchmark 97
9.6 Field Observations 97
9.7 McMaster Students Walk Over Assigned Kidd Copper Site Area 98
9.8 The Kidd Copper Property is Assayed 99
9.9 Graphic Log Constructed of the Topograpghy 99
9.10 Unearth the Layer of Top Soil 101
9.11 Soil Stripes Unearthed After Redeposition on Kidd Copper 102
9.12 Photographs from Field Observations show Stratified Layers 102

10.0 PRIMARY RESEARCH OF 120 BOREHOLE SAMPLES **104**

10.1 Data Acquisition of Soil Profiles 104
10.2 Raw Data of Boreholes Collected August 30, 2004 105
10.3 Kidd Copper Aggregate Base Map of Borings 132
10.4 Correlation of Lithographic Logs Defined 134
10.5 Lithographic Log West-North Transect 135
10.6 Lithographic Log North-South Transect 138

11.0 DATA ANALYSIS FROM FIELD LOGS **142**

11.1 Soil Taxonomy – Distinct Horizons 143
11.2 Eye-witness Ground Observations and Testimony 144
11.3 Quantitative Analysis of Kidd Copper Sediments 144
11.4 Matrix of Sediment Types and Descriptors on Kidd Copper 145
11.5 Matrix of Single Color Descriptors Found on Kidd Copper 147
11.6 Matrix of Multi-Color Descriptors 149
11.7 Origin of the Grain Sediments 150
11.8 Packing of the Grains 150
11.9 Matrix of Grain Size Descriptors 150
11.10 Kidd Copper Sediment Grain Size and Shape 151
11.11 Clayey Fabric Distribution and Orientation Patterns 152
11.12 Matrix of Grain Orientation Descriptors 153
11.13 Were the Layers Formed by Segregation? 155
11.14 Significance of Boreholes and Matrices 156
11.15 Correlation of Parallel and Perpendicular Cross-sections 156
11.16 Facies Analysis 157
11.17 Facies Change 158
11.18 Adjacent facies Deposition Changes Gives Clues to Time 158
11.19 Data Summarized from Field Logs 159
11.20 Significance of the Bouma Sequence and the Kidd Copper Property 160
11.21 Data Interpretation Techniques 160

12.0 FAST RATES OF DEPOSITION SEEN ELSEWHERE 162

12.1 MANY SUBENVIRONMENTS DISPLAY RAPID STRATIFICATION CHARACTERISTICS 162
12.2 DEBRIS FLOW CHARACTERIZE BOTH SLOW AND FAST SEDIMENTATION RATES 162
12.3 TURBIDITY CURRENTS RAPID DEPOSITION 163
12.4 MUDFLOW FAST RATES OF DEPOSITION 163
12.5 LANDSLIDES FAST RATE OF DEPOSITION 163
12.6 CASE OF HORGEN SLIDE 164
12.7 PRIMARY AND SECONDARY STRUCTURES 164

13.0 STRATIGRAPHIC ANALYSIS.. 165

13.1 NEW KNOWLEDGE AND IMPACT OF KIDD COPPER ANALYSIS.. 165
13.2 STENO'S FOUR PRINCIPLES 165
13.3 THE PRINCIPLE OF ORIGINAL HORIZONTALITY 167
13.4 FLOOD PLAINS ALMOST ENTIRELY HORIZONTAL LAYERS OF FINE SEDIMENT.................................... 168
13.5 THE LAW OF ORIGINAL HORIZONTALITY TESTED... 168
13.6 HORIZONTAL LAMINATION IS FOUND AT VARIOUS LEVELS ON FIRST DEPOSITS................................ 170
13.7 THE PRINCIPLE OF STRATIGRAPHIC SUPERPOSITION 170
13.8 ALL GEOLOGIC CHRONOLOGY IS BASED ON THE LAW OF SUPERPOSITION 171
13.9 THE LAW OF SUPERPOSITION DEFINED 172
13.10 IGNEOUS ROCK EXPRESSES LAYERS 173
13.11 BANDED STRUCTURES 174
13.12 THE LAW OF ORIGINAL HORIZONTALITY AND SUPERPOSITION ILLUSTRATED 174
13.13 APPLICATION OF THE LAW OF SUPERPOSITION 176
13.14 THE PRINCIPLE OF CROSS-CUTTING RELATIONSHIPS ON ROCKS 179
13.15 THE PRINCIPLE OF FOSSIL SUCCESSION 179

14.0 UNIFORMITARIANISM – FUNDAMENTAL PRINCIPLE 180

14.1 FATHERS OF MODERN GEOLOGY... 180
14.2 ACTUALISM – PRINCIPLE OF UNIFORMITY 181
14.3 CHEMICAL UNIFORMITARIANISM 182
14.4 LAW OF NATURE... 183
14.5 PRESENT-DAY PROCESSES 183
14.6 REPRODUCIBILITY.. 184
14.7 A FUNDAMENTAL ASSUMPTION.. 187
14.8 MODELS FOR FUTURE REPLICATION.. 187
14.9 TEST OF REPRODUCIBILITY 187
14.10 EXPERIMENTAL DESIGN OF A FUTURE FLUME STUDY... 188
14.11 SPIGOT DAM RE-CONSTRUCT MODEL... 192
14.12 ACCEPTANCE TESTING.. 198

15.0 THE STRATIGRAPHIC RECORD AND CORRELATION 199

15.1 STRATIGRAPHIC ANALYSIS OF THE ROCK RECORD 199
15.2 LAW OF THE CORRELATION OF FACIES 200
15.3 CORRELATION OF LITHOSTRATIGRAPHIC UNITS 202
15.4 CORRELATION OF SEDIMENTARY STRATA 203
15.5 WALTHER'S FACIES LAW 207

15.6 LAW OF ORIGINAL CONTINUITY 208
15.7 LATERAL CONTINUITY 208
15.8 LAW OF EQUAL DECLIVITIES 210
15.9 SAME SEQUENCE IS VISIBLE IN ROCK OUTCROPS 210
15.10 UNCONFORMITIES 211

16.0 ROCK DATING METHODS AND PRINCIPLES.. 213

16.1 GEOLOGIC TIME UNITS DEFINED 213
16.2 RELATIVE AGE DATING 213
16.3 HISTORICITY OF RELATIVE AGE DATING 214
16.4 RELATIVE TIME 214
16.5 UNITS OF GEOLOGICAL TIME 215
16.6 THE GEOLOGIC COLUMN 216
16.7 ABSOLUTE AGE 217
16.8 ABSOLUTE TIME 218
16.9 GEOLOGIC AGE 218
16.10 ABSOLUTE DATE 218
16.11 AGE EQUATION 219

17.0 GEOLOGY THE SCIENCE OF TELLING TIME 220

17.1 ROCK STRATA RECORD EARTH'S HISTORY 220
17.2 GEOLOGICAL TIME 221
17.3 CHRONOSTRATIGRAPHY 222
17.4 GEOLOGIC TIME SCALE 222
17.5 ASSUMPTIONS UNDERLYING THE GLOBAL CHRONOSTRATIGRAPHIC SCALE...... 226
17.6 THE NEW GEOLOGIC TIMESCALE, NAMED GTS 2004 227
17.7 LIMITATIONS OF GTS 2004 230
17.8 CALIBRATION OF LINEAR TIME OF EVENTS IN THE ROCK RECORD 230
17.9 STRATIGRAPHIC COLUMN AND THE STRATIGRAPHIC RECORD 231
17.10 GLOBAL STRATOTYPE SECTION AND POINT (GSSP) 232
17.11 DEVONIAN-CARBONIFEROUS DISCREPANCIES 232
17.12 PRECAMBRIAN POLARITY SEQUENCE IS POORLY KNOWN AND POORLY DATED..... 232
17.13 RADIOGENIC ISOTOPE GEOCHRONOLOGY HANDLED DISPARATE RESULTS 233
17.14 GEOMATHEMATICS ASSUMES A HYPOTHETICAL TRIAL AGE 233
17.15 THE CHI-SQUARE TEST 234
17.16 THE OPTIMUM CHOICE OF AGE AGTERBERG IMPROVED 234
17.17 METHODS FOR AGE ESTIMATIONS... 235
17.18 SPLINE CURVE FITTING DISADVANTAGES 237
17.19 CARBONIFEROUS AND PERMIAN COMPOSITE STANDARD 237
17.20 DECAY CONSTANTS AND THEIR UNCERTAINTIES 238
17.21 NEOPROTEROZOIC PALEONTOLOGY, ISOTOPES OR CLIMATE HISTORY 238
17.22 ARBITRARILY DEFINED GEOCHRONOMETY... 238
17.23 TRIASSIC & JURASSIC PERIODS 239
17.24 UNCONFIRMED DATA BASIS OF STRATIGRAPHIC RECORD 240

18.0 PROBLEMATIC DATING METHODS 241

18.1 DATING METHODS.. 241

18.2 PROBLEMS WITH DATING METHODS 241
18.3 THE RADIOMETRIC DATING TECHNIQUE 244
18.4 THE LEAD-LEAD METHOD 246
18.5 URANIUM-LEAD METHOD 247
18.6 THE POTASSIUM-ARGON METHOD 247
18.7 THE RUBIDIUM-STRONTIUM METHOD 247
18.8 THE CARBON-14 (RADIOCARBON) METHOD 247
18.9 FISSION-TRACK DATING 248

19.0 DATA ANALYSIS OF KIDD COPPER'S STRATIGRAPHY **249**
19.1 RATE OF SEDIMENTATION 249
19.2 RELEVANCE OF THE RATES OF SEDIMENTATION. 251
19.3 RATES OF EROSION 253
19.4 HEAVY PARTICLES FALL FIRST 253
19.5 CROSS LAMINATION 255
19.6 CROSS-STRATIFICATION 255
19.7 CROSS-BEDDING GEOLOGICAL MEASUREMENTS 255
19.8 PALAEOCURRENT DATA 257
19.9 PALAEOCURRENT PATTERNS 257
19.10 KEY INDICATORS 258
19.11 RHYTHMIC LAYERING UNITS IN DIFFERENT HIERARCHAL LEVELS 258
19.12 CLOSELY RESEMBLES RIBBON ROCK 261
19.13 EVENT STRATIGRAPHY 261
19.14 RECURRENCE TIME 262
19.15 KEY TIME FACTORS 262
19.16 DATA CONFIRMING RAPID STRATIFICATION 263
19.17 TILTING 263
19.18 EXTENSIVE COVERAGE 264
19.19 CONSISTENCY 264
19.20 RHYTHMIC SEDIMENTATION WAS NOT OBSERVED 264
19.21 MIGRATION OF TAILINGS 265
19.22 PEER-REVIEW PRESENTATION 266
19.23 RAPID STRATIFICATION PEER-REVIEW PRESENTATION QUESTIONS 269
19.24 AREAL DISTRIBUTION AND SIGNIFICANCE OF RAPID STRATIFICATION 269
19.25 SUMMATION OF FINDINGS AT KIDD COPPER 273

20.0 RAPID STRATIFICATION INTERPRETATIVE DATA **274**
20.1 STRATIFICATION DEFINED 274
20.2 ASSERTION THAT ONE-TIME DEPOSITION CONFIRMS RAPID STRATIFICATION 274
20.3 RAPID STRATIFICATION IN DEPOSITIONAL PROCESSES – INFERENCES 274
20.4 NOT INTERBEDDED OR INTERSTRATIFIED 275
20.5 LAMINATION COMPOSITION OF LAYERS IS NOT ANOMOLIES 275
20.6 START OF GEOLOGIC HISTORY 276
20.7 QUALITATIVE DATA INTERPRETATION 277
20.8 ADJACENT SEDIMENTARY STRATA 277
20.9 CONVENTIONAL INTERPRETATION FALSIFIED 278
20.10 OTHER CURRENT OBSERVATIONS SUPPORT YOUNG-EARTH STRATA DEPOSITS 278

20.11 Past Tendency to Interpret Each Laminae Separately 280
20.12 Structural Geology 281
20.13 Secondary Stratification 281
20.14 Complicated Geometries 281

21.0 STRATIFIED LAYERS AND TIME CONCLUSIONS.. 282

21.1 Young Age of Tailings 282
21.2 Stratigraphic Correlation Defined 283
21.3 Time-stratigraphic Facies Defined 283
21.4 Interpretation of Stratigraphic Unit Point 64/X1 283
21.5 Interpretation of Stratigraphic Unit Point 56/4 285
21.6 All Boreholes are Isochronous 286
21.7 Reconstruction of Palaeoenvironment of Kidd Copper Property 286
21.8 Extrapolations Not Needed at Kidd Copper 287
21.9 Rapid Stratification Opens the Door to a Younger Earth 287
21.10 Doorway to a Younger Earth Opens 288
21.11 Uniformitarianism Re-defined 290

22.0 OLD EARTH 291

22.1 Origin of the Elements 291
22.2 Origin of Man 292
22.3 Old Age Given to Date the Earth 292
22.4 Biological Evolution 293
22.5 Crustal Evolution 293
22.6 Planetary Evolution 294
22.7 Evolutionary Rate 294
22.8 Darwinism 295
22.9 A Geological Time Scale to Emphasize Mammalian Evolution 295
22.10 Young Data Is Not Readily Accepted 298
22.11 Geologic Timescale Calibration is a Sparse Reflection of Real Time 298

23.0 CLAIM FOR A YOUNGER EARTH 300

23.1 Age of the Earth <2 Billion Years Old 300
23.2 Time Correlation 301
23.3 Time-Series with Terrestrial Matter 301
23.4 Time Line 301
23.5 Geologic Time Scale GTS2004 Challenged 303
23.6 Timeline of Kidd Copper Events Re-constructed 304
23.7 Rapid Stratification Supports a 'Younger Age' for Planet Earth 305
23.8 Factors Underlying a Good Theory 307
23.9 Opening the Doorway Back to Special Creation 309

24.0 RAPID STRATIFICATION FINDINGS & IMPLICATIONS.. 310

24.1 Finding One: Historicity and Source Rock Well-established 310
24.2 Finding Two: A Cataclysmic Event Changes Landscape in Hours 311
24.3 Finding Three: Exogentic Processes are Minimal 312
24.4 Finding Four: A Base-line of Geological History at Kidd Copper 312

24.5 Finding Five: Heterogeneous Sediment Deposited 312
24.6 Finding Six: Primary Sedimentary Structure can be Heterogeneous 313
24.7 Finding Seven: Twisted Rocks are found at Initial Deposition 314
24.8 Finding Eight: No Uniformity in the Adjacent Structures 315
24.9 Finding Nine: Units Cycled are Not Proportional 316
24.10 Finding Ten: The Law of Equal Declivities Falsified 316
24.11 Finding Eleven: Lateral Continuity Not Found Kidd Copper Case 317
24.12 Finding Twelve: Some Major Geologic Laws Do Not Hold 317
24.13 Finding Thirteen: Concept of Uniformitarianism Further Refined 318
24.14 Finding Fourteen: The Principle and Law of Original Horizontality Falsified 318
24.15 Finding Fifteen: Falsification of the Law of Superposition 319
24.16 Finding Sixteen: Some Calibration Methods are Problematic 320
24.17 Finding Seventeen: Sedimentary Rock Order Falsified 320
24.18 Finding Eighteen: Supports All Scales Incomplete Geologic Record 321
24.19 Finding Nineteen: Assigning Ages to the Rock Layers is Unreliable 322
24.20 Finding Twenty: The Geologic Time Scale is Insufficient 322
24.21 Finding Twenty-one: A 'Younger Date' for Planet Earth is Reasonable 323
24.22 Finding Twenty-two: Old Age for Planet Earth Seriously Questioned 323
24.23 Finding Twenty-three: New Interpretation Given for Geologic Events.............. 325
24.24 Finding Twenty-four: New Time Parameters from Rock Readings................... 326

25.0 HYPOTHESIS SUPPORTED FOR RAPID STRATIFICATION **328**

25.1 Hypothesis Accepted that Stratification Equals a Fast Rate of Deposition 328
25.2 Weighting of Evidence Supports Claim for a Younger Earth 329
25.3 Conclusion: Strong Evidence to Support a 'Younger Date' for Planet Earth 330

REFERENCES **332**

APPENDICES **338**

APPENDIX A: DATA DICTIONARY **338**

A-1 Adjacency 338
A-2 Analysis 338
A-3 Architectural Elements 338
A-4 Attribute 338
A-5 Bedding Plane 338
A-6 Biofacies 338
A-7 Chronostratic Unit Interpretations 338
A-8 Clarity 339
A-9 Compaction 339
A-10 Competence 339
A-11 Compression 339
A-12 Conformable 339
A-13 Connectivity 339
A-14 Consolidation 339
A-15 Data Format 340
A-16 Datum 340
A-17 Deposition 340

A-18 Diffusion 340
A-19 Erosion 340
A-20 Expected Error 340
A-21 Face Method.. 340
A-22 Facies 340
A-23 Facies Associations 341
A-24 Facies Classification 341
A-25 Facies Maps 341
A-26 Facies Models 342
A-27 Facing 342
A-28 Formation 342
A-29 Geologic Log 342
A-30 Geologic Province 342
A-31 Geologic Section 343
A-32 Geologic Structure 343
A-33 Horizons 343
A-34 Humic Substances 343
A-35 Ichnofacies 343
A-36 Informal Unit 343
A-37 Inset 343
A-38 Interstitial 343
A-39 Isolith 343
A-40 Key Bed 343
A-41 Key Horizon 344
A-42 Laminations 344
A-43 Liquid Flow 344
A-44 Lithofacies 344
A-45 Lithology 344
A-46 Lithofacies Codes 344
A-47 Lithostratigraphy 344
A-48 Low-Angle Parallel Laminated Sand 346
A-49 Map Types: Base Map 346
A-50 Map Types: Dot Map 346
A-51 Map Types: Geologic Map 346
A-52 Map Types: Geotechnical Map 346
A-53 Map Types: Lithologic Map 346
A-54 Map Types: Real Map 346
A-55 Mask 346
A-56 Matrix 346
A-57 Offset 347
A-58 Page Co-ordinates 347
A-59 Plane Bed 347
A-60 Permeability 347
A-61 Pore space 348
A-62 Point Feature 349
A-63 Pore water 349
A-64 Porosity 349

A-65 PROFILE 350
A-66 PROFILE SECTION 350
A-67 QUANTITATIVE ANALYSIS 350
A-68 SATURATION 350
A-69 SHEET DEPOSIT 350
A-70 SILT COMPOSITION 350
A-71 SILTING 350
A-72 SILT LOAM 350
A-73 SILT SHALE 350
A-74 SILT SOIL 350
A-75 SILTSTONE 350
A-76 SLAG 350
A-77 SLICKENS 351
A-78 SOIL PROFILE 351
A-79 SOIL PROPERTIES AND BEHAVIORS OF SOIL SEPARATES 352
A-80 SOLE 352
A-81 STRATIGRAPHIC COLUMN 353
A-82 STRATOTYPE 353
A-83 STRATUM 353
A-84 THICK BEDS 353
A-85 THIN BEDS 353
A-86 TOPOLOGY 353
A-87 TURBULENT FLOW 353
A-88 VARIABLES TO RECONSTRUCT DAM AND FLOW 354
A-89 VERTICAL ACCRETION 354
A-90 VISCOSITY 354
A-91 YOUNGING 354

APPENDIX B: PARTICLES **355**

B-1 UDDEN-WENTWORTH GRAIN SIZE SCALE 355
B-2 VERY FINE SAND 356
B-3 RIPPLE-DRIFT 356
B-4 PROXIMAL OR DISTAL TURBIDITES 356
B-5 MECHANISMS THAT CAUSE SEDIMENT DEFORMATION 356
B-6 MECHANICAL SORTING 356
B-7 BEDDING DEFINITIONS 357

APPENDIX C: CLAY MINERALOGY **358**

C-1 CHLORINE CONTENT OF TERRESTRIAL ROCKS 358
C-2 NORMATIVE COMPOSITION 358
C-3 MINERAL COMPOSITION OF A ROCK 358
C-4 MINERAL GROUPS 358
C-5 CLAY DEFINED 359
C-6 CLAY IRONSTONE 359
C-7 CLAY MINERALS 359
C-8 CLAYSTONE 360
C-9 DESCRIPTION OF MUDSTONE, SHALE AND CLAY 360

C-10 FIELD RELATIONS 360
C-11 GEOCHEMICAL FACIES 360
C-12 HEAVY AND LIGHT MINERALS 361
C-13 ILLITE 361
C-14 LUTITES 361
C-15 MONTMORILLONITE – SILICATES (PHYLLOSILICATES, CLAY GROUP) 361
C-16 MONOMINERALIC 362
C-17 PYRITE 362
C-18 RESIDUAL CLAY 362
C-19 SECONDARY MINERALS 362
C-20 SENSITIVE CLAY 363
C-21 SULFIDES 363
C-22 THIXOTOPIC 363

APPENDIX D: COLORS **365**

D-1 ROCK COLOUR CHART – MUNSELL SYSTEM 365
D-2 CLAY 365
D-3 DETRITUS 365
D-4 FELSIC ROCKS 365
D-5 IRON 365
D-6 MAFIC 366
D-7 OCHRE 366
D-8 PIGMENTATION 366
D-9 RED BEDS 366
D-10 RED CLAY 366
D-11 RESIDUAL OCHRE 366
D-12 SIENNA 366
D-13 UMBER 366
D-14 WULFENITE 367
D-15 YELLOW MUD 367

APPENDIX E: FERRIC OXIDE **368**

E-1 FERRIC OXIDE DEFINED 368
E-2 FERRIC OXIDE 368
E-3 FERRIFEROUS 368
E-4 HAEMOSIDERIN 368
E-5 IRON 368
E-6 IRON FORMATION 368
E-7 IRON IN SEDIMENTATION 368
E-8 IRON ORE MINERALS 369
E-9 IRONSTONE 369
E-10 LITHIFICATION 369
E-11 OXIDE 369
E-12 OXIDATES 369
E-13 OXIDISED ZONE 369
E-14 OXIDIZATION 369
E-15 REACTING SYSTEMS 370

E-16 REACTING SYSTEMS: MINERAL REACTIVITY 370
E-17 RUST 370
E-18 RUSTING 370

APPENDIX F: EXPERIMENTAL FUTURE REPRODUCIBILITY PROJECTS **372**

F-1 COMPUTER MODELS 372
F-2 SLOPE-AREA MEASUREMENTS 372
F-3 CHOROPLETH MAP 372
F-4 DELAUNAY TRIANGULATION 372
F-5 DAM STRUCTURES 372
F-6 ENBANKMENT DAMS 372
F-7 GRAVITY DAMS 373
F-8 WIRELINE LOGS 373

APPENDIX G: SATELLITE SITE DATA **374**

G-1 COPPER CLIFF SATELLITE SITE DATA 374
G-2 RAINFALLS FOR VARIOUS DURATIONS 375
G-3 CRUDE DEPOSITIONAL RATE 376
G-4 CHANNEL SHAPE ON VELOCITY 376
G-5 DRAINAGE 376
G-6 MANNING EQUATION 377

APPENDIX H: PERMISSION STATEMENTS **378**

H-1 PERMISSION STATEMENTS COLLECTED 378
H-2 PERMISSION TO USE 32 STUDENTS' FIELD LOGS 378

APPENDIX I: KIDD COPPER 'DAM' FIELD ENTRIES **380**

I-1 BELAN 380
I-2 CARROLL 380
I-3 DE MEDEIROS 380
I-4 ENDERS 380
I-5 FARRUGGIA 380
I-6 GABRIEL 380
I-7 GRENIER 380
I-8 GYATT 380
I-9 HARPER 380
I-10 HYNES 381
I-11 JACK 381
I-12 KILMENKO 381
I-13 LENGERT 381
I-14 LI 381
I-15 MELYMUK 381
I-16 ROBERTSON 381
I-17 ROSAMOND 381
I-18 RUDD 381
I-19 SINGH 382
I-20 STUDDY 382

I-21 Van Pham.. 382
I-22 van Hengstum 382
I-23 Walker 382
I-24 Weller 382
I-25 Windus 382
I-26 Xiujuan 382
I-27 Zelek 382

APPENDIX J: GEOLOGY AND THE EARTH SCIENCES.......................... **383**

J-1 Biostratigraphy 383
J-2 Geology 383
J-3 Geochemistry 383
J-4 Geomorphology 383
J-5 Earth Sciences 383
J-6 Hydrology 384
J-7 Petrography 384
J-8 Lithology 384
J-9 Mathematical Geology 384
J-10 Mining Geology 384
J-11 Morphology 384
J-12 Physical Geology 384
J-13 Sedimentology 384
J-14 Science Stratigraphy 384
J-15 Strctural Geology 385
J-16 Structural Petrology 385

APPENDIX K: SEDIMENTARY PROCESSES AND TERMINOLOGY **386**

K-1 Sediment 386
K-2 Sedimentation 386
K-3 Science of Sedimentology 386
K-4 Sedimentation Delivery Ratio 386
K-5 Sedimentation Diameter.. 386
K-6 Sedimentary Differentiation 386
K-7 Sedimentary Facies 386
K-8 Sedimentary Insertions 386
K-9 Sedimentary Laccolith 387
K-10 Sedimentary Ore 387
K-11 Sedimentary Petrography 387
K-12 Sedimentary Petrology 387
K-13 Sediment-Production Rate 387
K-14 Sedimentation Radius 387
K-15 Sedimentation Rate 387
K-16 Sedimentary Rock 387
K-17 Sedimentary Structure 388
K-18 Sediment Transport 388
K-19 Sedimentary Trap 388
K-20 Sedimentation Trend... 389

K-21 SEDIMENTATION UNIT 389
K-22 SEDIMENT YIELD 389
K-23. SPATIAL DISTRIBUTION LAMINAE: CLOSE UP OF A BED: INTERNAL STRUCTURES 389

APPENDIX L: THE BIG-BANG THEORY **390**

APPENDIX M: THE GEOLOGIC TIMESCALE **391**

INDEX **393**

ABOUT THE AUTHOR, LOREEN SHERMAN, MBA 395

List of Figures

FIGURE 1 - SECTION 1.1: PHOTOGRAPH OF THE SUDBURY AREA MINING RAILWAYS, DALE WILSON 29
FIGURE 2 - SECTION 1.1: JORDAN SPREADER 5 IS ONE OF INCO'S AS PHOTOGRAPHED IN 1992 30
FIGURE 3 - SECTION 1.2: GEOGRAPHIC COORDINATES FOR KIDD COPPER PROPERTY, CITE GOOGLE EARTH, 2009. 31
FIGURE 4 - SECTION 1.3: KIDD COPPER IMPOUNDMENT IS SW OF PERCH LAKE, SUDBURY, ONTARIO 32
FIGURE 5 - SECTION 2.6: SLAG POTS PHOTO BY PAUL BINNEY 1994 (WILSON, 2003, P. ND.) 37
FIGURE 6 - SECTION 2.6: STOCKPILE OF TYPE II A.C.B.F. SLAG, www.sysco.ns.ca 38
FIGURE 7 - SECTION 2.7: PHOTOGRAPH OF THE MINETAILINGS SURFACE AT KIDD COPPER PROPERTY 39
FIGURE 8 - SECTION 3.2: 1970 AIR PHOTO REVEALING A SMALL LAND-BRIDGE ON THE WASTE SITE 42
FIGURE 9 - SECTION 3.3: SKETCH MAP SHOWING THE DAM ON THE SE CORNER OF KIDD COPPER PROPERTY 43
FIGURE 10 - SECTION 3.6: 1983 AIR PHOTO SHOWING EXTENDED TAILINGS AREA - 13 YEARS 46
FIGURE 11 - SECTION 3.9: 1992 AIR PHOTO OF THE KIDD COPPER MINE TAILINGS WASTE SITE 48
FIGURE 12 - SECTION 3.11: THE ROCK CYCLE. (DISCOVERY BOOKS, 1999, P. 21) 50
FIGURE 13 - SECTION 4.1: LOGARITHMIC GRAPH SHOWING STREAM VELOCITIES. 53
FIGURE 14 - SECTION 4.2: WATER FROM THE DAM IS THE DOMINANT SUPPORT MECHANISM 55
FIGURE 15 - SECTION 4.4: FLUID TURBULENCE 57
FIGURE 16 - SECTION 4.5: TURBIDITY CURRENT FEATURES; HEADED INTO A MINE TAILINGS DEPOSIT 59
FIGURE 17 - SECTION 4.9: COMMON SEDIMENTARY IRONSTONE MINERALS (NICHOLS, 2003, P.33) 64
FIGURE 18 - SECTION 4.10: TRANSPORT OF PARTICLES IN A FLUID (NICHOLS, 2003, P. 39) 65
FIGURE 19 - SECTION 6.2: TAILINGS NEST IN11 BETWEEN 0.01 AND 0.06 MM. (COGGANS, *ET AL.,* 1991, P.450) 80
FIGURE 20 - SECTION 7.2: GRADED BEDS - LARGEST PARTICLES SETTLE FIRST 84
FIGURE 21 - SECTION 7.4: BOUMA SEQUENCE – 5 DIVISIONS 87
FIGURE 22 - SECTION 7.4: THE IDEAL TURBIDITE SUCCESSION AND "BOUMA" INTERVALS 88
FIGURE 23 - SECTION 9.7: MCMASTER STUDENTS WALK OVER THE MINE TAILINGS SITE IN 2004 98
FIGURE 24 - SECTION 9.9: FIELD SKETCH FROM EDEN HYNES STUDENT LOG, AUGUST 30, 2004 100
FIGURE 25 - SECTION 9.10: STRATIFIED LAYERS UNEARTHED IN THE UPPER ZONE OF TRANSITION 101
FIGURE 26 - SECTION 9.12: EXPOSED STRATIFIED LAYER OF SEDIMENT FROM ONE EVENT 103
FIGURE 27 - SECTION 10.3: BASE MAP OF KIDD COPPER SITE, A COMPILATION OF BOREHOLES N = 120 133
FIGURE 28 - SECTION 10.4: CROSS-SECTIONS WEST-NORTH TRANSECT AND A NORTH-SOUTH TRANSECT 135
FIGURE 29 - SECTION 11.2: A TYPICAL FIELD OBSERVATION, SPOT X 144
FIGURE 30 - SECTION 11.11: CLAYEY FABRIC DISTRIBUTION 152
FIGURE 31 - SECTION 13.8: 'LAYER-CAKE' STRATIGRAPHY (NICHOLS, 2003, P. 232) 172
FIGURE 32 - SECTION 13.12: PLATE 1 PLANE-PARALLEL BEDDING: PANORAMIC VIEW 176
FIGURE 33 - SECTION 13.13: GRAND CANYON EXPOSED BEDS – UPPER PORTION 177

FIGURE 34 - SECTION 13.13 GEOLOGISTS PAINT A PICTURE OF GRAND CANYON'S HISTORY 178
FIGURE 35 - SECTION 14.1: A FATHER OF MODERN GEOLOGY, SIR CHARLES LYELL.............................. 181
FIGURE 36 - SECTION 14.11: CROSS SECTION ALONG A-A' 193
FIGURE 37 - SECTION 14.11: LOCATION OF INCO LTD. COPPER CLIFF TAILINGS IMPOUNDMENTS................... 194
FIGURE 38 - SECTION 15.2: FACIES ASSOCIATIONS, FACIES SEQUENCES & FACIES CODES (NICHOLS, 2003, P. 65) 201
FIGURE 39 - SECTION 17.4: MAGNITUDE OF GEOLOGICAL PROCESS IN SPACE AND TIME (NICHOLS, 2003, P.3) 225
FIGURE 40 - SECTION 18.1: MAJOR METHODS OF DATING ROCKS, (DISCOVERY BOOKS, 1999, P. 34) 241
FIGURE 41 - SECTION 18.3: PARENT AND DAUGHTER ISOTOPES WITH RESPECTIVE HALF-LIVES......................... 245
FIGURE 42 - SECTION 19.2: VERTICAL PROFILE OF FOUR ASSOCIATIONS, CROSS-SECTION B–B', AUGUST 30, 2004.......... 252
FIGURE 43 - SECTION 19.11: PLATE 59 RHYTHMIC STRATIFICATION IN FALLOUT DEPOSITS 259
FIGURE 44 - SECTION 19.12: BUNDLED UPBUILDING FOUND IN WAVE-FORMED RIPPLES.......................... 261
FIGURE 45 - SECTION 19.22: LOREEN RUDD POSES FOR FIELD OBSERVATIONS AND RECORDS, AUGUST 30, 2004 266
FIGURE 46 - SECTION 19.22: PEER-REVIEW POSTER PRESENTATION OF RAPID STRATIFICATION 267
FIGURE 47 - SECTION 20.10: SEVENTEEN HORIZONTAL NEW LAYERS IN THE ENBANKMENT, LOREN NELSON 279
FIGURE 48 - SECTION 21.4: STRATIGRAPHIC UNIT POINT 64/X1 284
FIGURE 49 - SECTION 21.5: STRATIGRAPHIC UNIT POINT 56/4 285
FIGURE 50 - SECTION 22.9: CHARLES DARWIN IN 1881, (LEAKEY, 1979, P. 8) 295
FIGURE 51 - SECTION 22.9: MAMMALIAN EVOLUTION (RETALLACK, 1990, P. 262).............................. 297
FIGURE 52 - SECTION 22.10: RELATIVE BIOSTRATIGRAPHIC POSITION TWO DATES REPRESENTED 298
FIGURE 53 - SECTION 23.4: TIME STRATIGRAPHIC FACIES BOREHOLE 64 AND BOREHOLE 56.......................... 302
FIGURE 54 - APPENDIX A: LAMINATIONS - SEDIMENTARY ROCK, (ALBIN, 2004, P. 19). 344
FIGURE 55 - APPENDIX A: LITHOFACIES CODES: (LEWIS, 1984, P.13).. 345
FIGURE 56 - APPENDIX A: PARALLEL-LAMINATED SAND.. 346
FIGURE 57 - APPENDIX A: BLAST FURNACE 351
FIGURE 58 – APPENDIX A: CLASSES 353
FIGURE 59 - APPENDIX B: SCALE 355
FIGURE 60 - APPENDIX B: GRAIN SIZE 355
FIGURE 61 - APPENDIX B: MEAN FLOW VELOCITY 356
FIGURE 62 - APPENDIX D: ROCK COLOR 365
FIGURE 63 - APPENDIX F: BLOCK DIAGRAM.. 373
FIGURE 64 - APPENDIX G: COPPER CLIFF SATELLITE SITE DATA (COGGANS, ET AL., 1991, P. 453) 374
FIGURE 65 - APPENDIX G: RAINFALLS FOR VARIOUS DURATIONS 375
FIGURE 66 - APPENDIX G: GRADIENT UPSTREAM 376
FIGURE 67 - APPENDIX G: RESERVOIR FORMS ON A DAM (TARBUCK & LUTGENS, 1999, P. 235) 376
FIGURE 68 - APPENDIX G: VALUES OF MANNING EQUATION 377
FIGURE 69 - APPENDIX K: CLASTIC AND NON-CLASTIC SEDIMENTARY ROCKS (NICHOLS, 2003, P. 11) 388
FIGURE 70 - APPENDIX K: GEOMETRY OF BEDDING AND SEDIMENTARY BODIES (LUCCHI, 1995, P. 28)........... 389
FIGURE 71 - APPENDIX M: THE TIME COLUMN SHOWS MILLIONS OF YEARS AGO 391
FIGURE 72 - APPENDIX M: TIME UNITS OF GEOLOGIC TIME SCALE (TARBUCK & LUTGENS, 1999, P. 205) 392

List of Tables

TABLE 1 - SECTION 5.2 SOURCES OF INDIVIDUAL IONS IN RAINWATER.. 69
TABLE 2 - SECTION 5.4: FACTORS THAT INFLUENCE WEATHER 73
TABLE 3 - SECTION 6.1: USGS LIBRARY MINERAL ENDMEMBERS USED WHEN UNMIXING AURORA DATA 79
TABLE 4 - SECTION 10.5: A SEQUENTIAL LITHOGRAPHIC COMPILATION OF CROSS-SECTION A-A' 136
TABLE 5 - SECTION 10.6: A SEQUENTIAL LITHOGRAPHIC COMPILATION OF CROSS-SECTION B-B' 139

TABLE 6 - SECTION 11.4: THE SEDIMENT COMPOSITION FOUND ON KIDD COPPER WHEN REDEPOSITED............................ 146
TABLE 7 - SECTION 11.4: THE MINERAL COMPOSITION FOUND ON KIDD COPPER AFTER REDEPOSITION............................ 147
TABLE 8 - SECTION 11.5: SINGLE COLOR DESCRIPTORS OF KIDD COPPER PROPERTY... 148
TABLE 9 - SECTION 11.6: MULTI-COLOR DESCRIPTIONS OF KIDD COPPER PROPERTY... 149
TABLE 10 - SECTION 11.9: SIZE OF GRAINS FOUND ON THE KIDD COPPER PROPERTY 151
TABLE 11 - SECTION 11.12: ORIENTATION OF THE GRAINS FOUND ON KIDD COPPER AFTER REDEPOSITION 153
TABLE 12 - SECTION 13.5: COMPARING THE LAW OF ORIGINAL HORIZONTALITY WITH OBSERVED FACTS 169
TABLE 13 - SECTION 14.10: POINT SOURCE LAYERING 189
TABLE 14 - SECTION 14.10: SHEET SOURCE LAYERING 190
TABLE 15 - SECTION 15.4: KIDD COPPER ARCHITECTURAL ELEMENTS 204
TABLE 16 - SECTION 15.4: SOME CHARACTERISTICS OF SEDIMENTARY ROCKS.. 208
TABLE 17 - SECTION 16.6: THE GEOLOGIC COLUMN (TARBUCK 1999; RITLAND 1981-INCOMPLETE)............................ 217
TABLE 18 - SECTION 17.4: 2009 GEOLOGIC TIME SCALE, THE GEOLOGIC SOCIETY OF AMERICA 223
TABLE 19 - SECTION 17.4: 1999 GEOLOGIC TIME SCALE, THE GEOLOGIC SOCIETY OF AMERICA 224
TABLE 20 - SECTION 17.17: CARBONIFEROUS – PERMIAN SPLINE-CURVE CROSS-VALIDATION....................................... 235
TABLE 21 - SECTION 17.17: CARBONIFEROUS – DEVONIAN – SILURIAN.. 236
TABLE 22 - SECTION 18.2: PROBLEMS WITH DATING METHODS 243
TABLE 23 - SECTION 19.9: PALAEOCURRENT PATTERNS... 258
TABLE 24 - SECTION 19.24: TABLE OF CONSIDERATIONS: IF – THEN PATTERNS 271
TABLE 25 - SECTION 21.10: PREDICTORS FOR DEPOSITIONAL OUTCOMES 289
TABLE 26 - APPENDIX A: MIALL FACIES CLASSIFICATION 342
TABLE 27 - APPENDIX A: THICKNESS SCALE 343
TABLE 28 - APPENDIX A: AVERAGE PERMEABILITY 348
TABLE 29 - APPENDIX A: SOIL PROPERTIES AND BEHAVIOR OF SOIL (BRADY &WEIL, 2002, P. 127)............................. 352
TABLE 30 - APPENDIX C: CHLORINE CONTENT OF TERRESTRIAL ROCKS... 358
TABLE 31 - APPENDIX C: DIAGNOSTIC PROPERTIES 359
TABLE 32 - APPENDIX E: RELATIVE REACTIVITY OF COMMON MINERALS AT PH = 5.. 371
TABLE 33 - APPENDIX E: ELEMENTAL COMPOSITIONS FOR SOIL HUMIC AND FULVIC ACIDS.. 371

Kidd Copper Property Report Based on 2004-August-30

The Kidd Copper property impoundment of tailings is an area located northwest of Sudbury, Ontario just south of Perch Lake. "In 1966, Kidd Copper commenced the construction of a 1000 tpd mill and commenced mining and milling in 1967 producing approximately 250,000 tons of ore prior to mining being concluded in 1968" (Article [2], 2008, *p. nd.*). The Kidd Copper property "man made dam broke" (Belan, 2004, p. 41). The historic authenticity of the property is well established.

On Monday, August 30, 2004, the McMaster University Geo3Fe3 field class was given Assignment #9, to "walk over the entire site" (McMaster University, August 2004, Assignment #9 Mine Tailings). Students exposed 120 boreholes, some to a depth of 50 cm. Core lithofacies logs were compiled from borehole observations and recorded in each student's Forestry Supplier's Field Book. The raw field data and analytical information obtained from this exploration has been incorporated into this study.

The sedimentary cycle is "a cycle which comprises the weathering of an existing rock, followed by the erosion of minerals, their transport and deposition, then burial" (Allaby & Allaby, 2003, p. 485). In a twenty-three year period, from 1960 -1983, three unique sedimentation depositions can be observed on this Kidd Copper disposal site. Three depositional processes are observed in the site area with definitive ages: (1) The original deposition of known parental material and subsequent deposits; (2) The upswelling like a turbulent flow caused by a dam breach; and (3) The redeposition of the sediment onto the major mine tailings area and a new extension. These depositions have been recorded in recent history eliminating speculation on their age. Time correlation and rapid stratification are the basis of this report showing that some geolgic units are formed in a short period of time.

A new discovery was made when the exposed strata revealed that rapidly stratified sediment deposits do not form uniform and even deposits for first-time deposition. This is the first report to indicate that under water as a dominant support mechanism the transported particles settle out of fluid as stratified layers. The new discovery is that the lithographic logs revealed heterogeneous deposits with unequal declivities, different slope angles and no symmetrical profile. The correlated logs did not continue laterally in all directions until it thins out as a result of nondeposition or until it abuts against the edge of the original basin of deposition. The rapid stratification of the sediment found at Kidd Copper revealed that the adjacent boreholes and the adjacent geologic units had no correlation, no pattern and were not laterally conformed to the marker. This was an unexpected outcome because all of the boreholes within this report had the same reference point of time and underwent the same depositional process. The boreholes were identical units deposited beside each other during the same depositional process but the facies correlations were not identical. The characteristics of each borehole were different from the adjacent or associated units, which should have reflected the condition of a different origin. The origin of the sediment material was identical and in this case each unit projected a different profile. The new discovery is that a time-stratigraphic correlation of heterogeneous material can be referenced by the same point in time even when the distribution and pattern are dissimilar.

This report is based on real-time observations and primary field work in a well explored mine tailings waste site in northern Ontario, with the collection of blind data to ascertain a depositional process for first layer strata that is deposited on Earth.

"Prediction: (1) the scientific ability to forecast the outcome of a process in advance. (2) The ability of a model to provide information beyond that for which measurements are available" (Clarke, 2003, p. 322). "Foresets may be termed heterogenous if the layering is due to variations in grain size, or homogenous if it is not" (Selley, 1982, p. 221). Based on a satellite study, the Grain size D_{50} (mm) of the sediment on the Kidd Copper property is between "0.01 and 0.06 mm" (Coggans, *et al.*, 1991, p. 450). With such a minute variation existing in grain size in the sediment on the Kidd Copper property it should be expected that the layering would be homogenous. The prediction is that the layering of the strata at Kidd Copper would be uniformly deposited. The predicted outcome from the textbooks will be cross-examined in relation to this real case scenario of original deposition throughout this report.

This report compares the attributes collected in field reports against resource material found in textbooks. This report is based on an application of real-time observations; current sedimentologic principles; and data from a well explored waste mining site in northern Ontario. The author endeavors to relate textbook and geologic principles and relationships with first-hand observations to depositional processes in the analysis of the sediment layering. The main methodology applied to this study is *in situ* (in place). Scientific report "validation by examination" (Clarke, 2003, p. 324).

The author will establish the fact that stratification is a rapid process. This is the first report to use rapid stratification as an application of the Genesis model in a real-world geologic situation. Field observations are used to compare time-based depositional processes and correlate to original layering of sediment on Earth.

What happens when sediment or sediments are originally deposited is the focus of this report. This is a detailed analysis of rapid stratification, which is observed in the layering of the mine-tailing deposits on the Kidd Copper Property. Over 120 boreholes at the mine tailings waste site support rapid stratification. On this fact, the author will establish the conclusion that rapid stratification is a primary depositional process; and the principles defined herein may assist future interpretations in the age of the earth "where lesser amounts of data are available" (Hughes, 1984, p. preface).

The author will develop this report into seven main facts:

(1) The historicity of the Kidd Copper property area, being well established and documented; (2) A dam breach resulted in the redeposition of a primary sediment

structure on the impoundment of tailings; (3) The redeposited sediment is examined in its diagenesis context and compared to the expected outcome(s) from the Geologic laws and principles; (4) Rapid stratification will be established as one-layer and can be a primary structure; (5) The Law of Original Horizonatality and some calibration methods will be questioned; (6) Time correlations will be examined; and (7) The door to a younger age for the Earth will be opened.

The scope of this report does not include: Universal time (astronomy to navigation, transits of stars), Ephemeris Time (comparing positions of the Sun, Moon, and the planets), (Gradstein, Ogg & Smith, 2004, p. 7) nor radiometric databases. "Astronomical time scale calibrations in the Paleogene face additional theoretical uncertainties in the orbital calculations (Laskar, 1999)" (Gradstein, *et al.*, 2004, p. 407) will not be discussed. Nor will this report endeavour to cover the topics on rainfall yields or "biogenic sedimentary structures" (Reineck & Singh, 1973, p. 74). Point data, including samples from the surface of the waste site for geochemical and mineralogical analysis are not discussed in this report.

Correlating and transferring data from the outdoor observatory where the boreholes are located on the Kidd Copper property south of Perch Lake (hereafter in this report referred to as the primary site, the site, or the site area) will be the basis for analysis of the hypothesis that rapid stratification supports the theory of a 'younger age' for planet Earth.

1.0 Historical and Geological Overview of Site

The primary study site for this scientific report is the Kidd Copper Property. "Kidd Copper is located southwest of Sudbury, Ontario, Canada. It was at one time a nickel-copper-cobalt-platinum mine" (Shang, *et al.*, 2004, p. 2). This specific site with a waste impoundment of tailings was chosen for this report because of how well known in both the scientific and mining community this established and recognized field site is. On this site a precise date can be given for the mine tailings. The initial deposition is known. The factors that relate to its sedimentation stages are known; and the aftermath of a catastrophic event are recorded. This makes the Kidd Copper property an ideal study site for dating sedimentation processes because time can be accounted.

1.1 The Kidd Copper Property Site and its History are Well Documented

The satellite property, Copper Cliff mine and "the region's first smelter was established nearby, beginning operation December 24, 1888 (Wilson, 2003, *p. nd.*). Comparisons between Copper Cliff and Kidd Copper will be examined throughout this report. Years after the Worthington mine collapsed on October 4, 1927, "two separate mines established themselves in the vicinity, the Totten Mine and Kidd Mines" (Charbonneau, 2003, *p. nd.*). "In 1965 the property was optioned by Sheridan Geophysics Ltd. and assigned to Kidd Copper Mines Ltd." (Article [2], 2008, *p. nd.*). The Kidd Copper property is well known and documented for its operations.

"Ore [GEOL]. (1) The naturally occurring material from which economically valuable minerals can be extracted. (2) Specifically, a natural mineral compound of the elements, of which one element at least is a metal. (3) More loosely, all metalliferous rock, though it contains the metal in a free state. (4) Occasionally, a compound of nonmetallic substances, as sulfur ore" (Licker, 2003, p. 231).

INCO Operations. INCO had an extensive mining operation, "one of the largest mining complexes in the world, cover 21 km^2 and constitue in excess of 10% of all mine tailings areas in Canada" (Coggans, *et al.*, 1991, p. 447). "Ore is mined at 11 mines in the area, sent to three mills, a smelter complex and finally three refineries producing nickel, copper and other metals along with sulphuric acid" (Wilson, 2003, *p. nd.*). Figure 1 - Section 1.1 is an aerial view of the massive INCO facility in the Sudbury area of INCO's mining railways, published by Bob Chambers/Kalmbach online showing the extensive mining operation. INCO's operation covers 21 km^2 and includes six tailings impoundments within the disposal area (Coggans, *et. al.*, 1991, p. 447).

Figure 1 - Section 1.1: Photograph of the Sudbury Area Mining Railways, Dale Wilson

"The company also has several advanced projects in Sudbury, which include the AER Kidd Project adjacent to Inco's Totten Mine (10.1 million tonnes @1.5% Ni, 2% Cu, 4.8 g/t PGE's)" (Bharti, November 2004, News Release).

More information and various records about the mine tailings can be found in the Township of Sudbury records and in the records of INCO. Other resources are the Ontario Ministry of Natural Resources and the Canandian Environmental Protection Act Association.

"The mine was originally operated by INCO Ltd. but it is currently managed by Crowflight International Inc." (Shang, *et al.*, 2004, p. 2). In 1965, Geophysicists drilled within the area to inspect covered minerals. Zachary Windus recalls "Crowflight Resources → remap geology of area, look for new areas … geophysicists → core sample" (2004, p. 33). "Crowflight is a Canadian junior mining exploration company listed on the TSX Venture Exchange focused on Nickel, Copper and Platinum Group Minerals exploration in the Thompson Nickel Belt and Sudbury Basin" (Bharti, November 2004, News Release). "Crowflight Resources decided to remap the geology of the area – bulldozed area and found new outcrops" (Gyatt, 2004, p. 54).

Figure 2 - Section 1.1 shows one piece of equipment, the "Jordan Spreader 5", that INCO used in their mining operation "after rebuild by new-parent Pandrol Jackson; *Thomas Sajnovic*" (Wilson, 2003, *p. nd.*).

Figure 2 - Section 1.1: Jordan Spreader 5 is One of INCO's as Photographed in 1992

Extensive historical data can be found on the abandoned waste site; for example in 1992, Inco and Falconbridge Noranda controlled the waste-site. As reported by Karl Belan on Geo3Fe3 Field Camp log entry on August 30, 2004 "FNX-geologists drill boreholes to determine mineral composition of rocks successfully" (page 34). The Kidd Copper property site and its history were well known. "Hyperspectral AURORA data were used for surface mineral mapping over a nickel-copper mine-waste site at Kidd Copper, Ontario. AURORA data were collected on July 13, 2002 for the Ontario Mineral Exploration Technologies (OMET) program" (Shang, *et. al.*, 2004, p. 1). Knowing the historical background establishes the conditions surrounding the mine's waste tailings.

1.2 Kidd Copper Impoundment Location and Coordinates

Photocopied maps are subject to slight distortion. "The space distortion of a map projection, consisting of warping of direction, area, and scale across the extent of the map" (Clarke, 2003, p. 310). Therefore, the photocopied maps are subject to "dropout [which is] the loss of data due to scanning at coarser resolution than the map features to be captured" (Clarke, 2003, p. 310).

Geographic Coordinates. The Atlas of Canada gives the "geographic coordinates the latitude and longitude coordinate system" (Clarke, 2003, p.313).

Perch Lake, Sudbury, Ontario recorded in Degrees, Minutes, Seconds: "(+)46° 23' 45" N (-)81° 25' 7" W" (Google Earth, 2009, Fairbank Lake, Ont.). The geographic coordinates for the Kidd Copper property mine tailings waste site are: "(+)46° 23' 30.59" N (-)81° 25' 38.00" W" (Google Earth, 2009, Fairbank Lake, Ont.). Refer to Figure 3 - Section 1.2.

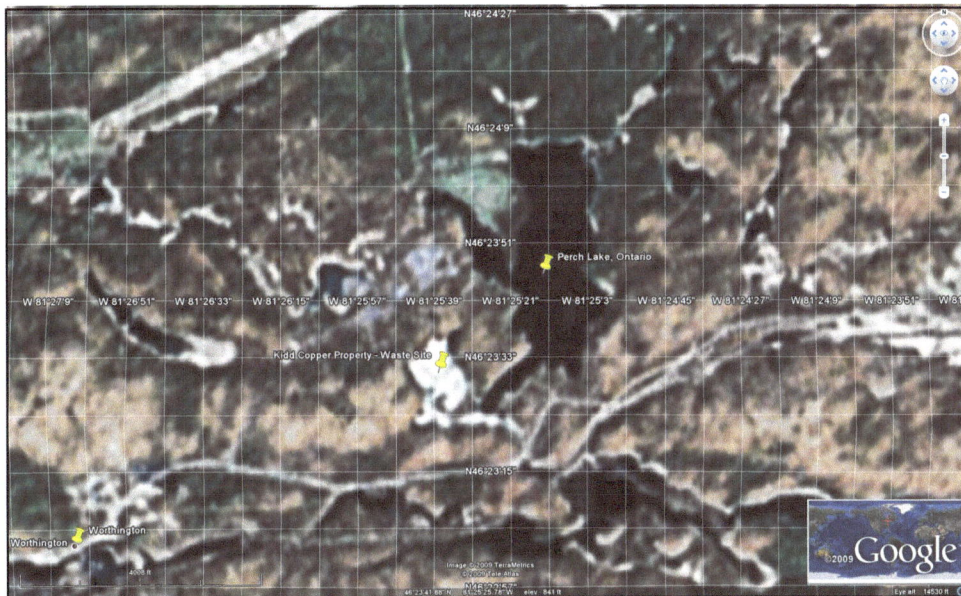

Figure 3 - Section 1.2: Geographic Coordinates for Kidd Copper Property, cite Google Earth, 2009.

1.3 Elevation of 850 Feet

UTM Coordinates. Reference maps from the National Air Photo Library and Natural Resources Canada are included in this report. A reference map is "a highly generalized map type designed to show general spatial properties of features" (Clarke, 2003, p. 323). For example, in Figure 1 - Section 1.1 the photograph of the mining railways shows the complexity of the railways as an aerial road map of the facility, while in Figure 4 - Section 1.3 a map is used to proximate the site of study's location. The absolute location is "a location in geographic space given with respect to a known origin and standard measurement system, such as a coordinate system" (Clarke, 2003, p. 304).

The Coordinate system is "a system with all the necessary components to locate a position in two-or three-dimensional space; that is, an origin, a type of unit distance, and two axes" (Clarke, 2003, p. 307). The UTM (Universal transverse Mercator) is "a standardized coordinate system based on the metric system and a division of the earth into sixty 6-degree-wide zones. Each zone is projected onto a transverse Mercator projection, and the coordinate origins are located systematically" (Clarke, 2003, p. 327). A coordinate pair is "an easting and northing in any coordinate system, absolute or relative. Together, these two values, usually termed (*x, y*), describe the location in two-dimensional geographic space" (Clarke, 2003, p. 307). GPS points are an example.

The Perch Lake UTM coordinates are "17 T 467248.0 5137841.0" (Converter, 2009, UTM). Perch Lake Coordinates in Degrees: "(+)46.393° N (-)81.426° W." (Converter, 2009, UTM).

Figure 4 - Section 1.3: Kidd Copper Impoundment is SW of Perch Lake, Sudbury, Ontario

Figure 4 - Section 1.3 is a Topographic Map (sheet number 41I6) from the Natural Resources of Canada, from her majesty the Queen in Right of Canada, 2007. Assessed from the National Topographic Database, Primary Map Data: 1:50 000 CanVec modified on 2006-02-06. Toporama – Location of the Kidd Copper property waste disposal site is marked at the place name waste. "Place-name: A text cartographic element that links the language given name to a feature by placing it close to the symbol to which it corresponds" (Clarke, 2003, p. 322). Scale 1: 40 000. In this case 1 inch on the map represents 40,000 inches, or about 0.631 miles on the ground (Williams, 1993, p. 13). The Worthington offset is located 30 miles SW of Sudbury, Ontario, Canada.

The site area is south-west of Perch Lake and the contour intervals, "the vertical difference in measurement units such as meters or feet between successive contour lines on a contour map" (Clarke, 2003, p. 307). A stream flows from Perch Lake downward to the train track. The stream flow direction is north to south following the path of least resistance and from higher elevation to lower. Refer to Figure 4 - Section 1.3, the waste site is mapped relatively level at an elevation of 850 feet where Kidd Copper is.

1.4 Parent Material is Goethite

The available data and geomorphological background of the parent material for the Kidd Copper site area is known. "Parent material is the starting point for a soil, and the degree of development of a soil is measured by the amount of change compared with its parent material" (Retallack, 1990, p. 255). Source rock defined as a geologic term is a "rock from which fragments have been derived which form a later, usually sedimentary rock" (Licker, 2003, p. 348).

"The Southern Province is a major source of iron ore in the Canadian shield" (Roberts, 1996, p. 64). "Northwest of Sudbury, Ontario, is a geological structure that has been a world leader in producing nickel ore, along with lesser amounts of copper, gold, and silver. It is a basin structure of Proterozoic rocks that includes a sill of igneous material. Since the structure is a basin, the sill outcrops all around it. There are mines along the margin of the sill, where its intrusion cracked the rock above it and mineral-rich solutions filled the cracks" (Roberts, 1996, p.64). "In the Sudbury Basin, mineral-rich solutions migrated into cracks in the bedrock from the intruding molten material" (Roberts, 1996, p. 64).

"From Massey to Sudbury, the bedrock and outcrops consist mostly of early Proterozoic sedimentary and metasedimentary rocks similar to those the highway traverses to the west, such as argillite, graywacke, quartzite, and conglomerate, with associated dikes and sills. There is a sill of metadiabase cut by a diabase dike .5 mile west of the Highway 6 intersection … that contains large feldspar grains" (Roberts, 1996, p. 95).

"The Kidd Copper site is located directly southwest of Sudbury, and the mine actually sits on a mineral intrusion. As a result, the bedrock is rich in goethite" (Shang, *et al.*, 2004, p. 9). On the Kidd Copper Property, the main geologic feature is the impoundment of mine tailings. A feature is "a single entity that composes part of a landscape" (Clarke, 2003, p. 311). Access to the site is from the north.

1.5 Three Steps in the Depositional Process on Kidd Copper Offers Answers

The significance of the Kidd Copper mine tailings (slag) site is that three steps of sediment deposits are recorded in its history from 1960-~1970. The *first step* is the initial slag deposits of hot liquid from the slag pots in the 1960's. During this time a daily accumulation of slag was deposited on the ground surface layer, in the upper and lower zones of transition. The *second step* is during the early 1970's, when the retaining dam at the southeast corner breaks, creating an upsurge of deposits water, which overflows the tailings. Sand, liquid water and the mine tailings (slag) mix in random order due to the velocity of the water overflow. The *third step* is approximately one week after the break in the dam. The water-mixed sediments settled onto the ground surface. The redeposited sediments are the sediments that were observed *in situ* on the Kidd Copper property and are the focus of this scientific report.

Environment of deposition is "an area within which a *sediment* is deposited under a particular set of physical, chemical and biological conditions … Depositional environments can be very broadly classified as *alluvial* (including deltaic), … *lacustrine*, *desert*, *shoreline* (*clastic* and *carbonate*), shallow marine, … deep clastic seas (*turbidite* basins), *pelagic* and *glacial* (continental and glacio-marine)" (MacDonald, *et al.*, 2003, p. 164). The depositional environment is known at the Kidd Copper Property, which is an accumulation of waste sediment that was deposited as large amounts of fine grained waste on the impoundment. When the dam broke the water swept up the mine tailings and redeposited them onto the site area. These facts are known making it easy for a geologist to determine the chronology of the sediment deposits. No guesswork is involved for the time of deposition or how the sediment was redistributed.

"To determine the relationship of geologic events to one another, four basic principles are applied. These are the principles of (1) original horizontality, (2) superposition, (3) lateral continuity, and (4) cross-cutting relationships" (Plummer, McGery & Carlson, 2003, p. 177). This report will consider each of these principles to determine whether the geologic events at Kidd Copper property follow conventional wisdom.

The importance of this report is that the deposit studied on the Kidd Copper property mine tailings site was formed in the aftermath of the dam breach; within recent history (at the time of this report < 40 years). This report will assemble text-book theory with geologic laws and compare the expectations of geologic principles with actual site observations of the aftermath. The contribution to the academic body of knowledge is a correlation of original excavations through three depositional processes to consider how the earth was formed. This correlation will consider speed, time and historical records.

2.0 Geomorphological Background – Step 1 Initial – Waste

Sedimentary rock is "formed by accumulation and consolidation of minerals and organic fragments that have been deposited by water" (Oldershaw, 2003, p. 221).

2.1 Step 1: Initial Deposition of the Mine tailings on the Waste Site

"Approximately 35,000 t/day of ore from the Sudbury Igneous Complex, a fine-to-medium-grained noritic to gabbroic rock locally containing up to 60 wt% sulfides, is mined by INCO Ltd. and processed at the Copper Cliff mill. Monoclinic and hexagonal pyrrhotite (Fe_{1-x} S), pentlandite ((Ni, Fe)$_9S_8$), chalcopyrite ($CuFeS_2$) , pyrite (FeS_2) , and cubanite ($CuFe_2S_3$) compromise the bulk of the sulfides. Minor concentrations of cobalite ($CoAsS$), nickelite ($NiAs$), galena (PbS), and millerite (NiS) are present (Hawley and Stanton, 1962; Naldrett, 1984)" (Coggans, *et. al.*, 1991, p. 447).

2.2 Tailings Analysis: Mineral Ore and Fine Sediment

"Approximately 95% of the copper and 80% of the nickel are recovered in the milling process, and the remainder is discharged to the impoundments with the bulk tailings (INCO Ltd., staff, personal commun., 1991)" (Coggans, *et al.*, 1991, p. 447).

Mine Tailings Chemical Composition. "Mine tailings: Silicates and sulphides, puritite, Pyrite when exposed to O_2 and H_2O quick reaction which generates sulphuric acid → dissolves minerals (ex U) and creates acid leacheate" (Jack, 2004, p. 31). "Unaltered tailings contain about 15 vol% pyroxene, primarily as augite ((Ca, Na)(Mg, Fe, Al, Ti)(Si,Al)$_2O_6$) and hypersthene ((Mg,Fe)$_2Si_2O_6$)" (Coggans, *et. al.*, 1991, 459).

"Reactive sulphides (e.g. pyrite and pyrrhotite) that exist in most mine waste are the agents responsible for AMD. AMD typically has pH values in the range of 2-4 and may contain high concentrations of metals and dissolved salts (Ash *et al.*, 1951; Nordstorm and Apers, 1999)" (Shang, *et al.*, 2004, p. 1).

2.3 Sediment Deposits on Kidd Copper Are Not Varves

"The sediment layer(s) deposited in a year constitute a varve" (Friedman-Sanders-Kopaska-Merkel, 1992, p. 175). "Varves, which are pairs of thin sedimentary layers deposited within a one-year period, have been used to correlate the lake deposits of many thousands of years, back into the Pleistocene and Holocene. Each varve pair consists of a fine winter layer and a coarse summer layer. The ages of deposits are determined by counting the pairs in a sample sediment core" (MacDonald, *et al.*, 2003, p. 131). Varves are "annually formed sediment couplets" (Eyles & Boyce, September 2004, GEO3E03 Courseware, p. 81).

2.4 Pre-Cambrian Crust Banded Ore Formations Occur

Banded ore formations occur in the Pre-Cambrian crust of all continents. "Banded iron formation [GEOL] a sedimentary mineral deposit consisting of alternate silica-rich (chert or quartz) and iron-rich layers formed 2.5–3.5 billion years ago; the major source of iron ore" (Licker, 2003, p. 34).

"Banded iron formation (BIF) – sedimentary rocks that are typically bedded or laminated and composed of at least 25 per cent iron, mostly as oxides (*Hematite*, *Magnetite*) and microcrystalline quartz (chert, chalcedony, jasper). Banded iron formations occur in the Precambrian crust of all continents, from 3800 Ma onwards, with a peak at 2500 Ma, and a scatter of occurrences in the late Neoproterozoic. They are commonly used as low-grade iron ore. Because banded iron formations have not been formed since Precambrian time, it is thought that special conditions may have existed contemporaneously with their formation. Their origin has been ascribed primarily to a marine biochemical process in which oxygen from photoautotrophic bacteria produce seasonally rhythmic deposition of oxidised iron from ferrous iron in solution. Other factors, including volcanic activity and the chemical oxidation of ferrous iron, may be involved" (MacDonald, *et al.*, 2003, p. 50).

Banded iron formations are not the result of early deposits neither from the Precambrian crust nor from a marine biochemical process as ascribed above; rather the banding of iron is a result of the redeposited settlement from the upswelling of the water in recent history on the Kidd Copper Property.

2.5 Mine Tailings Smelting and Flotation Process

"Ore in a 30,000-ton facility is fed through screens and crushers until it is the size of a marble, ground to powder to which water is added to make slurry. Flotation removes much of the rock, producing the waste material 'tailings. At this stage nickel and copper ores are separated" (Wilson, 2003, INCO operations). There are "two types of nickel tailings (slag from smelting and tailings from flotation process)" (Munoz, 2009, Abstract ID 2093). Mine tailings are large amounts of fine grained waste, produced as a result of the enrichment of metal ores.

"The nickel ore, called 'nickel concentrate' goes two places. Some will have its sulphur content reduced at the sulphuric acid plant. This by-product is shipped in tank cars in unit trains. The somewhat purified ore is called nickel calcine and returned to the smelter where it is mixed with nickel concentrate and the mixture thickened through removal of most of the water. This mixture is put into the roasters where sand is added to help separate out impurities into slag, which is carried by rail to the dump" (Wilson, 2003, INCO operations). Further details about slag can be found in the Data Dictionary. Refer to Appendix A-76.

"Molten nickel-copper is made up of 25-ton ingots, and the cooling process forces the copper content to gather together. The ingots are ground up and put through another flotation process, separating the copper and nickel" (Wilson, 2003, INCO operations).

"Copper ore is dried to powder, mixed with sand and put through a flash furnace, fired by natural gas (coal was used for many decades); this produces liquid copper and slag, the latter being skimmed off and dumped into 'slag pots'; molten copper at 2100 degrees F goes into 120-ton capacity, torpedo shaped, firebrick lined railway cars for its journey, done five or six times a day, to the refinery. There the copper is cast into anodes and further refined through electrolsis" (Wilson, 2003, INCO operations).

2.6 Mechanical-Human-Transport to Dump Molten Slag

The Worthington offset is located 30 miles SW of Sudbury. "Molten slag was removed from the plant in special slag pot cars and taken about 3 miles to be dumped as waste. Slag temperatures range from 1,200 degrees F for nickel to 1,600 for cobalt. This operation took place about every 2 to 3 hours, 24 hours a day using 10 car trains requiring two electric locomotives" (Wilson, 2003, Railway Operations). "CNR used 80-ton drop bottom dump cars" (Wilson, 2003, Railway Operations).

Figure 5 - Section 2.6: Slag Pots photo by Paul Binney1994 (Wilson, 2003, *p. nd.*)

Slag pots carried the mine tailings composed of nickel and cobalt; then dumped the slag deposits of hot liquid on the ground. Refer to Figure 5 - Section 2.6. The slag accumulated over the ground thickly on the Kidd Copper Property. "Mining operations must contain and treat the water runoff from their operations, even when the mines are no longer active" (Fresques, 2004, p. 1). The mine tailings produced sulfuric acid because the high concentration of metal in the mine tailings reacts with rain water and oxygen in the atmosphere contaminating surface and ground water.

Figure 6 - Section 2.6: Stockpile of Type II A.C.B.F. slag, www.sysco.ns.ca

The slag looks like colluvium (slope deposit) deposited by sheet flows running down face of the tailings pile. The dumping of the slag pots over the hillside developed a "slope wash – earth material moved down a slope, principally by the action of gravity aided by non-channelled running water" (MacDonald-Burton-Winstanley-Lapidus-Coates, 2003, p. 402). Colluvium is "unconsolidated material at the bottom of a cliff or slope, generally moved by gravity alone. It lacks stratification and is usually unsorted" (MacDonald, et al., 2003, p. 99). An example is found in Figure 6 - Section 2.6 showing the mounding of the slag accumulations. An important point here is that when the slag was deposited on the Kidd Copper Property from the mining operations it lacked stratification.

2.7 Tailings Waste Deposit at Kidd Copper Property

"In 1966 Kidd Copper commenced the construction of a 1000 tpd mill and commenced mining and milling in 1967 producing approximately 250,000 tons of ore prior to mining being concluded in 1968" (Article [2], 2008, *p. nd.*). The Kidd Copper property continued its operation until ~1970's when the dam broke and the waste site was stopped. "Abandoned mine sites can be extremely dangerous places" (Umpherson, Bennett, & Webb, 1991, p. 34).

On August 30, 2004 students take a photograph while they observe the surface of the Kidd Copper Property. Refer to Figure 7 - Section 2.7. This photograph is of the surface where mine tailings redeposited. "Tailings deposited within the impoundments average 6 wt% total sulfides, predominantly pyrrhotite" (Coggans, *et al.*, 1991, p. 447).

Figure 7 - Section 2.7: Photograph of the Minetailings Surface at Kidd Copper Property

2.8 Duration of 'Silty' Tailings Deposits 1936-1970

"To understand the earth's history, it is necessary to have some sequential framework in which to place our observations; that is, the relative or absolute ages of the rock units must be known" (Garrels & Mackenzie, 1971, p. 63). This is field geology. "That part of *Geology* that is practiced by direct observation in the field" (MacDonald, *et al.*, 2003, p. 182).

The Kidd Copper mine tailings "dam broke ~1970" (Enders, 2004, p. 49). And "no more tailings added after early 1970's" (Rudd, 2004, Tailings Exercise Marking Sheet). "The Kidd Copper site has been inactive for over 20 years" (Shang, *et al.,* 2004, p.2). "Repetitively monitoring a waste site using both ground observations and remote sensing will help us understand changes at the site over time" (Shang, *et al.*, 2004, p. 2)

The Kidd Copper site area establishes an accurate time clock in an outdoor observatory where observations at the immediate sub-surface can be observed and sampled directly. "The geologist interprets ancient events by observing the effects that they have had on natural materials preserved in sequences of rocks" (Garrels & Mackenzie, 1971, p. 63).

The Kidd Copper property was well known for its operations. The abandoned waste site is the primary target of this report. The known rock classification of the mine tailings eliminates guesswork. This report focuses on the mine tailings deposits left on the Kidd Copper Property. An inactive site provides a framework for time observations.

2.9 Closed System

The geographic area of the Kidd Copper mine tailings site is "a closed system in which there is no transfer of matter or energy into or out of the system during a particular process" – relatively young (MacDonald & Burton, 2003, p. 96). This area once was a mine tailings pond with waste products of iron oxides. Now the impoundment of waste tailings is abandoned.

3.0 Geomorphological Background – 2 Middle – Upswelling

On hand were "a series of four air photos and one hyperspectral image showing changes at the Kidd Copper property through the period 1970 to 2002. These photos show changes in the dimension of the tailings and can be used to interpret changes in processes occurring on the property through time" (Map, Kidd Copper Propery Mine Tailings). "A synoptic overview of the waste site" area was "achieved using remote-sensing data" (Shang, *et al.*, 2004, p. 2). This report includes: 1970, 1983 and 1992.

3.1 Surface Mineral mapping of the Kidd Copper Mine tailings

The Kidd Copper study site for over the last six decades has both aerial data and photographs using AURORA hyperspectral imagery (Shang, *et al.*, 2004, p. 3). "Hyperspectral AURORA data were used for surface mineral mapping over a nickel-copper mine-waste site at Kidd Copper, Ontario. AURORA data was collected on July 13, 2002 for the Ontario Mineral Exploration Technologies (OMET) program" (Shang, *et al.,* 2004, p. 1), "from an altitude of about 2340 m above sea level (Quantec Geoscience, 2002)" (Shang, *et al.*, 2004, p. 3).

"Traditional panachromatic aerial photographs covering the study site were obtained from the National Air Photo Library of Natural Resources Canada. Historical air photos are valuable for providing information on temporal changes in the geographical extent of the tailings. Scales of the air photos obtained in 1970, 1983 and 1992 are approximately 1:50,000. The air photos were scanned at 300 dpi so that they could be more readily compared in a digital environment; hence the pixelated appearance of some of the photos" (Shang, *et al.*, 2004, p. 3).

3.2 1970 A Small Land Bridge

Sedimentary structure is "the external shape, the internal structure, or the forms preserved on bedding surfaces, generated in sedimentary rocks by sedimentary processes or contemporaneous biogenic activity" (Allaby & Allaby, 2003, p. 485). The impoundment area is the preservation of the activity on the mine tailings site from the original deposition to the flood and redeposition of mixed sediment.

The landscape features on the Kidd Copper property are distinct characteristic forms. The light tones in Figure 8 - Section 3.2 indicates a major mine-tailing deposit. A small land-bridge is visible at the south-east corner of the Kidd Copper waste tailings site in 1970. The landscape feature in 1970 is recognizable as the land bridge. However, in 1983 the landscape feature of the land bridge has been replaced with the rectangular-shaped tailings deposit on the Kidd Copper property, refer to Figure 10 - Section 3.6.

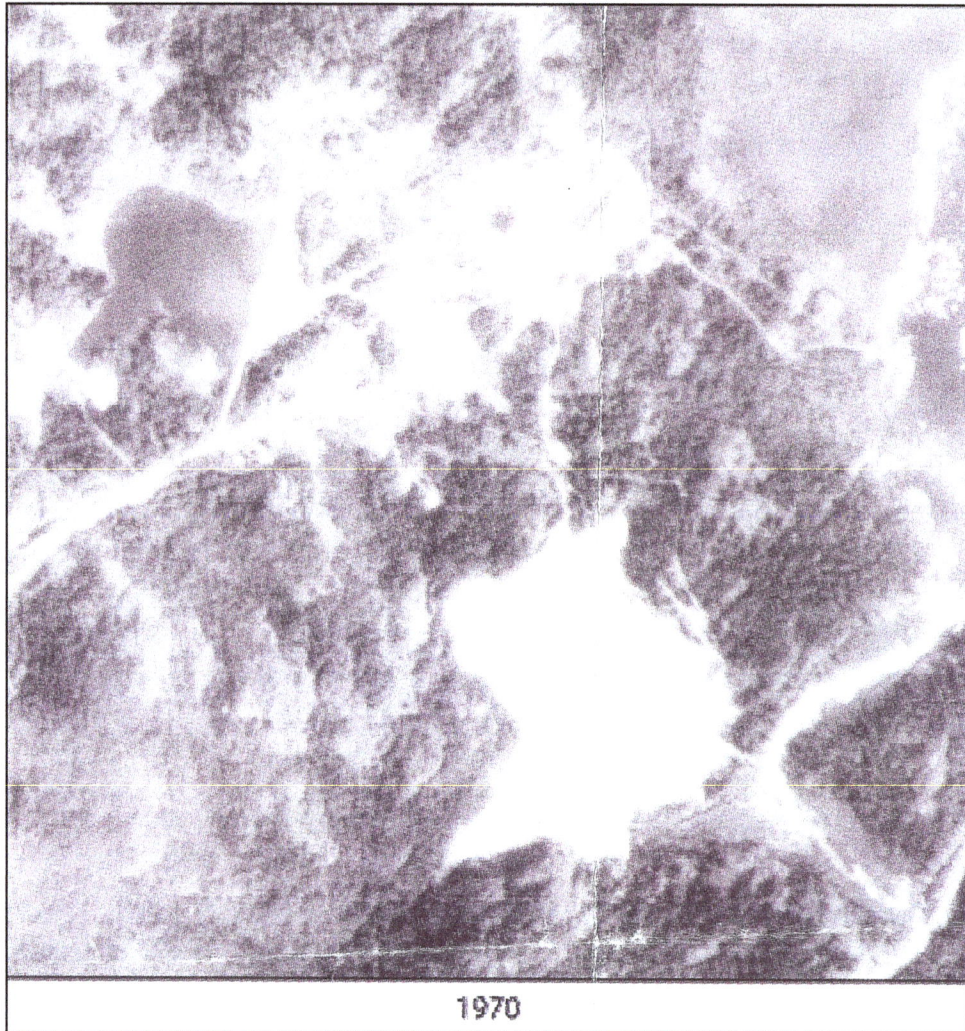

1970

Figure 8 - Section 3.2: 1970 Air Photo Revealing a Small Land-Bridge on the Waste Site

3.3 Historical Context of Kidd Copper Dam

Mine tailings "are usually disposed of in a tailings dam" (McMaster University, August 2004, Geo3Fe3, Assignment #9 Mine Tailings). One reason the tailings dam was constructed may have been to lessen the toxic wastes' acidity by dilution. "The high concentration of metal in mine tailings reacts with rain water and oxygen in the atmosphere, producing sufuric acid which contaminates surface and ground water" (Fresques, 2004, p. 1). "All mining operations produce a significant volume of waste products that are usually disposed of in a tailings dam. Depending on the ore mineralogy and mining process most mine waste comprise a variable mixture of benign silicates, reactive sulphides, and precipitates. Exposure of the reactive sulphides to oxygenated waters in the tailings pond results in a chemical reaction that eventually leads to the formation of an acid-rich leachate (acid mine drainage – AMD)" (McMaster University, 2004, Geo3Fe3, Assignment #9 Mine Tailings).

"Some time between 1977 and 1983, the retaining dam at the southeast corner of the site gave way and tailings deposits were carried into a lower-lying valley" (Shang, *et al.*, 2004, p. 4, 5). The velocity of the water created an upsurge of deposits when the breach occurred. Threshold velocity is "the minimum velocity at which wind or water in a specified location, and under given conditions, will begin to move particles of sand, soil or other material" (MacDonald, *et al.*, 2003, p. 430). The water overflowed the tailings. Sand, liquid water and the mine tailings (slag) mix in random order due to the velocity of the water overflow. The mine tailings that were sequentially deposited through the 1960's were uplifted and redeposited in 1970's because of a natural dam break on the site. The waste-site has been inactive for more than the last 20 years. The mine tailings sediment has remained untouched for the last 20 years because the mining production has stopped. Too little profit to fix a broken dam (Studdy, 2004, p. 86).

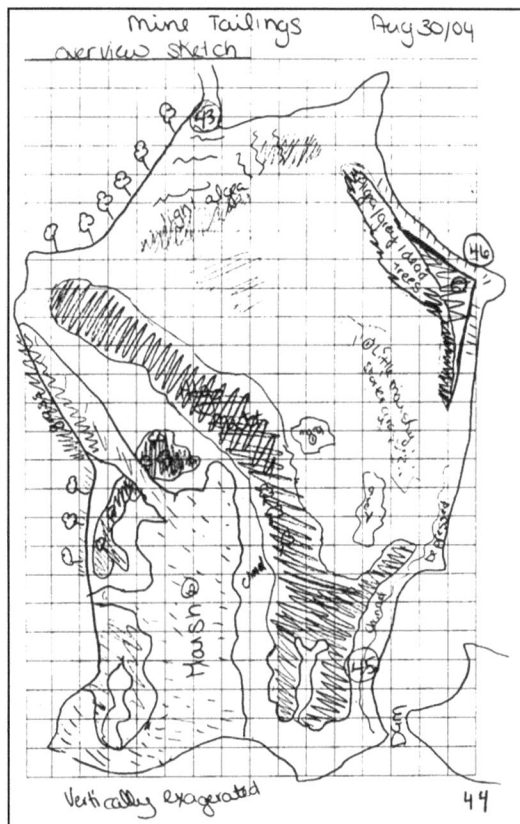

Figure 9 - Section 3.3: Sketch Map showing the Dam on the SE corner of Kidd Copper Property

Figure 9 - Section 3.3 is a sketch map drawn by Karen De Medeiros on August 30, 2004 showing the location of the dam on the SE corner of Kidd Copper property   (De

Medeiros, 2004, p. 44). Karen de Medeiros was one of the thirty-two students present at Kidd Copper property on August 30, 2004 to complete a Mine Tailings Exercise assigned by Dr. Bill Morris from McMaster University. She records the dam's coordinates on her sketch map as 137200 546300. Karen's coordinates of the waste site are as follows: NW corner: 0467113; NE corner: 5137600; SW corner: 467000 and SE corner: 546300.

History of the Dam in the early 1970's is found in Appendix I where a detailed list of field book references to the dam on the Kidd Copper property are recorded. These records are from McMaster University Geology students describing the dam from their observations recorded in the field entries. The composition of the dam was likely "blocks of rock", "an anthropogenic dam" see Zelek's and Rosamond's entries respectively in Appendix I – Section 27, 17. "Miners built dam to seal tailings pond" is referenced in Hynes'entry in Appendix I – Section 10. Note the retaining dam at the southeast corner breaks on the Kidd Copper property in the early 1970's after heavy rainfall for two weeks and the outflow lasted one hour (McMaster University, 2004, staff, *personal commun.*).

3.4 Retaining Dam at the Southeast Corner Broke

"Dam broke ~1970s" (Enders, 2004, p. 49). Heather Robertson another student provides these coordinates for the dam in her field book entry for August 30, 2004: "0467187 5137453" (p. 51, 53). She writes "dam broke and area exposed to air (30 years ago)" (p.50). "Some time between 1977 and 1983, the retaining dam at the southeast corner of the site gave way and tailings deposits were carried into a lower-lying valley" (Shang, *et al.,* 2004, p. 5). This extended area was well photographed refer to Figure 10 - Section 3.6.

Catastrophic Event. Some floods occur rapidly and die out just as quickly. "A flash flood is "a brief but powerful surge of water … over a surface" (Allaby & Allaby, 2003, p. 208). "*Flash floods* are local, sudden floods of large volume and short duration, often triggered by heavy thunderstorms" (Plummer, McGery & Carlson, 2003, p. 246). Or floods can be "caused by heavy convectional rainfall of short duration" (Allaby & Allaby, 2003, p. 208). Flashflood has a "very high discharge" (Plummer, McGery & Carlson, 2003, p. 557).

Flooding is the "inundation of the land surface by water" (Bates & Jackson, 1984, p. 187). When the dam broke the water would have swept the mine tailings up, there would have been an upswelling from the increased velocity of the water. "As the dam breaks it clogs with sediment and flows more slowly → a floodplain is created. A change occurs in the base level (gets lower) and creates steeper slumping channels within the other channels" (Jack, 2004, p. 38). When the retaining dam on Kidd Copper broke the water flooded the mine tailings waste site.

Information outlined for a possible dam construction including a composite of a spigot model for later experimental and reproducibility projects are included in Sections 14.9-11 and in Appendix F such as computer models and information on embankments.

3.5 Step 2: Upsurge of Mine tailings Waste Sediments

"During the processing of metal ores the waste material (called *tailings*) is pumped to holding ponds. The result is a tailings pile that may contain significant amounts of sulfide minerals. During, or after cessation of, mining operations rainwater percolating through the tailings can become acidified" (Eby, 2004, p. 353). 'Some materials may be removed from soils and rocks by simple dissolution and transport in pore water" (Retallack, 1990, p. 139). "In abandoned mining districts spoils and tailings are significant sources of acid drainage. The impact, treatment, and control of acid drainage from these abandoned sites is in part a function of the site history" (Eby, 2004, p. 353).

The impoundment area was a "tract of land covered by water during the highest known flood" (Bates & Jackson, 1984, p. 187). "During floods both capacity and competency increase. Therefore, the greatest erosion and sediment transport occur during these high-water periods" (Tarbuck & Lutgens, 1999, p. 244). Many geologic word definitions can be found in the data dictionary, refer to Appendix A. "During floods, flood plains may be covered with water carrying suspended silt and clay. When the floodwaters recede, these fine-grained sediments are left behind as a horizontal deposit on the flood plain" (Plummer, McGery & Carlson, 2003, p. 237). Refer to Figure 10 - Section 3.6 where a significant extended area was noted on the Kidd Copper property impoundment area in 1983. A significant amount of sedimentation was deposited in a very short period of time, roughly 13 years.

3.6 1983 Topography Has Substantially Changed

"Landform [GEOGRAPHY] All the physical, recognizable, naturally formed features of land, having a characteristic shape; includes major forms such as a plain, mountain, or plateau, and minor forms such as a hill, valley, or alluvial fan" (Parker, 1997, p. 198). The major landform is the impoundment of mine tailings which substantially changed in geomorphologic shape due to the dam breaking and carrying the tailings into "a lower-lying valley" Shang, *et al.*, 2004, p. 5). The rectangular-shaped tailings deposit, in1983, shows a relatively 'new' landscape feature on the Kidd Copper Property. Refer to Figure 10 - Section 3.6 an "area qualitative map… that shows the existence of a geographic class within areas on the map. Colors, patterns, and shades are generally used. Examples are geology, soil, and land-use maps" (Clarke, 2003, p. 305).

In 1970, the noticeable landform was the land bridge. Thirteen years later (a short period in the geologic time scale) the topography of the area had dramatically and quickly changed due to a catastrophic event when the dam broke. This dramatic change was not due to a slow erosional process over time; in fact, the "erosion surface [is] a land surface shaped and subdued by the action of erosion, esp. by running water" (Bates & Jackson, 1984, p. 170).

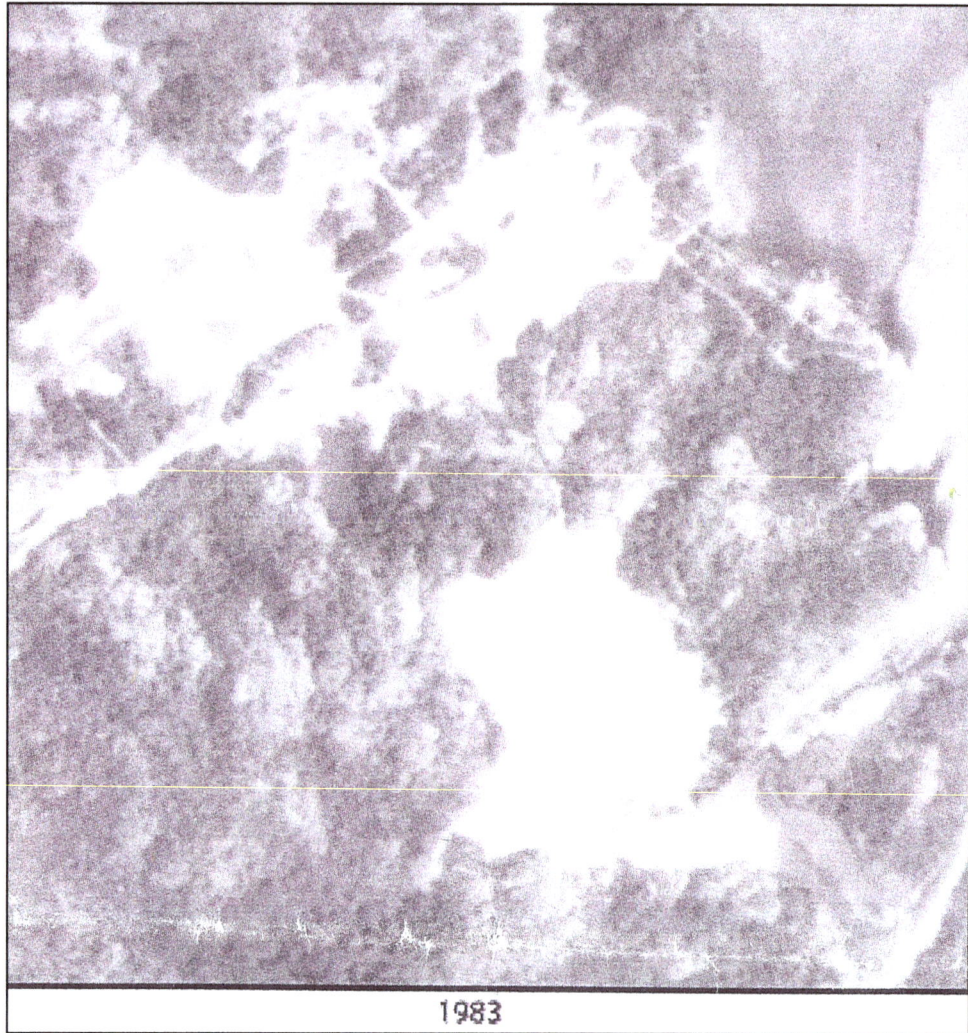

1983

Figure 10 - Section 3.6: 1983 Air Photo Showing Extended Tailings Area - 13 years

The topography of Kidd Copper changed when the dam broke significantly. Figure 10 - Section 3.6 shows the site area on Kidd Copper where the dam broke over the mine tailings. "Noticeably, from 1970 to 1983 a small land bridge was replaced with a larger, rectangular-shaped tailings deposit at the SE corner of Kidd Copper Property" (Rudd, 2004, Tailings Exercise Marking Sheet), "extended area of the tailings deposit" (Shang, *et., al.*, 2004, p. 5).

3.7 Stream Erosion was a Minor Contributor

Topographic texture is "the disposition or general size of the topographic elements composing a particular typography. It usually refers to the spacing of drainage lines in stream-dissected areas" (MacDonald, *et al.*, 2003, p. 432). "By 1992, the extended area of tailings had increased in size. Field observations indicate that this was probably due to stream erosion of the tailings in the main accumulation area followed by deposition of the eroded materials in the lower part of the the valley. Given the relatively porous nature of much of the tailings deposit, the majority of stream erosion probably occurs during snowmelt when large volumes of water are available and the tailings deposits are partially frozen" (Shang, *et al.*, 2004, p. 5).

"Ground water, the water that lies beneath the ground surface, filling the pore space between grains in bodies of sediment and clastic sedimentary rock, and filling cracks and crevices in all types of rock" (Plummer, McGery & Carlson, 2003, p. 260).

"The erosive process that is usually most effective on a rocky stream bed is *abrasion*, the grinding away of the stream channel by the friction and impact of the sediment load. Sand and gravel tumbling along near the bottom of the stream wear away the stream bed much as moving sandpaper wears away wood. The abrasion of sediment on the stream bed is generally much more effective in wearing away the rock than hydraulic action alone. The more sediment a stream carries, the faster it is likely to wear away its bed" (Plummer, McGery & Carlson, 2003, p. 230). "A stream flowing over limestone, for example, gradually dissolves the rock, deepening the stream channel. A stream flowing over other sedimentary rocks, such as sandstone, can dissolve calcite cement, loosening grains that can then be picked up by hydraulic action" (Plummer, McGery & Carlson, 2003, p. 230).

In the section on *Data Analysis* of the aerial photographs, Shang, Morris and Howarth record that the aerial photograph "in 1977, the major mine-tailing deposit (outlined in red) displays almost the same distribution as the present day" (Shang, *et al.*, 2004, p. 4). Thus, erosional processes like stream erosion were not as significant for the depositional changes visibly noted on Kidd Copper property as the catastrophic event.

3.8 1992 Data No Major Distribution Change After 1983

The variance of change is relatively small from 1983-1992 when compared to the magnitude extension of the tailings from 1970-1983 because of the dam break.

Comparing Figures 8 – Section 3.2; 10 - Section 3.6 and 11 - Section 3.9 of air photos from 1970-1992, "we can see the evolution of the Kidd Copper tailings site over time" (Shang, *et al.*, 2004, p. 4). The "images display the presence of a number of distinct spatial regions within the tailings. The reflectance data were used to characterize the mine-tailing surface at the Kidd Copper site" (Shang, *et al.*, 2004, p. 4). Unlike the dramatic change in 1983 to the Kidd Copper property the site area of the impoundment showed virtually no change in 1992. As earlier stated by Shang, Morris and Howarth the major mine tailing deposit was almost the same in 2004 as in 1977.

3.9 Landscape Changes as a Result of the Flood

"Although catastrophism had wide acceptance in the scientific community, it was short-lived and was supplanted by Hutton's principle of *Uniformitarianism*. A revived form of *new catastrophism* has emerged in recent years with the recognition of the intense global environmental effects of major impacts of extraterrestrial bodies and eruptions from *supervolcanoes*" (MacDonald, *et al.*, 2003, p. 84). "*Cataclysm* – Any of a number of geologic events, such as an exceptionally violent earthquake, that causes sudden and extensive changes in the earth's surface. An overwhelming flood of water (deluge) that spreads over a wide area of land also is sometimes referred to as the cataclysm" (DePree & Axelrod, 2003, p. 121).

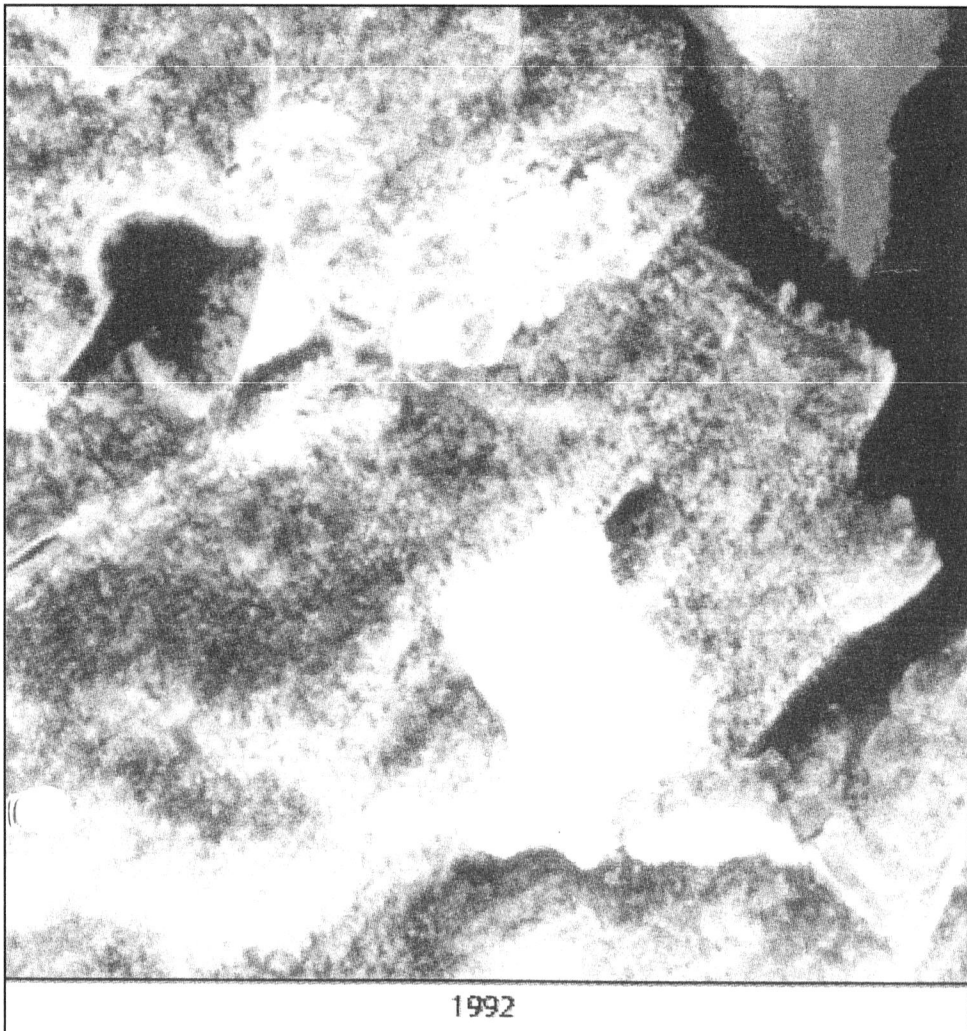

1992

Figure 11 - Section 3.9: 1992 Air Photo of the Kidd Copper Mine Tailings Waste Site

A stage defined in geology is "(1) a developmental phase of an erosion cycle in which landscape features have distinctive characteristic forms. (2) a phase in the historical development of a geologic feature" (Licker, 2003, p. 352). The process on Kidd Copper to accumulate the rectangular-shaped tailings deposit was not slow, nor was it erosional; rather the accumulation was due to a catastrophic event. The dam broke and in a short period of time (<13 years) the larger rectangular-shaped deposit accumulated. "Cuvier, a French biologist, supported the doctrine, explaining most extinctions and landscape changes as consequences of the Biblical deluge" (MacDonald, *et al.*, 2003, p. 84).

Without knowing the history of the Kidd Copper site one could mistakenly interpret the *stage* as "a time-stratigraphic unit ranking below series and above chronozone, composed of rocks formed during an age of geologic time" (Licker, 2003, p. 352). Instead, the facts surrounding Kidd Copper support the "Catastrophism – theory that the geological features of the Earth were formed by a series of sudden, violent 'catastrophes' beyond the ordinary workings of nature" (Machines Plc., 2003, p. 100). "Water can change our landscape in the span of just a few hours" (Albin, 2004, p. 34). Figure 11 - Section 3.9 shows that in 1992 the rectangular-shaped deposit did not continue to grow, as it suddenly formed in 1983.

3.10 Evolution of Mine tailings established During a Short Period in Time

A clear evolution of mine tailings is established during a short period in time: (1) mine tailings waste accumulates from 1936-1970 in the six impoundments, of which the primary site south of Perch Lake on Kidd Copper property was one; (2) the retaining dam breaks; (3) the water overflows the impoundment area of mine tailings; and (4) redepositing the sediment again onto the impoundment area and onto a lower valley. This dramatic change took place in a relatively short period of geologic time.

All operations on the mine disposal site stopped. There was no folding of rocks or intrusions detected on the Kidd Copper property site area. Garrels & Mackenzie note that the lack of folding of rocks contributes to a closed system "because of the assumption of equal probability of exposure to erosion of all equal masses, the same source materials of the total mass would always be available to erosion, so that the result of deposition of a constant ratio of rock types, derived from a different ratio of source rock types, results in changes in the proportions of rock types as a function of age" (1971, p. 273). Observed proportional differences were not a function of long-age on Kidd Copper but rather occurred rapidly in time.

3.11 Rock Cycle and Word Definitions

A few definitions are introduced to provide more evidence and explanations of the geologic findings at Kidd Copper. Refer to Appendix K for more on sedimentary rocks. "A rock may be thought of as an aggregate of minerals" (Knopf, 1979, p. 15).

"Primary rocks [PETR] rocks whose constituents are newly formed particles that have never been constituents of previously formed rocks and that are not the products of alteration or replacement, such as limestones formed by precipitation from solution" (Licker, 2003, p. 274). "Rework [GEOL] any geologic material that has been removed or displaced by natural agents from its origin and incorporated in a younger formation" (Licker, 2003, p. 301). More information on particles found in Appendix B.

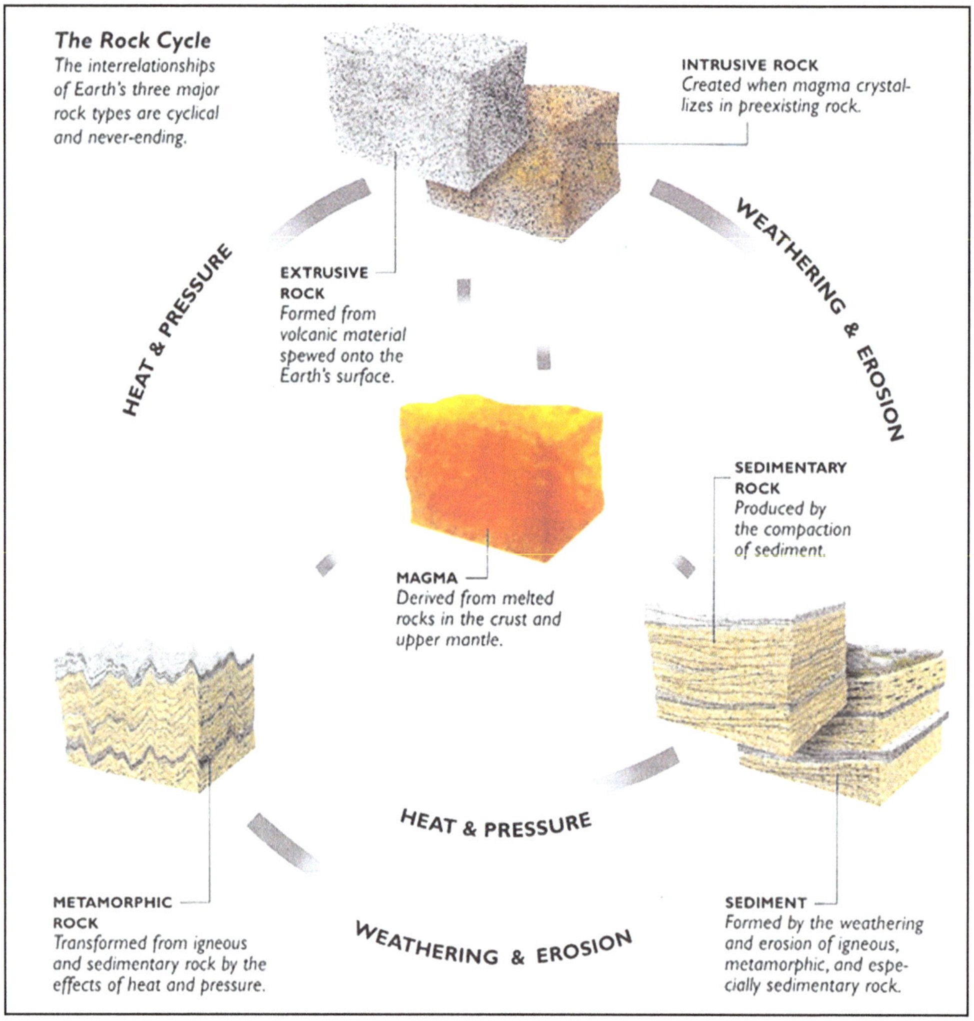

Figure 12 - Section 3.11: The Rock Cycle. (Discovery Books, 1999, p. 21)

"Rock cycle [GEOL] the interrelated sequence of events by which rocks are initially formed, altered, destroyed, and reformed as a result of magmatism, erosion, sedimentation, and metamorphism" (Licker, 2003, p. 308).

"Sedimentary rock [PETR] a rock formed by consolidated sediment deposited in layers. Also known as derivative rock; neptunic rock, stratified rock" (Licker, 2003, p. 329).

> "*Sedimentary rocks* are formed by a process that we could actually see, if we could stay long enough to watch it happen. Wind and water flowing across Earth's surface can weather both igneous and metamorphic rocks. This process breaks these rocks into small pieces known *as sediment*. As sediment accumulates and compacts into a solid mass, a sedimentary rock is formed. Again, if the sedimentary rock is subjected to extreme heat and pressure, it can be converted into igneous or metamorphic rocks. All three rocks are linked through this continuous process known as the *rock cycle*" (Albin, 2004, p. 14).

There are three types of rock in the rock cycle. The "interrelationships of Earth's three major rock types are cyclical and never-ending" (Discovery Books, 1999, p. 21). "The rock cycle is a consequence of plate tectonics. One complete cycle takes about 100 million years" (Kump, Kasting & Crane, 1999, p. 121). The three major rock types are: (1) Intrusive/Igneous; (2) Sedimentary; and (3) Metamorphic rocks. Rapid stratification is found in the major rock types. Layers are found in magma, which is igneous rock, as evidenced by the example in Figure 12 - Section 3.11.

3.12 Kidd Copper Depositional Setting Answers Sediment Bed Question

In 2004, mainstream geologists remained perplexed as to why a sediment bed had various geometries during sediment transport. The Kidd Copper property research provides a solid answer to their question.

"Why a sediment bed is molded into complicated geometries instead of simply remaining flat during sediment transport" (Harms, *et al.*, 2004, p. 8). "The basic problem in analyzing the development of some initial disturbance into an equilibrium bed configuration, whether in a formal stability analysis or in a physical model ... is that in order to obtain a differential equation for bed height as a function of time and position, which might then be solved to account for development of the bed configuration, a relation for local sediment transport rate in terms of flow and sediment characteristics must be known or assumed" (Harms, *et al.*, 2004, p. 10).

Up to now, "no very satisfactory answer to the intriguing question of why a sediment bed is molded into complicated geometries instead of simply remaining flat during sediment transport" has been given (Harms, *et al.*, 2004, p. 8). The reason for their perplexity is that their reasoning was flawed.

Now a very satisfactory answer can be given to the intriguing question of why a sediment bed is molded into complicated geometries instead of simply remaining flat during sediment transport. From the information acquired from Kidd Copper the sediment beds are known to be folded – distorted – layered – heterogeneous; rather than flat – homogeneous.

The answer is found upon the quickness that the event occurs. The events causing this disfiguration are catacalysmic. The sediment beds settle rapidly, which causes the layers. The origin of the layers of sediment was not a slow build up through time but rather a fast deposit of sediment from a catastrophic event. Insufficient time is available for the sediment to find a smooth, flat surface to rest upon. The sediment falls quickly into place and stays in the position, scattered, and disheveled. This is observed by the randomness of the distribution pattern on all of the lithographical logs taken on Kidd Copper Property.

The observations found at Kidd Copper confirm the upswelling of sediment was because of a flood event. A sediment bed is molded into complicated geometries because of the randomness of the mixing process of the Great Deluge that flooded the Earth and settled its many layers as random deposits of sediment. "Shells which were once crawling on the bottom of the sea, now standing nearly 14, 000 feet above its level" (Darwin & Jones, 2001, p. 286).

In the case researched at Kidd Copper the sediment resettled in similar amazement, totally unexpected from what would be expected without knowing the mechanism and force of the transport. The tailings were mixed and distributed in random order. Kidd Copper mine tailings did not fall in even layers because stratification is a primary structure. The stratified features on Kidd Copper formed shortly after the flood water mixed the tailings and redeposited them as beds on the impoundment. This first deposit of the re-mixed tailings sediment is a primary structure.

The Kidd Copper research answers why a sediment bed is molded into complicated geometries because primary structures do not have to be horizontally layered; instead the research shows that primary structures can be molded into complicated geometries. Mixing of sediment does not equate with flat transportation or an equal distribution of sediment. In the case of Kidd Copper, the mixing of sediment re-distributed irregular layers rather than flat.

4.0 Mechanism for Uplift

The dominant support mechanism for upswelling and transportation of the sediment particles on Kidd Copper Proptery is the water released from the dam breaking.

4.1 Velocity of Water

"A flood wave travels with a velocity about 1.67 times that of the water itself" (Dingman, 2002, p. 427). "During floods, even in large rivers, flow velocities can become great enough for the upper flat bed phase to be present" (Harms, Southard, Spearing & Walker, 2004, p. 32). "The distance water travels in a stream per unit time is called the stream velocity. A moderately fast river flows at about 5 kilometres per hour (3 miles per hour). Rivers flow much faster during flood, sometimes exceeding 25 kilometers per hour (15 miles per hour)" (Plummer, McGery & Carlson, 2003, p. 227).

"During floods a stream's discharge and velocity increase, usually as a result of heavy rains over the streams drainage basin. Flood discharge may be 50 to 100 times normal flow. Stream erosion and transportation generally increase enormously as a result of a flood's velocity and discharge. Swift mountain streams in flood can sometimes move boulders the size of automobiles. Flooded areas may be intensely scoured, with riverbanks and adjacent lawns and fields washed away. As floodwaters recede, both velocity and discharge decrease, leading to the deposition of a blanket of sediment, usually mud, over the flooded area" (Plummer, McGery & Carlson, 2003, p. 230).

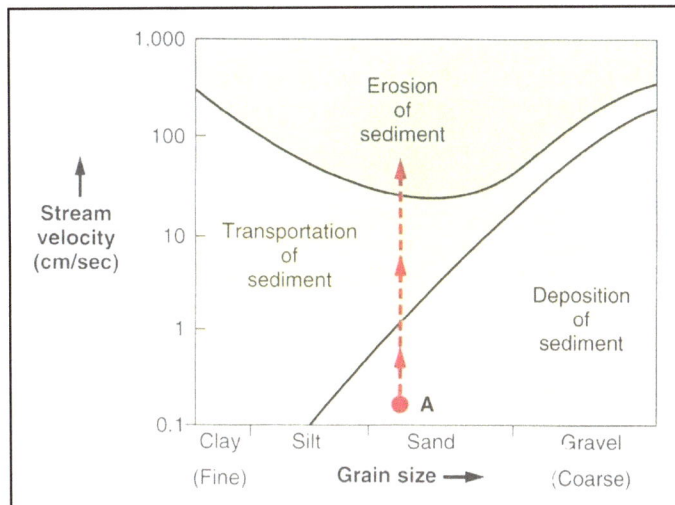

Figure 13 - Section 4.1: Logarithmic Graph Showing Stream Velocities.

In Figure 13 - Section 4.1"a logarithmic graph showing the stream velocities at which erosion and deposition of sediment occur" (Plummer, *et al.*, 2003, p. 228). "For each grain size, these velocities are different. The upper curve represents the minimum velocity needed to erode sediment grains. This curve shows the velocity at which previously stationary grains are first picked up by moving water" (Plummer, *et al.*, 2003, p. 227). Site observations at Kidd Copper, compiled in Table 10 - Section 11.9 show the clay to have a light and fine-grained texture, which would be easily picked up and transported.

"Velocity is the key factor in a stream's ability to erode, transport, and deposit. High velocity (meaning greater energy) generally results in erosion and transportation; low velocity causes sediment deposition" (Plummer, *et al.*, 2003, p. 227).

Sturzstroms are Examples of High Velocity. "To conclude, rocks slide and fall under gravity and the broken debris form sturzstroms. The unusually high velocity of sturzstroms occurs when debris volumes exceed $10^6 - 10^7$ m^3, and the runoff length tends to increase with the rockfall volume" (Hsü, 2004, p. 85).

Grand Banks Sediment. "After the termination of the Grand Banks sediment-gravity flow at the foot of the continental slope, sands and muds in suspension flowed downslope to break the cables which had been laid across the abyssal plain, as far as 800 km distant from the origin of the slide. At the most distal site of the cable break, the speed was reduced to less than 10 m s^{-1}. The deposition of a meter-thick turbidite at that site has been confirmed by coring" (Hsü, 2004, p. 90).

Further investigation into the exact velocity can be carried out by applying the Manning equation, which is beyond the scope of this report. Of interest, see Section 14.11 and Figure 68 - Appendix G where more information on dam structures and a mathematical model for assimilating a replication for a spigot dam is available. This information will provide the basis to run future tests to replicate the stream velocity and other factors of interest but are beyond the scope of this report. The provisions for these projects have been included only for the consideration of future reproducibility.

4.2 Fluid Lift Force

"The dominant support mechanism during flow is the strength of the continuous phase (water plus clays plus other relatively fine sediment that forms the matrix of the final deposit)" (Lewis, 1984, p. 29). The fluid turbulence dispersed clasts.

In Principles of Sedimentary Deposits, Chapter 7–
Sedimentary Processes: Physical, Biological and Chemical; Gerald
Friedman, John Sanders and David Kopaska-Merkel write: "In order
to move a particle that is locked in a bed among other particles and
hence not free to roll easily, a fluid must transfer enough energy to
overcome the downward force exerted by gravity on the net weight of

the particles. In the case of water, this net weight is known as the *immersed weight*. A fluid can energize a particle directly and acting alone in one of four ways: (1) *fluid lift force*, (2) changes in fluid pressure and surges related to wave action, (3) fluid impact, and (4) support from flow within turbulent eddies. Fluid lift force: The flow of a fluid above a stationary particle creates a fluid lift force which is related to the airfoil principle that enables airplanes to fly. Careful observation of sand in wind tunnels has shown that the lift force of the air can raise quartz particles of sand size. Three comments need to be made: (1) Not only are sand particles lifted by a strong-enough air current, but at the same time they are subjected to a rotational couple so that they spin vigorously. (2) The lift force can be shown to stop operating by the time the clearance beneath the lifted particle is about half the diameter of the particle. (3) Despite the effects of item (2), lifted particles can be observed to rise to heights equal to several particle diameters. This is interpreted to mean that the lift force accelerates the particles and that once they are moving, the inertia of the particles keeps them going for a short distance after the effectiveness of the lift force stops and before the drag created by viscosity becomes totally effective. Once the effect of viscosity overtakes the inertia, the particle stops rising. If nothing else happens, fluid impact will push the particle in the direction of local flow, and gravity will pull it back to the bed" (1992, p. 239).

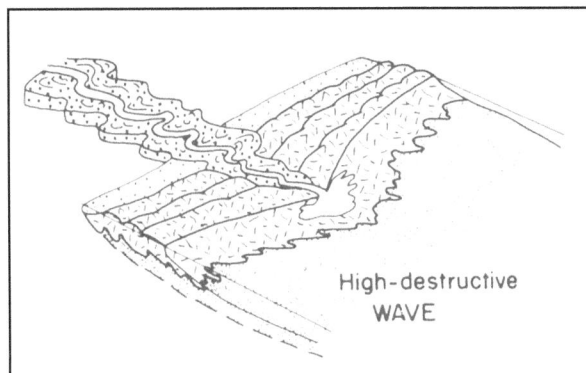

High-destructive
WAVE

Figure 14 - Section 4.2: Water from the Dam is the Dominant Support Mechanism

The inertia of water in a flood can create highly destructive waves as illustrated in Figure 14 - Section 4.2. "Transport of particles in water is by far the most significant of all transport mechanisms. Water flows on the land surface in channels and as overland flow" (Nichols, 2003, p. 37). "These flows may be strong enough to carry coarse material along their base and finer material in suspension" (Nichols, 2003, p. 37).

4.3 The Rate of Water Movement through Soils

Soil is "a layer of organic and inorganic weathered material that accumulates at the Earth's surface. The soil plus subsoil together make up the *regolith*.... Soils reflect the influence of climate, relief and nature of the parent material as well as the physical, chemical and biological processes involved" (MacDonald, *et al.*, 2003, p. 404).

"Two types of water vapor movement occur in soils, *internal* and *external*. Internal movement takes place within the soil, that is, in the soil pores. External movement occurs at the land surface, and water vapor is lost by surface evaporation" (Brady & Weil, 2002, p. 204). "Water tends to move from areas of high (less negative) water potential to areas of low (more negative) water potential" (Isaacs, *et.al.,* p. 835). The following example of Sturtzoms and the Grand Banks sediment demonstrate the factors of the volume of water moving through the soil and what resultants occur.

4.4 Strength of Turbidity Currents

"Sediment is transported in two fashions, depending on the type of flow. If it is carried smoothly and evenly across the bottom of the streamed by a laminar flow, it is known as the bed load. If it is carried in short spurts, in a bounding manner across the river bottom by a turbulent flow, it is known as saltation" (Albin, 2004, p. 34). "The transition from a laminar to a turbulent flow is difficult to describe" (Abulnaga, 2002, 2-3-2). "Ground waters exhibit laminar flow; turbulent flow is characteristic of most other natural fluids – wind, water in streams, lake and ocean water currents" (Garrels & Mackenzie, 1971, p. 177).

"Turbulent flow takes place when the velocity of the fluid is great enough to overcome the viscous forces that maintain continuity of the fluid" (Garrels & Mackenzie, 1971, p. 177). "That turbidity currents might be responsible for cutting submarine canyons and producing the associated submarine channels" (Blatt, Middleton & Murray, 1972, p. 168). The force of the dam breach uplifted a substantial amount of the initially deposited mine tailings waste. The sediment was uplifted from the accumulated impoundment on the Kidd Copper Property. "Sandstone has visible grains like coarse sugar, up to almost 1/10 inch (2mm). To move particles of this size, the water must have some force" (Busbey III-Coenraads-Willis-Roots, 1999, p. 51). Figure 15 - Section 4.4 taken from Lewis, 1984, p. 27 shows the fluid turbulence. Mixtures of the matrix of sediments are carried by the velocity of the turbidity current. The force was strong and as the sediments bounced around the clasts dispersed.

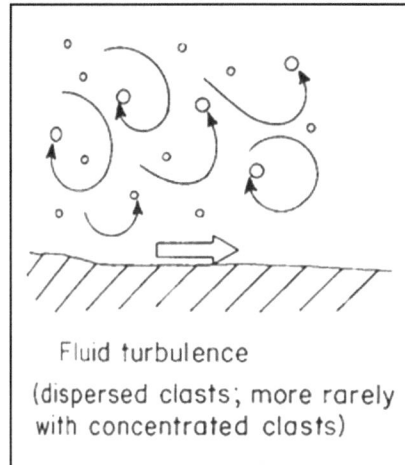

Fluid turbulence
(dispersed clasts; more rarely
with concentrated clasts)

Figure 15 - Section 4.4: Fluid Turbulence

Turbulent flow is "the motion of a liquid or gas turbulent when the velocity at any point changes direction and magnitude in a random manner. Its diffusive action distinguishes turbulence from an irregular wave motion, i.e. turbulence tends to increase rates of transfer of momentum, heat and water vapour. When the velocity of a fixed-size flow is increased, a smooth or laminar motion will become unstable and turbulent" (MacDonald, *et al.*, 2003, p. 438). Turbulent flow is an "eddying, swirling flow in which water drops travel along erratically curved paths that cross the paths of neighboring drops" (Plummer, McGery & Carlson, 2003, p. 565). "*Turbulent flow* with three-dimensional motion in which the velocity at any point fluctuates with time and *streamlines* may not be parallel. Occurs at *Reynolds number* > 200." (Kearey, 2001, p. 278).

"Turbulent flow is characteristic of most other natural fluids – wind, water in streams, lake and ocean water currents. Turbulence spreads throughout the fluid if the Reynolds number of the flow is exceeded. The Reynolds number is a dimensionless quantity defined as:

$$R_e = dvh/n$$

Where *d* is the density of the fluid in grams per cubic centimeter, *v* is the velocity of the fluid in centimeters per second, *h* is the depth in centimeters, and *n* is the viscosity in poises. Reynolds numbers for large bodies of water range from about 300–600; to exceed these values requires only small velocities; thus turbulent flow is characteristic of most water bodies" (Garrels & Mackenzie, 1971, p.177). *Turbulence intensity* is "the ratio of *turbulence* to mean flow velocity in *turbulent flow*" (Kearey, 2001, p. 278).

"Solid particles, when set in motion in stream waters or other fluids, follow laws of fluid flow. Fluids flow either laminarly or turbulently. Laminar flow occurs when fluids are moving relatively slowly, whereas turbulent flow takes place when the velocity of the fluid is great enough to overcome the viscous forces that maintain continuity of the fluid" (Garrels & Mackenzie, 1971, p. 177). "Ground waters exhibit laminar flow" (Garrels & Mackenzie, 1971, p. 177). "In laminar flow, all the fluid particles move in parallel streams; turbulent flow is characterized by complex swirls and eddies" (Murck, 2001, p. 127). "A particle suspended in a stream flowing laminarly would follow a smooth path, whereas if the stream were flowing turbulently, the particle path would describe a series of eddies.

"*Autosuspension* the condition in a fluid, only approached in *turbidity currents*, in which the hydraulic forces of motion provide sufficient energy to maintain a suspended load in the flow" (Kearey, 2001, p. 20). "Small silt and clay particles can be suspended in the water, and some sediment is so small that it dissolves completely" (Albin, 2004, p. 34).

4.5 Turbidity Currents Features

"As distance from a boundary and velocity increase, the inertia of the moving water increasingly overcomes viscous friction and the flow paths of individual water 'particles' increasingly deviate from the parallel layers of laminar flow. At relatively modest distances and velocities, all semblance of parallel flow disappears and the water moves in highly irregular eddies. This is the phenomenon called *turbulence*" (Dingman, 2002, p. 545). "*Turbulence* a series of quasi-random motions at different scales in a fluid; the root mean square of the instantaneous fluctuating velocity component in *turbulent flow*" (Kearey, 2001, p. 278). "*Bedform* the shape of the surface of a bed of granular sediment produced by fluid flow over it" (Kearey, 2001, p. 26, 27).

Erosion in the trough of a bedform is depicted in Figure 16 - Section 4.5 from the section of processes and sedimentary structures in Nichols' Sedimentology & Stratigraphy, (p. 55). This figure is the pictorial "shape of the head of a density (or turbidity) current, showing the pattern of flow within the head" (Blatt, Middleton & Murray, 1972, p. 167).

"In the *head*, the flow pattern … that sediment will be swept off the bottom and circulated round to the back of the head, where it may either settle back into the head (presumably the fate of the coarser grains) or be lost in eddies torn from the rear of the head. In turn these eddies will be added to the zone of entrainment of the body of the flow and swept down current. The coarsest sediment may be held in suspension in the head until there is a general deceleration produced by flattening of the slope or dilution of the current" (Blatt, Middleton & Murray, 1972, p. 169).

Copyright 2010 Star-Ting, Inc. All rights reserved. Worldwide.   1.877.896.7292   http://www.star-tingpublishing.com

A current is "the concentrated flowing of water" (Bates & Jackson, 1984, p. 124). Whereas, turbidity currents or "density current in which the excess density is created by suspended sediment" (Lucchi, 1995, p. 249).

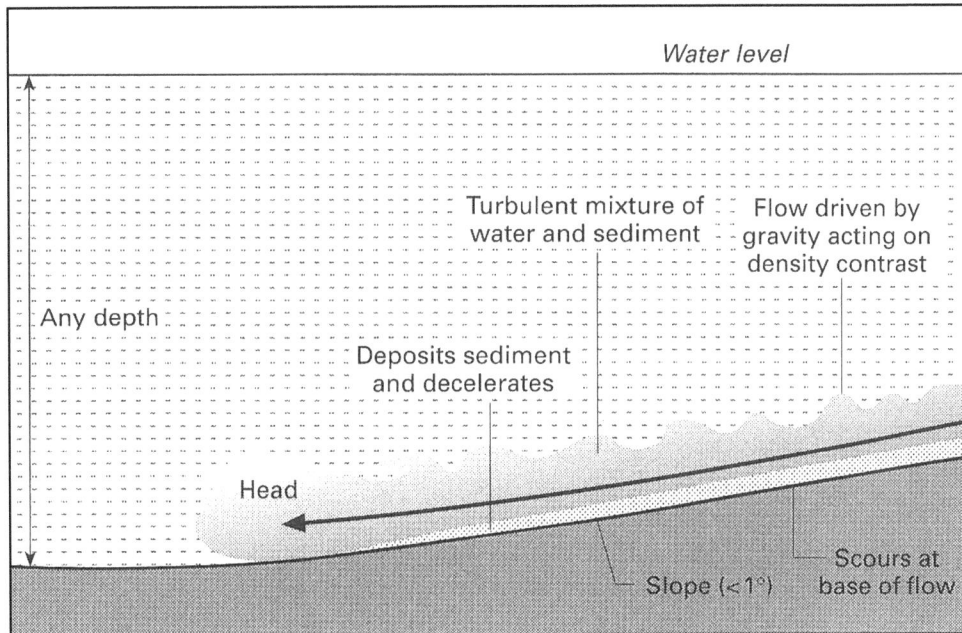

Water level

Turbulent mixture of water and sediment

Flow driven by gravity acting on density contrast

Any depth

Deposits sediment and decelerates

Head

Slope (<1°)

Scours at base of flow

Figure 16 - Section 4.5: Turbidity Current Features; Headed into a Mine Tailings Deposit

A general belief is "that *submarine canyons* are at least partially eroded by turbidity currents" (MacDonald, *et al.*, 2003, p. 438). A suspended load defined by Lucchi, 1995 is 'the bulk of particles carried in suspension by a flow" (p. 248). Turbidity "currents flow downslope at very high speeds and spread along the horizontal. The distance covered is determined by the turbulence and velocity of the current" (MacDonald, *et al.*, 2003, p. 438). Suspension current is "gravity driven (density) current in which the density contrast with surrounding fluid is created by suspended particles; cf. turbidity current, pyroclastic surge flow" (Lucchi, 1995, p. 248). See Figure 61 – Appendix B for further information about mean flow velocity.

A turbidity current is "a flowing mass of sediment-laden water that is heavier than clear water and therefore flows downslope along the bottom of the sea or a lake" (Plummer, McGery & Carlson, 2003, p. 565). "*Turbidity currents* are underwater floods. A river flood entering a lake with a suspension which has a density greater than that of lake water will, for example, flow along the lake bottom until its load is deposited" (Hsü, 2004, p. 47). In Figure 16 - Section 4.5 the features of a turbidity current show that as sediments deposit the current decelerates, that a turbulent mixture of water and sediment results by a flow driven by gravity of a slope less than 1°.

"Turbidity current is also a fluid-gravity flow, but the mixing of debris flow and ambient water produces a suspension less dense and less viscous than mud, so that the current flows turbulently. Turbidity currents are not only derived from debris flows, they may be derived directly from a submarine sediment-gravity flow. The sand, silt, and clay particles 'boiled' out of a stream of avalanche debris from the Grand Banks mixed with seawater to form a suspension and this kind of suspension current in water, as postulated by Kuenen and others, flows downslope as a turbidity current" (Hsü, 2004, p. 90).

"Fluid flow over sediment which has already been deposited can result in the partial and localized removal of sediment from the bed surface" (Nichols, 2003, p. 59). Sometimes "these turbulent eddies erode into the bed and create a distinctive erosional scour called a *flute cast*" (Nichols, 2003, p. 59). "Turbulent water flow over a bed surface results in localized eddies even when the bed surface is smooth and flat" (Nichols, 2003, p. 59).

4.6 Flow Slope

Minimum Slope for Flow Downslope. "It is possible to determine that the minimum slope for subaqueous flow of a grain-water mixture is 18 degrees in the inertial regime. The slope might be decreased (a) by presence of a denser interstitial fluid than water, for example, a mud matrix; (b) by the existence of an excess pore pressure acting against the gravity normal stress; (c) by addition of an additional downslope shear stress imposed from a flow above; and (d) by turbulence" *(Blatt, Middleton & Murray, 1972, p. 163)*.

"Density current can flow for many miles without becoming completely mixed with the surrounding water and so losing its identity" (Blatt, Middleton & Murray, 1972, p. 166). "Downstream from the point of separation, at a distance that is commonly 6 to 8 times the height of the step, the zone of diffusion has expanded to the extent that the flow becomes *reattached* to the bottom (Blatt, Middleton & Murray, 1972, p. 127).

In Figure 16 - Section 4.5 a turbulent mixture of water and sediment results by a flow driven by gravity of a slope less than 1°. Some of the mine tailings at Kidd Copper property were swept down current. The "tailings deposits were carried into a lower-lying valley" (Shang, *et al.*, 2004, p. 5). A noticeably large deposit of mine tailings was redeposited downstream, Refer to the 1983 photo, Figure 10 - Section 3.6

4.7 The Load of Streams

"The total material, other than water, that is transported by a stream is called its load. The materials may be moved in solution (*solution load*), in suspension (*suspension or wash load*), or by traction along the bed (*traction* or *bed load*). The rate at which sediment is moved is called the *sediment discharge* and is defined as the sediment load moved past a given cross section of the stream in unit time" (Blatt, Middleton & Murray,

1972, p. 21). "Sediment discharge, i.e. the amount of sediment carried by a stream, measured in kg s^{-1}. Determination of the *bed load* discharge is difficult, however, so estimates of total discharge may be inaccurate" (MacDonald, *et al.*, 2003, p. 143).

"Measurement of sediment load and rate of movement are relatively easy for solution and suspended load but are very difficult for bed load. The only accurate values for bed load transport rates are those obtained by measuring the sediment trapped in reservoirs and those obtained at a few specially constructed 'flumes' on rivers, where sufficient turbulence is generated to take the entire bed load into suspension. Estimates of bed load good to within an order of magnitude may also be obtained by use of theoretical 'bed load formulas,' by use of sediment traps, or by monitoring the movement of subaqueous dunes in rivers. In most studies of denudation rates, however, the bed load is ignored or is conventionally considered to compromise a certain fraction (commonly 10%) of the total load" (Blatt, Middleton & Murray, 1972, p. 21).

"The *sediment load* transported by a stream can be subdivided into *bed load*, *suspended load*, and *dissolved load*. Most of the stream's load is carried in suspension and in solution. The *bed load* is a large or heavy sediment particles that travel on the stream bed. Sand and gravel, which form the usual bed load of streams, move by either *traction* or *saltation*. Large, heavy particles of sediment, such as cobbles and boulders, may never lose contact with the stream bed as they move along in the flowing water. They roll or slide along the stream bottom, eroding the stream bed and each other by abrasion" (Plummer, McGery & Carlson, 2003, p. 231).

"Silt and clay move mainly as wash load in continuous suspension and are either transported rapidly to the mouth of the river or stored on floodplains where they have been deposited by overbank flooding" (Blatt, Middleton & Murray, 1972, p. 102, 103). Overbank flooding would be characteristic of a dam breach as seen on the Kidd Copper Property.

"Coarse to fine sand moves mainly by traction and intermittent suspension, with local and temporary storage in large structures such as dunes and point bars" (Blatt, Middleton & Murray, 1972, p. 102). "Sand grains move by traction, but they also move downstream by *saltation*, a series of short leaps or bounces off the bottom" (Plummer, McGery & Carlson, 2003, p. 232).

"Saltation [GEOLOGY] Transport of a sediment in which the particles are moved forward in a series of short intermittent bounces from a bottom surface" (Parker, 1997, p. 332).

> "*Saltation* begins when sand grains are momentarily lifted off the bottom by *turbulent* water (eddying, swirling flow). The force of the turbulence temporarily counteracts the downward force of gravity, suspending the grains in water above the stream bed. The water soon slows down because of velocity of water in an eddy is not constant;

then gravity overcomes the lift of the water, and the sand grain once again falls to the bed of the stream. While it is suspended, the grain moves downstream with the flowing water. After it lands on the bottom, it may be picked up again if turbulence increases, or it may be thrown up into the water by the impact of another falling sand grain. In this way sand grains saltate downstream in leaps and jumps, partly in contact with the bottom and partly suspended in the water" (Plummer, McGery & Carlson, 2003, p. 231).

"The *suspended load* is sediment that is light enough to remain lifted indefinitely above the bottom by water turbulence. The muddy appearance of a stream during a flood or after a heavy rain is due to a large suspended load. Silt and clay usually are suspended throughout the water, while the coarser bed load moves on the stream bottom. Suspended load has less effect on erosion than the less visible bed load, which causes most of the abrasion of the stream bed. Vast quantities of sediment, however, are transported in suspension" (Plummer, McGery & Carlson, 2003, p. 231, 232).

"Soluble products of chemical weathering processes can make up a substantial *dissolved load* in a stream. Most streams contain numerous ions in solution, such as bicarbonate, calcium, potassium, sodium, chloride, and sulphate. The ions may precipitate out of water as evaporate minerals if the streams dries up, or they may eventually reach the ocean. Very clear water may in fact be carrying a large load of material in solution, for the dissolved load is invisible. Only if the water evaporates does the material become visible as crystals begin to form" (Plummer, McGery & Carlson, 2003, p. 232).

From the impoundment of tailings at Kidd Copper, the load of streams would be ions in solution. These are light enough to remain above the bottom by water turbulence. Findings in Table 10 - Section 11.9 shows the average sediment size found on the Kidd Copper property is very fine sand; therefore sufficient turbulence is generated to take the entire bed load into suspension.

Whenever it rains on the Kidd Copper property an insignificant amount of sediment is eroded or transported to contribute to the tailings heap because from 1983 to 1992 no significant changes were noted in the site area, as Figure 10 - Section 3.6 and Figure 11 - Section 3.9 reveal. No substantial amount of additional strata is being deposited to the mine tailing's ground surface.

4.8 Hydraulic Gradients at the Copper Cliff Impoundment

"Movement of water is driven by a pressure gradient, commonly gravitational, and is retarded by friction with the grains of the rocks" (Garrels & Mackenzie, 1971, p. 52). "The elevated tailings impoundments are a regional ground-water recharge zone where precipitation infiltrates downward through the tailings and migrates radially away from the impoundments. The elevated water table … slopes gently to the southeast but hydraulic gradients increase sharply near the tailings dam" (Coggans, *et al.*, 1991, p. 450). See Appendix G.

"The northern portion of the tailings has a slightly higher elevation than the other parts of the tailings. Increased concentration of sulfides is near the input to the tailings site" (Shang, *et al.*, 2004, p. 9). "Gradient [GEOL] The rate of descent or ascent (steepness of slope) of any topographic feature, such as streams or hillsides" (Licker, 2003, p. 141). "The permeability can be tested by determining the rate at which the water will flow through a given cross-sectional area of a rock sample of a given length" (Garrels & Mackenzie, 1971, p. 52). Permeability equation can be found in Appendix A-60 for future investigation.

"One factor that controls a stream's velocity is the *stream gradient*, the downhill slope of the bed (or of the water's surface, if the stream is very large). A stream gradient is usually measured in feet per mile in the United States… A gradient of 5 feet per mile means that the river drops 5 feet vertically for every mile that it travels horizontally… A stream's gradient usually decreases downstream. Typically, the gradient is greatest in the headwater region and decreases toward the mouth of the stream" (Plummer, McGery & Carlson, 2003, p. 228). On the Kidd Copper Property, the velocity or the force of the flood was strong enough to create an upsurge of sediments when "the retaining dam at the southeast corner of the site gave way" (Shang, *et al.*, 2004, p. 5). Since the retaining dam was not re-built it can be concluded that the impoundment of tailings on the site area was only flooded once.

4.9 Solubility

Solubility is the "measure of the amount of solute (usually a solid or gas) that will dissolve in a given amount of solvent (usually a liquid) at a particular temperature" (Research Machines Plc., 2003, p. 550). "The degree of solubility is the concentration of a solute in a saturated solution at any given temperature" (DePree & Axelrod, 2003, p. 686). "Solubility may be expressed as grams of solute per 100 grams of solvent" (Research Machines Plc., 2003, p. 550). "The degree of solubility of most substances increases with a rise in temperature ,but there are cases (notably their organic salts of calcium) where a substance is more soluble in cold than in hot solvents" (DePree & Axelrod, 2003, p. 686). Solubility is "a property of a substance by virtue of which it forms mixtures with other substances that are chemically and physically homogeneous throughout" (DePree & Axelrod, 2003, p. 686).

"Solvents – The term solvent generally denotes a liquid that dissolves another compound to form a homogeneous liquid mixture in one phase. More broadly, the term is used to mean that component of a liquid, gaseous, or solid mixture present in excess over all other components of the system" (DePree & Axelrod, 2003, p. 686).

Figure 17 - Section 4. 9 is a list of some common sedimentary ironstone minerals. "The least soluble compounds are precipitated first" (Nichols, 2003, p. 31). "Most oxides of non-metals are acidic (dissolve in water to form an acid)" (Research Machines Plc., 2003, p. 423). Although outside this report's scope, more information in Appendix E.

Common sedimentary ironstone minerals.		
Oxides	Haematite	Fe_2O_3
	Magnetite	Fe_3O_4
Hydroxides	Goethite	FeO.OH
	Limonite	$FeO.OH.H_2O$
Carbonate	Siderite	$FeCO_3$
Sulphide	Pyrite	FeS_2
Silicates	Glauconite	$KMg(FeAl)(SiO_3)_6.3H_2O$
	Chamosite	$(Fe_5Al)(Si_3Al)O_{10}(OH)_8$

Figure 17 - Section 4.9: Common Sedimentary Ironstone Minerals (Nichols, 2003, p.33)

"The lake is divided into three vertical layers: (1) the *epilimnion*, which is the "well-mixed surface layer" (Eby, 2004, p. 335). The epilmnion is seen in the "uppermost layer of water in a lake" (Bates & Jackson, 1984, p. 168).

"– stream and ocean water currents and wind – when the forces exerted by the drag of the fluid overcome the gravitational and cohesive forces that hold the particles at rest" (Garrels & Mackenzie, 1971, p. 178). "Lighter particles move down-current with steep upward trajectories and gentler downward glide paths. This process is known as saltation. At the same velocity the lightest particles are borne along by the current in suspension" (Selley, 1982, p. 179).

"Sands from many different aqueous environments show this correlation and the fine sands are generally better sorted than very fine sands or silts" (Blatt, Middleton & Murray, 1972, p. 104). Often the "orientation parallel to the current is the most common orientation for sand grains" (Blatt, Middleton & Murray, 1972, p. 108).

The water displaced the original mine-tailing deposits as the turbulent water lifted and thoroughly mixed the mine tailings; see Figure 18 - Section 4.10 shows the "mechanisms of transport of particles in a flow" (Nichols, 2003, p. 39). "The same particle might be moved either by sliding, rolling, saltation, intermittent suspension, or continuous suspension, depending on the intensity of the flow" (Blatt, Middleton & Murray, 1972, p. 102). "Traction [GEOL] Transport of sedimentary particles along and parallel to a bottom surface of a stream channel by rolling, sliding, dragging, pushing, or saltation" (Licker, 2003, p. 377). Saltation is "a type of *sediment transport* in air or water in which particles are moved forward in short, abrupt leaps. This process is intermediate between *suspension* and *traction*" (MacDonald, *et al.*, 2003, p. 377, 378).

4.10 Fluids are Thoroughly Mixed

"In this process [saltation], small silt and clay particles can be suspended in the water, and some sediment is so small that it dissolves completely" (Albin, 2004, p. 34). An important consideration is to note that as the particles are suspended in water they dissolve. This mixing of sediment would become a blended mixture, smoothing out and dissolving the sediment particles like chocolate in milk. Accordingly, the minerals on Kidd Copper would have fully dissolved in the water.

As discussed earlier, the current was most likely to be turbulent from the dam breach. "Water flows are only laminar at very low velocities or very shallow water depths, so turbulent flows are much more common in aqueous sediment transport and deposition processes" (Nichols, 2003, p. 39). "Most flows in water and air which are likely to carry significant volumes of sediment are turbulent" (Nichols, 2003, p. 39). "In turbulent flows, molecules in the fluid move in all directions but with a net movement in the transport direction.

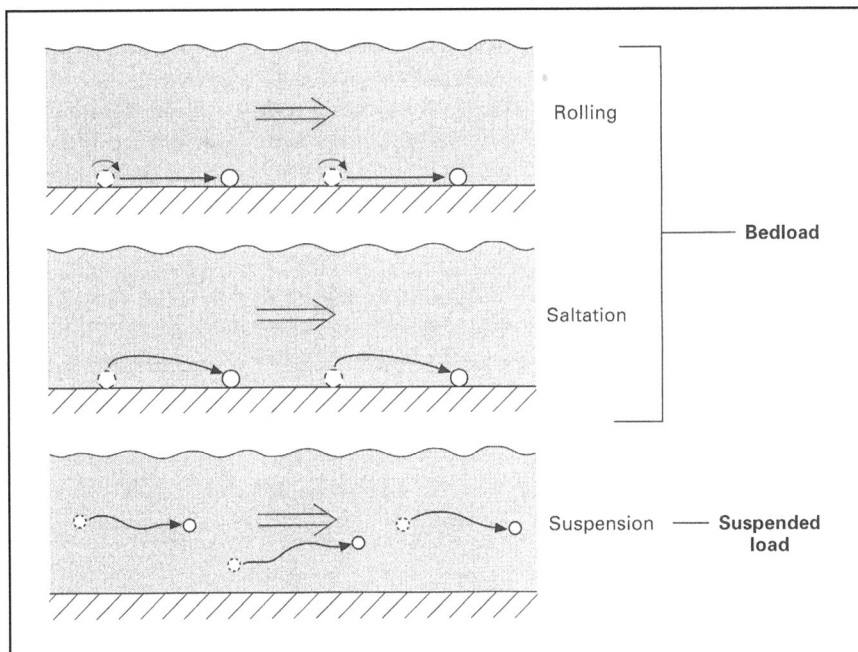

Figure 18 - Section 4.10: Transport of Particles in a Fluid (Nichols, 2003, p. 39)

Kidd Copper's diagenesis context, "heterogeneous fluids are thoroughly mixed in turbulent flows" (Nichols, 2003, p. 38). The mine tailings (slag) mixed thoroughly due to the velocity of the water overflow. This mixture of sediment and water was quickly redeposited onto the site area as heterogeneous sediment layers. These stratified layers of sediment when examined in the bore holes revealed they are a primary structure.

Contrary to conventional wisdom, the expectations on Kidd Copper's redeposited sediment, even with a thorough mixing of fine-grained sediment a homogeneous layer of sediment was not detected. The redepositional process happened in a very short period of time and so after fully mixing, the Kidd Copper property sediment should have been detected as an original layer of horizontal, fully mixed and homogeneous sediment when redeposited. A uniform layer was to be expected because all of the Kidd Copper sediment was deposited at the same time from the receding floodwaters. Instead, Kidd Copper sediment resettled its sediment deposits onto the site area in a rapid depositional process as a heterogeneous mixture. The records provided in the description of the matrix of sediment types and descriptors on Kidd Copper property, see Section 11.4, reveal that a flat deposit of fine-grained sediment was layered randomly and its distribution pattern was unordered.

5.0 Exogenetic Processes

The Kidd Copper property area was once a mine tailings pond with waste products. "*Waste* the unwanted material which is removed during mining and quarrying" (Kearey, 2001, p. 291). The surface cover looks reddish – iron, orange-brown because of the oxidation of the minerals. Lisa Melymuk records that Borehole 49 has "rust coloured (ferrihydrite) crust on surface" (2004, Field Book 18 page 44). The desiccation is evidence of water once being there because the dried out sediment shows water had once been prevalent. Algae mats are noted on the Kidd Copper property in 2004 (Enders, 2004, p. 48).

"Those processes which operate on or close to the surface of the Earth and which involve weathering, mass movement, fluvial, aeolian, glacial, periglacial, and coastal processes" are called exogenetic processes (Allaby & Allaby, 2003, p. 197). "The term is normally used in contrast to the endogenetic processes, whose origin is within the Earth" (Allaby & Allaby, 2003, p. 197). These processes are not the dominant factor for change at Kidd Copper because the change of the impoundment distribution is notably caused by the outflow of water from the dam breach. The overtopping of the dam was the dominant force that provided enough water to move the fine tailings quickly in a short period. Other exogenetic processes do not establish the same mass movement of the waste tailings; so, that these other exogenetic processes play a very minor degree of change.

5.1 Biochemical Activity

Some evidence exists for Bog iron ore to exist on the Kidd Copper Property. Bog iron ore is "a general term for a soft, porous deposit of impure hydrous iron oxides formed in bogs, swamps, and shallow lakes by precipitation from iron-bearing waters and by the oxidizing action of algae, iron bacteria, or the atmosphere. Composed principally of limonite impregnated with plant debris, clay, and clastic material; … an iron ore of poor quality" (Bates & Jackson, 1984, p. 60). Bog iron ore (*mineral*) is "a soft, spongy, porous deposit of impure hydrous iron oxides formed in bogs, marshes, swamps, peat mosses, and shallow lakes by precipitation from iron-bearing waters and by the oxidation action of algae, iron bacteria, or the atmosphere. Also known as lake ore; limnite; marsh ore; meadow ore; morass ore; swamp ore" (Licker, 2003, p. 48).

> "The upper part of the soil is a zone of intense biochemical activity. The bacterial population near the surface is large but decreases downward with a steep gradient. One of the major chemical processes is the oxidation of organic material, which releases CO_2. Samples of soil gases obtained above the zone of water saturation commonly show CO_2 contents 10–40 times that of the free atmosphere and in some instances CO_2 makes up as much as 30 percent of the soil gases as opposed to 0.03 percent of the free

atmosphere. In addition to the acid effects of CO_2 produced by oxidation, a highly acid microenvironment is created by the roots of living plants. Values of pH as low as 2 have been measured immediately adjacent to root hairs" (Garrels & Mackenzie, 1971, p. 141, 142).

Some biochemical activity is seen on the Kidd Copper site area but is beyond the scope of this report to verify further.

5.2 Precipitation

The erosion that occurs on the Kidd Copper property after 1983 is due mainly to the impact of raindrops. Refer to Appendix G-2 for a time series of rainfall on the ground in selected world locations. Note at 20 mins – 8.10 inches has been recorded and at 42 mins – 12.00 inches fell and at 2 hr 10 mins – 19.00 inches (Dingman, 2002, p. 151). "The Copper Cliff area in Northern Ontario has a cool climate that averages 78.2 cm of annual precipitation (van der Leeden, *et al.*, 1991) " (Coggans, *et al.*, 1991, p. 450).

"Average pH of rain water is 5.7 and that the H_2O is in equilibrium with the P_{CO_2} ($10^{-3.5}$ atm) of the atmosphere" (Garrels & Mackenzie, 1971, p.106). "The pH for water in equilibrium with atmospheric CO_2. This calculation gave a value of pH = 5.7. By definition, acid rain is any rainwater that has a pH of less than 5.7. The two most important components of acid rain are sulfuric and nitric acid. Hence, there is a direct relationship between the SO_4 and NO_x content of rainwater and rainwater acidity" (Eby, 2004, p. 283).

"Rainwater becomes acidified by dissolving carbon dioxide from the air, dissolving carbon dioxide from decaying organic matter in the regolith, or interacting with anthropogenic sulfur and nitrogen compounds" (Murck, 2001, p. 125). "Rain is not pure water. It is slightly acid, because it dissolves chemicals from the air. This means that it can slowly dissolve rock. Acid rain, which has extra acid in it from polluted air, can eat the nose off a stone statue in just a few years" (Claybourne, 2002, p. 13).

Table 1 - Section 5.2 provides the sources of the individual ions in rainwater as they react with different marine, terrestrial and pollution inputs. Kidd Copper had the potential to have pollutants as a result of the dumping of the slag but this report will not deliberate about these detrimental effects to the environment.

Table 1 - Section 5.2 Sources of Individual Ions in Rainwater

Ion	Origin		
	Marine inputs	Terrestrial inputs	Pollution inputs
Na^+	Sea salt	Soil dust	Biomass burning
Mg^{2+}	Sea salt	Soil dust	Biomass burning
K^+	Sea salt	Biogenic aerosols Soil dust	Biomass burning Fertilizer
Ca^{2+}	Sea salt	Soil dust	Cement manufacture Fuel burning Biomass burning
H^+	Gas reaction	Gas reaction	Fuel burning
Cl^-	Sea salt	None	Industrial HCl
SO_4^{2-}	Sea salt DMS from biological decay	DMS, H_2S, etc. from biological decay Volcanoes Soil dust	Biomass burning
NO_3^-	N_2 plus lightning	NO_2 from biological decay N_2 plus lightning	Auto emissions Fossil fuels Biomass burning Fertilizer
NH_4^+	NH_3 from biological activity	NH_3 from bacterial decay	NH_3 fertilizers Human, animal waste decomposition (Combustion)
PO_4^{3-}	Biogenic aerosols adsorbed on sea salt	Soil dust	Biomass burning Fertilizer
HCO_3^-	CO_2 in air	CO_2 in air Soil dust	None
SiO_2, Al, Fe	None	Soil dust	Land clearing

*From Berner and Berner (1996).

Source: Eby, 2004, p. 279

"The mechanics of soil erosion by water is fundamentally a three-step process:

(1) *Detachment* of soil particles from the mass;

(2) *Transportation* of the detached particles downhill by floating, rolling, dragging and splashing; and

(3) *Deposition* of the transported particles at some place lower in elevation" (Brady & Weil, 2002. p. 750).   However, the breach in the dam was more significant in changing the landform at Kidd Copper.

"Major storms that bring heavy rains over a large region, cause most floods" (Tarbuck & Lutgens, 1999, p. 260). A look at the hydrogeological setting can be found in Appendix G-2. "During floods, even in large rivers, flow velocities can become great enough for the upper flat bed phase to be present" (Harms, *et al*., 2004, p. 32). "A severe storm surge… can result in… flooding" (Nichols, 2003, p. 140). The increased precipitation caused a break in the natural walls of the dam.

A plausible explanation is as follows: "… pulsating drive of percolating rain followed by dry weather" (Garrels & Mackenzie, 1971, p. 51, 52). "The most effective water-carriers are sandstones" (Garrels & Mackenzie, 1971, p. 52).

"Rainwash [GEOL] (1) The washing away of loose surface material by rainwater after it has reached the ground but before it has been concentrated into definite streams. (2) Material transported and accumulated, or washed away, by rain water" (Licker, 2003, p. 290). Water table is "upper level of ground water (water collected underground in porous rocks). Water that is above the water table will drain downwards; a spring forms where the water table meets the surface of the ground. The water table rises and falls in response to rainfall and the rate at which water is extracted" (Research Machines Plc., 2003, p. 636).

> "The pore spaces of the rocks tend to fill up to the ground surface during the rain, when addition is faster than flow through the rock; this increase in potential energy beneath the upland tends to disappear between rains, causing continuous flow. The surface of saturation (water table) is clearly a function of the amount and frequency of rainfall and of the water-carrying characteristics of the rocks" (Garrels & Mackenzie, 1971, p. 52).

"Some rocks can be dissolved by water. *Solution*, although ordinarily slow, can be an effective process of weathering and erosion (weathering because it is a response to surface chemical conditions; erosion because it removes material)" (Plummer, McGery & Carlson, 2003, p. 230).

"In arid areas, evaporation may exceed rainfall, and no zone of saturation may exist; in areas of heavy continuous rain the saturation surface remains at the topographic surface" (Garrels & Mackenzie, 1971, p. 52). For example, "As a generalization, the zone of saturation in the United States, exclusive of the desert areas, is encountered at an average depth of 6– 2 m and appears as a subdued image of the topography" (Garrels & Mackenzie, 1971, p. 52). Precipitation has a small influence upon the dramatic landform change in 1970-1983-1992 as noted previously. The dramatic change in the landform did not result from the erosional effects caused by precipitation over a long period of time but rather the force of water changed the landform quickly in 1983.

5.3 Air Erosion

Wind erosion is "the *deflation* of surfaces and *abrasion* of consolidated and unconsolidated material by wind action … particularly important during dry periods in areas lacking vegetation cover. Deflation is very effective in removing the finer sands and silts from unconsolidated sediments, leaving coarser material to form a *desert pavement*" (MacDonald, *et al.*, 2003, p. 457). "Wind erosion [GEOL] Detachment, transportation, and deposition of loose topsoil or sand by the action of wind" (Licker, 2003, p. 396). "Flowing air erodes the land surface in two ways. The first, *abrasion*, results from the impact of wind-driven grains of sand. Airborne particles act like tools, chipping small fragments off rocks that stick up from the surface. When rocks are abraded in this way they acquire distinctive, curved shapes and a surface polish. A bedrock surface or stone that has been abraded and shaped by windblown sediment is called a *ventifact* ('wind artifact'). The second wind erosional process, *deflation* (from the Latin word meaning 'to blow away'), occurs when the wind picks up and removes loose particles of sand and dust. Deflation on a large scale takes place only where there is little or no vegetation and loose particles are fine enough to be picked up by the wind" (Murck, 2001, p. 131). "NE side is a gypsum rich area wind erosion creates hoodoos" (Gyatt, August 2004, p. 60). Air erosion effects surface not sub-surface topography; therefore air erosion had little effect on the boring samples taken at Kidd Copper.

"*Ablation (geomorphology)* – essentially, the wasting away of rocks; the separation of rock material and formation of residual deposits, as caused by wind action or the washing away of loose and soluble materials" (DePree & Axelrod, 2003, p. 2). "The sandblasting action of the particles carried by the wind abrades *boulders* and *pebbles* to produce *ventifacts*, while preferential erosion of softer or weaker rock surfaces may form *yardangs* or *zeugen*" (MacDonald, *et al.*, 2003, p. 457). Some examples of the strength of wind erosion are as follows: (1) atmospheric dust flux; (2) yardangs; and (3) zeugens. Kidd Copper was not affected to a noticeable affect by ablation because the sediment on Kidd Copper was fine-grain and not boulders.

Atmospheric Dust Flux. "Atmospheric transport of materials into the oceanic system is the least important pathway" (Garrels & Mackenzie, 1971, p. 111).

"Yardang – An elongated ridge carved mostly from relatively weak earth materials by wind *abrasion* in a desert region. Yardangs appear as sharp-edged topographical forms up to 15 metres in height but some in Iran stand over 200 metres. The ridges are separated by steep-sided troughs, both the ridges and troughs lying parallel to the prevailing winds. Yardangs are common in the deserts of Central Asia and on the Peruvian coast" (MacDonald, *et al.*, 2003, p. 459).

"Zeugen – tabular erosional remnants, often 30 metres or more in height, formed in desert regions by the erosional effect of sand-laden winds. Zeugen are masses of resistant rock resting on supports of softer rock, having become separated by *erosion* from the once continuous horizontal stratum" (MacDonald, *et al.*, 2003, p. 460).

Summation. The exogenic processes noticeable on the Kidd Copper property include: air erosion, stream erosion and surface weathering, which minimally affected the topography or crustal geology of the site area. Surface area looks reddish is evidence of water once being there, highly probable that the "surface layers of rust coloured oxidated tailings" (Belan, 2004, Field Book 1 p. 36). Stream erosion caused by snowmelt had not substantially affected the overall distribution on the impoundment area. The chips and small fragments of rock sticking up from surface were unaffected by air and barely noticeable. Air erosion virtually left the landform untouched.

5.4 Physical Weathering

"Hutton argued that the scale of time for Earth's history was much greater than previously thought, because the weathering of rock from one form to another takes far longer than just a few thousand years" (Discovery Books, 1999, p. 32). "Sedimentary rocks provide us with evidence that for at least a billion years – and perhaps more – conditions on the earth's surface were very much as they are today. An atmosphere contained the sun's heat, rain, wind, running water, and the chemicals of the air changed the surface rocks. This process of weathering and erosion began as soon as the first rocks formed, and it continues today" (Zim, 1961, p. 30).

What then is weathering? "Weathering is the physical and chemical breakdown of rock. Erosion is the combination of weathering and the transport of weathered particles" (Murck, 2001, p. 121).

"Weathering [GEOL] Physical disintegration and chemical decomposition of earthy and rocky materials on exposure to atmospheric agents, producing an in-place mantle of waste" (Licker, 2003, p. 394). "During the course of weathering, ferrous iron is released from the minerals and oxidized by O_2 in the soil water to ferric iron which hydrolyzes to form ferric oxides. "The minerals include the hydrated ferric oxide dimorphs goethite ($\alpha FeO \bullet OH$) and lepidocrocite ($\gamma FeO \bullet OH$) and the anhydrous dimorphs hematite (αF_2O_3) and maghemite (γFe_2O_3, magnetic hematite" (Garrels & Mackenzie, 1971, p. 150).

"Because these alteration products are yellow and orange-brown to red in color, the weathering of iron-bearing rocks usually results in vividly colored soil zones that contrast sharply with the drab fresh rock from which the soils are derived. A few percent total iron in a rock is all that is necessary to impart strong reddish coloration to a soil" (Garrels & Mackenzie, 1971, p. 150). On the Kid Copper Property according to Heather Robertson, Borehole 54 has "5 m strip reddish 3" (2004, Field Book 19 page 50). It was reported that at Borehole Point 33, the "surface is a light sandy color, medium grain (like sand)" (Jack, 2004, Field Book 13 p. 32). In another spot (Borehole Point 35) the "upper surface is dark brown and black due to weathering (Jack, 2004, Field Book 13 p. 33). Table 2 - Section 5.4 is a chart of factors that influence the rate of weathering.

Table 2 - Section 5.4: Factors That Influence Weather

Rate of Weathering			
	Slow	**→**	**Fast**
Mineral resistance	High (e.g., quartz)	Intermediate (e.g., mica)	Low (e.g., calcite)
Frequency of joints	Few (meters apart)	Intermediate (~0.5-1.0 m apart)	Many (centimeters apart)
Steepness of slope	Gentle	Moderate	Steep
Vegetation	Dense	Moderate	Sparse
Temperature	Cold (avg. ~5° C)	Temperate (avg. ~15° C)	Warm/Hot (avg. ~25° C)
Rainfall	Low (>40 cm/yr)	Intermediate (40-130 cm/yr)	High (>130 cm/yr)
Burrowing animals	Rare	Frequent	Abundant
Source: Murck, 2001, p. 125.			

"Surfaces exposed by erosion, landslides or human activity, form *outcrops* and may range from almost plane and even to irregular and rugged; it depends on various factors, such as the type of sediment or rock (lithology), the type and duration of weathering, the orientation of the surface with respect to the wind and sun, the erosional agent (wind, water, ice), and the chemical or mechanical action of organisms (plant roots, lichens, etc)" (Lucchi, 1995, p. 2). The rate of weathering varies with mineral resistance being slow and burrowing animals to be more abundant. Kidd Copper property did not report burrowing animals as a factor. The sediment cores on Kidd Copper were under the initial ground surface and covered until exposed on August 30, 2004.

Weathering and erosional processes were insignificant in the correlation of the boreholes. Loreen Rudd's presentation "dismissed the stratification due to erosional or surface run-off processes because the entire area of the mine-tailings site exposed consistent data the stratification was noted through-out the entire area" (Rudd, December 2, 2004, McMaster University Presentation Rapid Stratification). Each borehole was subject to the same processes as recorded in the history of the site area.

5.5 Chemical Weathering

"*Chemical weathering* is the decomposition of rocks and minerals as a result of chemical reactions that change them to new minerals that are stable at the Earth's surface" (Murck, 2001, p. 122; Bates & Jackson, 1984, p. 85). "Chemical weathering is primarily caused by water that is slightly acidic" (Murck, 2001, p. 124). "Chemical weathering involves a chemical transformation of rock into one or more new compounds" (Tarbuck & Lutgens, 1999, p. 120). "*Chemical weathering* produces new minerals. The main chemical reactions are oxidation, hydration and solution. The main products of weathering are resistant minerals such as quartz, new minerals such as clay minerals produced through the weathering of feldspar and mica, and soluble substances carried away in solution" (MacDonald, *et al.*, 2003, p. 455).

"Chemical reactions between rock minerals and soil waters produce dissolved constituents and solid residues" (Garrels & Mackenzie, 1971, p. 136). "The marked tendency of iron to form oxide minerals in the weathering zone and the remarkable persistence of these minerals in soils can be explained in terms of their vanishingly low solubilities under oxygenated conditions" (Garrels & Mackenzie, 1971, p. 150). "The dissolved constituents enter the ground-water system and eventually move into streams" (Garrels & Mackenzie, 1971, p. 136). "Delayed runoff – Water from precipitation that sinks into the ground and discharges later into streams through seeps and springs" (Bates & Jackson, 1984, p. 131). "Many ground waters have compositions near those predicted entirely from rock mineral-soil water reactions. The solid residues, from the moment of their formation, are subject to dispersal by the agents of erosion, primarily running water" (Garrels & Mackenzie, 1971, p. 136).

"The chemical weathering of iron-bearing minerals and rocks … most iron mineral-soil water interactions involve oxidation-reduction reactions and may also involve the formation of several residual solid phases. Most of the major iron minerals, e.g. magnetite (Fe_3O_4), siderite ($FeCO_3$), pyrite (FeS_2), pyrrhotite (FeS), and iron silicates of various compositions, weathering at the earth's surface, contain ferrous iron in their structure" (Garrels & Mackenzie, 1971, p. 149). "These minerals usually occur dispersed throughout the rock but may be in discrete bands or segregations. Ferrous and ferric iron also are found in rock minerals, other than those in which iron is a major element, substituting for ions of similar size and charge, e.g. Fe^{2+} for Mg^{2+} in dolomite and Fe^{3+} for Al^{3+} in montmorillonite" (Garrels & Mackenzie, 1971, p. 150).

> "The chemical weathering of iron-bearing minerals and rocks results in the formation of ferric oxides because of their wide range of environmental stability. Ferrous iron is the primary iron species produced in solution as a result of weathering, but it is found only at the lower part of the soil zone and is oxidized when the soil waters eventually get back to the surface. Consequently, the iron initially dissolved in acidic ground waters is transported in streams as discrete ferric oxide particles or as ferric oxide coatings on mineral grains rather than in solution" (Garrels & Mackenzie, 1971, p. 153).

How much does chemical weathering play on the Kidd Copper Property? Vividly colored soil zones under oxygenated conditions can result from chemical weathering of the exposed area but would not be expected in the underlying protected areas, such as is visible in Figure 26 - Section 9.12 two feet below the surface layer. Therefore, chemical weathering over a short period of time has little influence in the layers below the initial surface as they had not been exposed to the direct atmosphere. Recall weathering occurs "when the soil waters eventually get back to the surface" (Garrels & Mackenzie, 1971, p. 153). Like silverware wrapped in a blanket the chemicals in the lower zone of transition on the Kidd Copper property are protected from tarnishing.

Four "major kinds of chemical reactions in paleosols using molecular weathering ratios" are: (1) hydrolysis; (2) oxidation; (3) hydration and (4) salinization (Retallack, 1990, p. 67). Chemical reactions are beyond the scope of this report except to note three chemical ratios of oxidization: (1) ferric/ferrous iron = Fe_2O_3 / FeO; (2) total iron/alumina = $(Fe_2O_3 + FeO)$ / Al_2O_3; and (3) total iron & manganese / alumina = $(Fe_2O_3 + FeO + MnO)$ / Al_2O_3" (Retallack, 1990, p. 67).

"The formation of new minerals in place is called authigenesis" (Retallack, 1990, p. 138) and will not be discussed in this report. However, information is provided in Appendix C on *Clay Mineralogy* for the reader's interest.

5.6 Mechanical Weathering

Weathering of Minerals and Rocks. "Mechanical weathering is accomplished by physical forces that break rock into smaller and smaller pieces without changing the rock's mineral composition" (Tarbuck & Lutgens, 1999, p. 120). "*Mechanical weathering* is the physical breakdown of rock with no change in mineralogy or chemistry" (Murck, 2001, p. 121). Weathering is the "destructive natural processes by which rocks are altered with little or no transport of the fragmented or altered material. *Mechanical weathering* occurs with the freezing of confined water and the alternate expansion and contraction caused by temperature changes" (MacDonald, *et al.*, 2003, p. 455).

Detrital Sedimentary Rocks – "sediment may be an accumulation of material that originates and is transported as solid particles derived from both mechanical and chemical weathering" (Tarbuck & Lutgens, 1999, p. 145). "When these dissolved substances are precipitated by either inorganic or organic processes, the material is known as chemical sediment and the rocks formed from it are called *chemical sedimentary rocks*" (Tarbuck & Lutgens, 1999, p. 145). Refer to Appendix K to expand on sedimentary rocks and their processes.

Migration of Surface Waters. "In addition to the processes … relief and climate also play a role in the chemistry of surface waters. Chemical weathering occurs at a maximum rate in wet and warm climates and at a minimum rate in dry and cold climates. The degree of relief determines the rapidity of the erosional process. For example, an area of low relief subjected to wet and warm climatic conditions would undergo extensive chemical weathering and most of the mobile components would be leached from the rocks and soil" (Eby, 2004, p. 331).

Chemical Weathering from the Rain. "Total water at any one time in the atmosphere as water vapor is approximately 0.13×10^{20} g, and this water turns over approximately every 3 or 4 weeks. The average rainfall on the land areas of the earth is about 66 cm/year" (Garrels & Mackenzie, 1971, p. 137).

Frost Heaving. "Water in small cracks and crevices in rocks may freeze at night and thaw during the heat of the day. Changes in the specific volume of H_2O as it alternately freezes and thaws may force the cracks to enlarge, an eventually the rock is weakened and broken apart. This mechanism is operative particularly in semiarid regions or mountainous areas where temperature extremes occur frequently" (Garrels & Mackenzie, 1971, p. 136). "Striped ground [GEOL] a pattern of alternating stripes formed by frost action on a sloping surface. Also known as striated ground; striped soil" (Licker, 2003, p. 357).

Reaction of Rock Minerals with Soil Waters. This subject matter is beyond the scope of this report, further study is required. Refer to Table 30 - Appendix C, where the chlorine content of terrestrial rocks is identified in ppm.

"Chemical weathering of rocks is the integration of the reactions of soil waters with the individual minerals composing rocks. As a result of these reactions, dissolved constituents are produced and added to the reacting aqueous solutions" (Garrels & Mackenzie, 1971, p. 142). "Thus, water percolating through a salt bed or a rock containing NaCl can dissolve a great deal of NaCl before the solution becomes saturated, and beds of salt are rapidly dissolved away. Any relatively insoluble constituents in the salt bed such as quartz and clay minerals remain at the source as a residue" (Garrels & Mackenzie, 1971, p. 143).

The Kidd Copper site area has virtually remained untouched; therefore the exogenetic processes discussed here play a relatively small factor in the reshaping of its landform. Exogenetic processes barely influenced the change in major distribution of sediment and erosional processes had little impact on the impoundment in this report. Kidd Copper provides valuable information on temporal changes in the geographical extent of the tailings and orientation of the sediment layers as they have virtually remained intact since redeposition after the flood.

6.0 Ore Mineralogy

Clay mineralogy is a complex subject which is described in more detail in Appendix C but is outside the scope of this report. No boulders or very coarse grain sized sediment was studied on the Kidd Copper Property. This report specifically inspects Kidd Copper mineralogy endmembers to verify that the size of grains is small and is easily mixed.

6.1 Mineralogy Defined

"The difference between a rock and a mineral should be clearly understood. Rocks are the essential building materials of which the earth is constructed, whereas minerals are the individual substances that go to make up the rocks. Most rocks, therefore, are aggregates of two or more minerals" (Pearl, 1955, p. 17). "A *rock* is a naturally formed, coherent aggregate of minerals, which may be mixed with other solid materials such as natural glass or organic matter. The essential distinction between minerals and rocks is that rocks are *aggregates*. This means that rocks are collections of mineral grains (and sometimes other types of particles, such as organic material) stuck together. Rocks usually consist of several different types of minerals, but sometimes they are made of just one type of mineral. In either case, a rock will contain many grains of the constituent mineral or minerals. Sometimes the grains are held together by naturally formed cement. In other cases, the grains of the rock stick together because they have grown as crystals with interlocking grain boundaries" (Murck, 2001, p. 42). "Rock [PETR] (1) A consolidated or unconsolidated aggregate of mineral grains consisting of one or more mineral species and having some degree of chemical and mineralogic constancy. (2) In the popular sense, a hard, compact material with some coherence, derived from the earth" (Licker, 2003, p. 308).

"Minerals are pure, solid substances made up of crystals. Some minerals, like gold and carbon, are made of one element. Others, like salt and quartz, are made of a combination of elements. Only natural substances are true minerals" (Claybourne, 2002, p. 5). "The main difference is that a mineral is 'homogeneous' – it is the same all the way through. If you look at a lump of salt under a microscope, every part of it has the same structure. But rock is not homogeneous, it is a mixture. For example, granite, a rock, is made up of three minerals, mica, feldspar, and quartz" (Claybourne, 2002, p. 4).

"Once a metallic ore has been taken from the ground, it has to be processed to extract the metal it contains. This is often done right at the mine site. Huge smelting ovens melt down ore and separate metal from the waste, or gangue, which is then discarded" (Discovery Books, 1999, p. 60). "*Ore* – A metalliferous *mineral* or an *aggregate* of metalliferous *minerals* mixed with *gangue* that can be exploited at a profit" (Kearey, 2001, p. 189). "Recovery of valuable metal pollutants was actually profitable for industries" (Brady &Weil, 2002, p. 821). "Mine tailings, the waste material left over from processing ore, are essentially tons of crushed rock with zero percent organic matter" (McGinley, 2002, p. 18).

"Most mine waste comprise a variable mixture of benign silicates, reactive sulphides, and precipitates. Exposure of the reactive sulphides to oxygenated waters in the tailings pond results in a chemical reaction that eventually leads to the formation of an acid-rich leachate" (McMaster University, 2004, Geo3Fe3, Assignment #9 Mine Tailings). "In natural waters sulfur exists as four aqueous species: HSO_4^-, SO_4^{2-}, $H_2S_{(aq)}$, and HS^-. The latter two species only exist under strongly reducing conditions. $H_2S_{(aq)}$ is the dominant species under acidic conditions and HS^- is the dominant species under alkaline conditions" (Eby, 2004, p. 371).

"Another example of an important reacting system is that of soil formation. In this situation minerals are continuously exposed to a new supply of CO_2^- rich oxygenated water. The CO_2 reacts to dissolve minerals incongruently, and the presence of oxygen further complicates the picture by changing the valence of various species, such as ferrous iron or manganous manganese, usually causing the formation of slightly soluble higher valence oxides or hydroxides" (Garrels & Mackenzie, 1971, p. 371).

"More than 99 percent of the dissolved solids in river water can be accounted for by the constituents chlorine, sodium, sulphate, magnesium, calcium, potassium, bicarbonate, and silica" (Garrels & Mackenzie, 1971, p. 107). "Some of the clay particles, especially those in the coarser clay fractions, are composed of the minerals already cited as quartz, hematite, and gibbsite. And even more important group, however, is the complex aluminosilicates. Three main minerals types are at present recognized although others are known to occur in significant quantities – *kaolinite*, *illite*, and *montmorillonite*. It will suffice at this time to say that these groups vary markedly in plasticity, cohesion, and adsorption, kaolinite being lowest in each case and montmorillonite highest" (Buckman & Brady, 1968, p. 43, 44).

"Soil organic matter and high-surface-area clays tend to be the strongest adsorbents for some compounds, while oxide coatings on soil particles strongly adsorb others. The presence of certain functional groups, such as – OH, –NH_2, – NHR, – $CONH_2$, –COOR, – and +NR_3, in the chemical structure encourages adsorption, especially on the soil humus" (Brady &Weil, 2002, p. 802, 803).

"The waste materials at Kidd Copper are derived from the same mineral deposit environment as the more well studied Copper Cliff deposit. The Kidd Copper site has been inactive for over 20 years. It has a similar geological setting and mineral composition as the Copper Cliff site which makes it an ideal test for a comparative study" (Shang, *et al.,* 2004, p. 2). "Reflectance spectra for the minerals found at the Copper Cliff tailings site of Sudbury (Shang *et al.*, 2002) were used as endmembers. This is logical since the Kidd Copper tailings site displays a similar geolgocial setting to the Copper Cliff site" (Shang, *et al.*, 2004, p. 5).

Table 3 - Section 6.1 shows "the spectra of the mineral endmembers were extracted from the USGS spectral library (USGS, 2003). It must be noted, however, that the USGS mineral endmembers represent specific mineral phases which may not be fully representative of the mineralogy of the present site" (Shang, *et al.,* 2004, p. 5).

Table 3 - Section 6.1: USGS Library Mineral Endmembers Used when Unmixing AURORA Data

Mineral Endmember	Formula
Calcite	$CaCO_3$
Chlorite	$(Mg,Fe)_3(Si,Al)_4O_{10}(OH)_2-(Mg,Fe)_3(OH)_6$
Dolomite	$CaMg(CO_3)_2$
Goethite	alpha-$FeO(OH)$
Gypsum	$CaSO_4 \cdot 2H_2O$
Hematite	alpha-Fe_2O_3
Illite	$(K,H_3O)(Al,Mg,Fe)_2(Si,Al)_4)_{10}[(OH)_2,H_2O]$
Natro-Jarosite	$(Na)Fe_3^{+3}(SO_4)_2(OH)_6$
K Jarosite	$KFe^{+3}_3(SO_4)_2(OH)_6$
Hydro K Jarosite	$(K,H_3O)Fe^{+3}_3(SO_4)_2(OH)_6$
Maghemite	gamma-Fe_2O_3
Magnetite2	$Fe^{+2}Fe^{+3}_2O_4$
Pyrite3	FeS_2
Pyrrhotite	$Fe_{1-x}S$
Quartz3	SiO_2

Source: Shang-Morris-Howarth, 2004, p.5

The mineral signatures left from the dumping of slag determine that "pyrite and pyrhotite were the primary sulphide waste minerals deposited at the Copper Cliff tailings and at the Kidd Copper site" (Shang, *et al.*, 2004, p. 7), refer to Table 3- Section 6.1; however the mineralogy, concentration of the metals and salts as well as the effects of the reactive sulphides are beyond the scope of this report.

6.2 Fine-Grain Sized Waste Products on Kidd Copper Property

Texture is "the general character or appearance of a rock as indicated by relationships between its component particles, specifically grain size and shape, degree of crystallinity and arrangement … The physical nature of a soil expressed in terms of relative proportions of sand, silt and clay" (MacDonald, *et al.*, 2003, p. 428).

C.J. Coogans, D.W. Bowes, W.D. Robertson, and J.L. Jambor took detailed grain-size analysis of tailings of the Copper Cliff area in Northern Ontario, which they classified from 0.01 to 0.06, on the Udden-Wentworth scale of grain size, refer to Figure 59 - Appendix B this would translate between very fine sand to coarse silt (Nichols, 2003, p. 12). "The tailings deposited in the CD and M impoundments along cross-section A-A` (Figure 19 - Section 6.2) are typically a mixture of silt to medium-grained sand" (Coggans, *et al.*, 1991, p. 450).

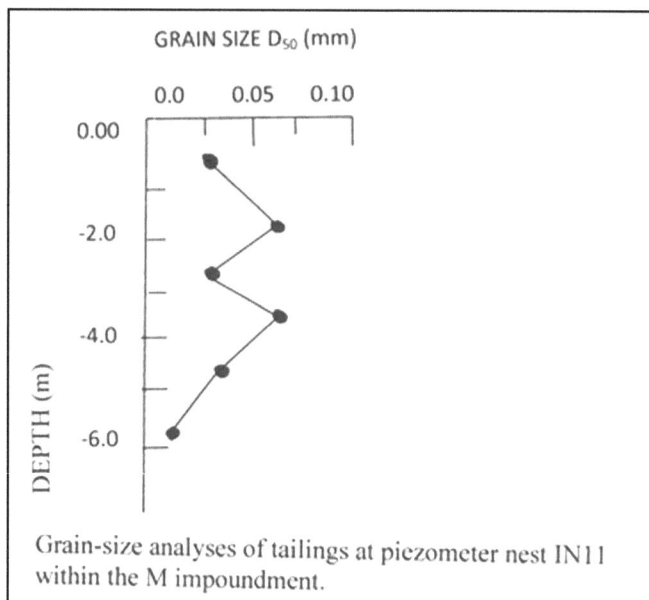

Grain-size analyses of tailings at piezometer nest IN11 within the M impoundment.

Figure 19 - Section 6.2: Tailings Nest IN11 between 0.01 and 0.06 mm. (Coggans, *et al.*, 1991, p.450)

Homogeneous Prediction Based on Conventional Wisdom. The average sediment size found on the Kidd Copper property is *Very Fine Sand* according to the findings in Table 10 - Section 11.9. "Foresets may be termed heterogenous if the layering is due to variations in grain size, or homogenous if it is not" (Selley, 1982, p. 221). Since no variations in grain size appear in the sediment on the Kidd Copper property the expected finding would be homogenous layering and to predict the strata layers at Kidd Copper to be uniformly deposited.

6.3 Satellite Site – Impoundment at Copper Cliff, Ontario

"The waste materials at Kidd Copper are derived from the same mineral deposit environment as the more well studied Copper Cliff deposit" (Shang, *et al.*, 2004, p. 2). Refer to Appendix G-1 for Copper Cliff data. "The Kidd Copper tailings site displays a similar geological setting to the Copper Cliff site" which makes it "an ideal test site for a comparative study" (Shang, *et al.*, 2004, p. 5, 2). Copper Cliff provides useful information for the classification of the grain size and the hydraulic gradient for the Kidd Copper waste-site mine tailings south of Perch Lake, Ontario.

Copper Cliff. "The modeled flow system along a 3 km cross section through tailings deposited from 1940 to 1960… Maximum concentrations of 1060 mg/l Fe, 8700 mg/l SO_4, 641 mg/l Ni, and 23.3 mg/l Co are observed in the shallow tailings. The Ni and Co concentrations are probably representative of tailings in the Sudbury area" (Coggan, *et. al.*, 1991, p. 447).

"Dissolved Fe, Ni and Co exhibit similar mobilities within the tailings water flow system. Although amorphous ferric hydroxide and goethite seem to be acting as a solid phase controls on dissolved Fe concentrations, no discrete secondary Ni and Co minerals were observed. However, appreciable amounts of Ni were detected by energy dispersive electron probe microanalysis within goethite precipitates and minor amounts were detected in altered biotite. The mobilities of dissolved Zn, Cr, Pb, and Cu in the tailings water are strongly dependent upon the pH and are controlled through a series of precipitation, dissolution, adsorption, and desorption reactions with secondary minerals and by solid-solution reactions… In the absence of an effective remedial program, it is anticipated that the observed concentrations of SO_4, Fe and other metals will continue to move downward and laterally through the Copper Cliff impoundments and will ultimately be discharged from the impoundments toward Copper Cliff Creek or into the underlying geologic material" (Coggans, *et al.*, 1991, p. 464).

"A vegetation program, initiated in 1957, has developed self-sustaining grass covers on the C-D, M, and M1 tailings areas … Revegetation has not measurably lessened the oxidation of sulfide minerals or the transport of dissolved constituents within the mine tailings" (Coggans, *et al.*, 1991, p. 447).

"The Copper Cliff tailings disposal area consists of several valleys filled to depths of up to 45 m with silty tailings that have been deposited sequentially since 1936" (Coggans, *et al.*, 1991, p. 447). The tailings on Kidd Copper have the same "mineral composition as the Copper Cliff site" (Shang, *et al.*, 2004, p. 2).

Homogeneous Prediction Based on Conventional Wisdom. The Kidd Copper silty tailings are small and the minerals dissolved, which supports the prediction that the mine tailings sediment would distribute itself as uniformly deposited.

6.4 Clastic Sedimentation

Sediment is the "solid material, organic or inorganic in origin, that has settled out from a state of *suspension* in a fluid and had been transported and deposited by wind, water or ice. It may consist of fragmented rock material, products derived from chemical action or from the secretions of organisms. Loose sediment such as sand, mud and till may become consolidated and/or cemented to form coherent sedimentary rock" (MacDonald, *et al.*, 2003, p. 385, 386). The sedimentation revealed in the boreholes on the Kidd Copper property revealed loose sediment deposits. Eden Hynes writes in her Forestry Supplier's Field Book, the surface "very powdery fine grained texture" (2004, p.

45). Caitlin Vanderkooy adds "chalk like fine grain" (2004, p. 117) to describe the sediment on the Kidd Copper property.

"Clastic sedimentation occurs when particles suspended and transported by water or wind are deposited and the energy source for transport decreases … clastic sedimentation is solely caused by physical forces" (Albin, 2004, p. 19). "Nonclastic sedimentary rocks are formed by chemical precipitation from water in a process called chemical sedimentation" (Albin, 2004, p. 19), which is beyond the scope of this report.

6.5 Spherical and Evenly Distributed

Homogeneous Prediction Based on Conventional Wisdom. The water flow released from the dam breach on Kidd Copper property can be assumed to carry "a substantial concentration of dissolved material" (Garrels & Mackenzie, 1971, p. 361). "Further assumptions are that the sulfide grains are evenly distributed throughout the tailings and the grains are spherical and of one size" (Coggans, *et al.*, 1991, p. 457). On Kidd Copper the sulphide grains are expected to be evenly distributed throughout because the grains are spherical and of one size.

6.6 Colloidal Homogeneous Appearance

Colloid is "a mixture of extremely small particles of a substance dispersed in another in which it does not dissolve. The particles (groups of atoms, molecules or ions) are smaller than in suspension" ie. milk. (Stockley, Oxlade &Wertheim, 2001, p. 145). "*Colloid* – A dispersion of extremely fine particles in suspension, possible because the ultra-small size of the particles (1-10μm) allows the supporting forces to exceed the gravitational forces promoting settling" (Kearey, 2001, p. 53). "*Colloids* are particles that occupy the size range between true solutions and suspensions. This size range is not precisely defined but is generally considered to encompass particles between 10 μm and 0.01μm in diameter. In this size range Brownian motion plays an important role in maintaining particle suspension. In natural waters colloids are usually clay minerals, silica, oxyhydroxides, organic matter, and bacteria. The colloidal size range and amount vary as a function of the hydrologic environment" (Eby, 2004, p. 343).

The water mineral interactions are such that "ions that have low ionic potential more readily lose their attached water molecules and, hence, form inner-sphere complexes" (Eby, 2004, p. 342). "Define … the actual fraction of the soil which is in the colloidal state. The upper limit in size of the mineral colloidal particles is less than 0.001 mm., or one micron (μ), values as low as 0.5 or even 0.2 microns been commonly accepted. Since the maximum size limit of the clay fraction of a soil is considered as 0.002 mm., or 2μm, by no means is all of the clay strictly colloidal" (Buckman & Brady,1968, p. 69).

"Flocculation is a term applied to a coagulation of the
dispersed particles. A very good example of flocculation is afforded

by treating a colloidal clay suspension (wherein the particles are dispersed) with a small amount of calcium hydroxide. The tiny clay particles almost immediately coalesce into floccules. Because of their combined weight, these sink to the bottom of the containing vessel, leaving the supernatant liquid clear. The phenomenon is called *flocculation* because of the peculiar appearance of the aggregates. The same action apparently takes place in the soil itself, but, of course, much less rapidly" (Buckman & Brady, 1968, p. 103).

"Colloidal clay particles are so small that they cannot be seen even with the aid of an ordinary microscope. An electron microscope must be used" (Buckman & Brady, 1968, p. 10). Colloid "substance composed of extremely small particles of one material (the dispersed phase) evenly and stably distributed in another material (the continuous phase). The size of the dispersed particles (1–1,000 nanometres across) is less than that of the particles in the suspension but greater than that of molecules in true solution" (Research Machines, 2003, p. 133).

Homogeneous Prediction Based on Conventional Wisdom. If one could have seen the sediment mixture when the dam flooded the mine tailings site, it would be highly probable the Kidd Copper sediment was a coagulation of the particles leaving the supernatant liquid clear because of the combining and mixing aspects of the light grained sediment. Refer to Table 10 - Section 11.9. The average grain sized detected on the Kidd Copper property is *Very Fine Sand* 0.06 – 0.135 mm; therefore these particles are small enough to be dispersed evenly after mixing in solution and would be expected to maintain a homogenous appearance.

7.0 Geomorphological Background – 3 Final – Tailings Redeposited

"Resedimentation [GEOL] (1) sedimentation of material derived from a preexisting sedimentary rock, that is, redeposition of sedimentary material. (2) mechanical deposition of material ... (3) the general process ... such as the formation of a turbidity-current deposit" (Licker, 2003, p. 298).

7.1 Step 3 Observable Sediment Redeposited on Kidd Copper Property

"A geologist's ability to describe an outcrop is closely related to his understanding of the observable phenomena" (Blatt, Middleton & Murray, 1972, p. 5). The observed facts confirmed the grains on the Kidd Copper property were not mainly deposited by original settling; washed in by percolating groundwater; or blown in from wind (aeolian factors); rather, the resultant grains were redeposited as a result of the flood waters subsiding as supported by the evidence noted earlier in Section 3.4. Recall, when the dam broke the water would have swept the mine tailings up, there would have been an upswelling from the increased velocity of the water. "As the dam breaks it clogs with sediment and flows more slowly→ a floodplain is created. A change occurs in the base level (gets lower) and creates steeper slumping channels within the other channels" (Jack, 2004, p. 38).

The water drains off the region and redeposits the mine tailings. Then as the washout settled, the particles were redeposited. "The faster and harder the water flows, the more sediment will be deposited" (Albin, 2004, p. 34). "Redeposition [GEOL] formation into a new accumulation, such as the deposition of sedimentary material that has been picked up and moved (reworked) from the place of its original deposition, or the solution and reprecipitation of mineral matter" (Licker, 2003, p. 293).

7.2 Sediment Deposits

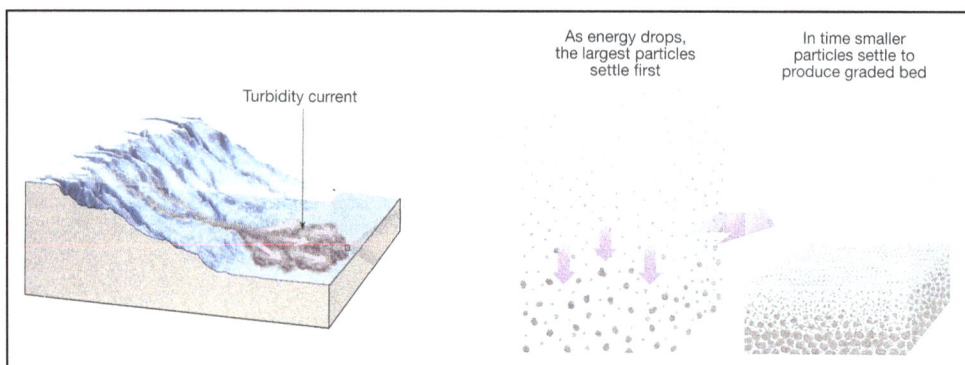

Figure 20 - Section 7.2: Graded Beds - Largest Particles Settle First

"The sedimentary load is gradually dropped as the current slackens and the water comes to rest. Such deposits are called *turbidites* and are characterised by moderate sorting, graded bedding and well-developed primary structures" (MacDonald, *et al.*, 2003, p. 438; Nichols, 2003, p. 55). "Sediment discharge [is] the amount of sediment moved by a stream in a given time, measured by dry weight or by volume; the rate at which sediment passes a section of a stream" (Bates & Jackson, 1984, p. 455).

In Figure 20 - Section 7.2 taken from Tarbuck & Lutgens, 1999 on page 161 is an illustration of a turbidity current showing that "as energy drops, the largest particles settle first. In time smaller particles settle to produce a graded bed. Each layer grades from coarse at its base to fine at the top" (Tarbuck & Lutgens, 1999, p. 161).

> Larger sediment particles "tend to congregate at the bottom of the current and are dragged along by the fluid stress exerted on the grain flow by the overlying flowing suspension. The finer silt and clay particles remain in suspension, been uplifted by the upward component of the turbulence. The energy loss by a turbidity-current flow is, therefore, not only related to the resistance to fluid flow such as that considered by Chezy, additional energy has to be expended to maintain the falling particles in suspension. In order to maintain sediment particles in suspension, not only the competence, but also the capacity of the fluid force must be considered" (Hsü, 2004, p. 91).

> "The total capacity of sediment transport is related to the power of current, i.e., available energy per unit time. Taking air transport as an example, we know that an airplane losing power has to dump part of its load to keep flying at the same speed. When a current loses its power, part of its suspension settles. The deposition of turbidite is, therefore, an indication that the current is no longer accelerating, but is losing power" (Hsü, 2004, p. 91).

Stream load is "all the material that is transported by a stream. The *dissolved load* consists of material in ionic solution; the ions are part of the fluid and move with it. They are derived from groundwater that has filtered through weathering soil and rocks. The *suspended load* or *wash load* comprises clay, silt and sand particles so fine that they remain in *suspension* almost indefinitely. Material that is too coarse to be lifted by the stream water is bounced or rolled along the bottom; this is the *bed load*" (MacDonald, *et al.*, 2003, p. 415). "Because of their small size and platy shape, clay minerals remain in suspension in quite weak fluid flows and only settle out when the flow is very sluggish or the fluid still. Clay particles are therefore present as suspended load in most currents of water and air, and are only deposited when the flow ceases" (Nichols, 2003, p. 19). For example, "as a flooding river spreads over a flood plain, it slows down. The velocity of the water is abruptly decreased by friction as the water leaves the deep channel and moves in a thin sheet over the flat valley floor" (Plummer, *et al.*, 2003, p. 239). "*Flood deposits* are usually silt and clay" (Plummer, *et. al.*, 2003, p. 243).

7.3 The Dam Drains

Drainage is "the manner by which the waters of an area flow off in surface streams or subsurface conduits" (Bates & Jackson, 1984, p.149). On Kidd Copper property the water drained off the region and redeposited the mine tailings over the impoundment area and to a distance in the lower lying valley. "Valley is basin, traps tailings" (Rosamond, 2004, p. 77). Figure 10 - Section 3.6 testifies to the fact that the mine tailings sediment was lifted up in a turbulent flow, mixed and redeposited causing an extension of the mine tailings deposit in the aftermath. "*Flood erosion* is caused by the high velocity and large volume of water in a flood" (Plummer, *et al.*, 2003, p. 243).

"The clastic sedimentary rocks–those deposited mechanically as grains – contain from 30 to 90 percent by volume of water at the time of deposition; as they are compacted by the weight of subsequent deposits, the water is squeezed out until the grains make sufficient contact with each other to form a load-supporting structural framework. In the absence of any later cementing material, sandstones and shales average about 30 percent porosity" (Garrels & Mackenzie, 1971, p. 49, 51) …"the original pore water may be in part retained, apparently with selective loss of water and salts with increasing compaction, or it may be completely displaced by fresh waters percolating downward from the surface or by waters moving from adjacent rocks. The original porosity may be increased by solution of grains by moving waters, or it may be eliminated by precipitation and cementation from such waters" (Garrels & Mackenzie, 1971, p. 51).

"By providing areas for storage of water as a flood wave moves downstream, floodplains reduce the velocity (and the height) of the flood peak" (Dingman, 2002, p. 427). Waning current is a "current related to a catastrophic event, which looses energy after an initial peak" (Lucchi, 1995, p. 249). The waning current left behind the sediment on the impoundment area. As Toni Jack recorded, "as the dam breaks it clogs with sediment and flows more slowly, a flood plain is created" (2004, Fieldbook 13 p. 38).

This was a "single-event" in geologic time; the Kidd Copper flood "simulates time-varying response of a system to an isolated input" (Dingman, 2002, p. 29).

7.4 Sediment Deposition like That of a Turbidity Current

Refer to Appendix K-2 for a definition of sedimentary processes.

"A turbidity current is driven by the gravity of the suspension, which is related to the effective density of the suspension" (Hsü, 2004, p. 90). Suspension is "a type of *sediment transport* in which sediment particles are held in the surrounding water by the upward component of eddy currents associated with turbulent flow" (MacDonald, *et al.*,

2003, p. 421). "Larger sediment particles settle rapidly from suspension" (Hsü, 2004, p. 90, 91).

"When a [turbidity] current loses its power, part of its suspension settles" (Hsü, 2004, p. 91). "When the discharge exceeds the capacity of the channel water flows over the banks and out on to the floodplain where *overbank* or *floodplain* deposition occurs" (Nichols, 2003, p. 120). "As the discharge decreases, the load of sediment is gradually deposited" (Plummer, *et al*., 2003, p. 230).

"The deposition of turbidite is, therefore, an indication that the current is no longer accelerating, but is losing power" (Hsü, 2004, p. 91).

"The internal characteristics of a turbidite show more than just a simple grading: a pattern of textures and sedimentary structures in these deposits was first noted by Bouma (1962) after whom the *Bouma sequence* of internal characteristics named …

Bouma Division 'a'… little sorting in this basal layer and no sedimentary structures form …

In Bouma Division 'b' laminated sand characterizes this layer, the grain size is normally finer than in the 'a' layer and the material is better sorted …

In Bouma Division 'c' cross laminated medium to fine sand, sometimes with climbing ripple lamination, forms the middle division of the Bouma sequence …

Bouma Division 'd' fine sand and silt in this layer are the products of waning flow in the turbidity current …

Bouma Division 'e' the top part of the turbidite consists of fine-grained sediment of silt and clay grade. This material is deposited from suspension as the turbidity current comes to rest" (Nichols, 1999, p. 55-56).

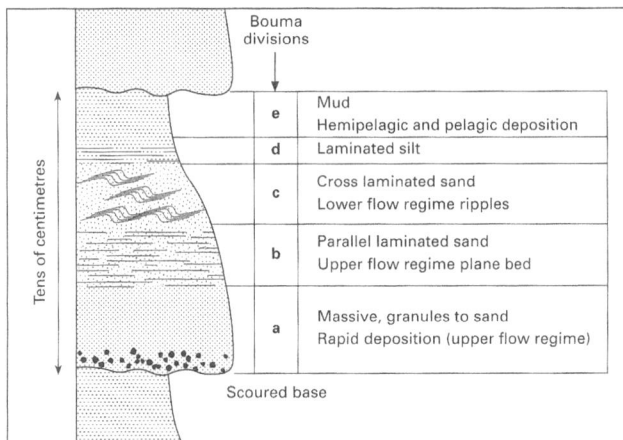

Figure 21 - Section 7.4: Bouma Sequence – 5 Divisions

"An ideal turbidite deposit contains five divisions ('a' – 'e') in the Bouma scheme … although most turbidites do not contain all five divisions" (Nichols, 2003, p.56), refer to Figure 21 - Section 7.4. "Turbidite beds range from a few centimetres to a metre or more in thickness"; and "in many turbidites, not all divisions are developed" (Reineck & Singh, 1973, p. 107).

In the geologic rock record many findings have supported differentiations in the settling of the sediment. Frequently geologists will find that the ideal Bouma sequence is not standard in every turbidite deposit.

The redeposited units on the Kidd Copper property are compared to the Bouma sequence in terms of flow regimes because "this material is deposited from suspension as the turbidity current comes to rest" (Nichols, 1999, p. 56). Recall the dam broke and the water flooded the mine tailings which lifted the sediment up and caused a mixing.

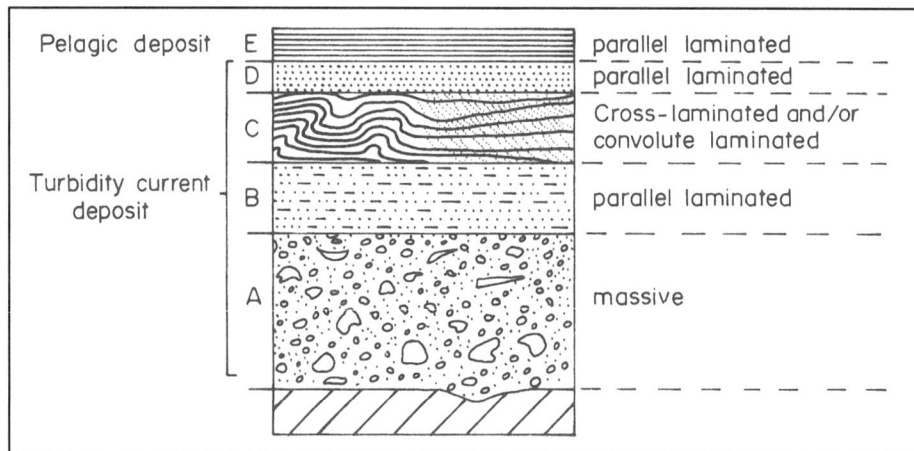

Figure 22 - Section 7.4: The Ideal Turbidite Succession and "Bouma" Intervals

Figure 22 - Section 7.4 is an ideal turbidite succession and depicts 'Bouma' intervals from D.W. Lewis (1984) on page 28. This figure shows that "any sediment or rock transported and deposited by a turbidity current, generally characterized by graded bedding, large amounts of matrix, and commonly exhibiting a Bouma sequence" (Licker, 2003, p. 381).

The Kidd Copper deposits fulfill the definitions of redeposition and resedimentation because the transport mechanism was like that of the turbidity currents seen in marine climates. The Kidd Copper sediment was formed into a new accumulation as a result of the deposition of sedimentary material that had been picked up and reworked.

The fact "in many turbidites, not all divisions are developed: AE, BCE, CDE, and CE sequences are common" (Reineck & Singh, 1973, p. 107). Mixed layers are seen elsewhere in the geologic rock record. The grading found in the rock records as the turbidite succession and the "Bouma" intervals supports that the varied stratification found at Kidd Copper is a geologic process. This is important because it validates that the findings on Kidd Copper are repeatable.

The suggested experiment in Table 13 - Section 14.10 on point source layering is another proposed method to expand and confirm the resedimentation process as it uses turbidity currents as a transport mechanism for future examination. An alternative experiment is suggested in Table 14 – Section 14.10 on sheet source layering that split-testing variables could be done to further develop the question as to final deposition: patterned or unpatterned or heterogeneous distribution and as to proportional amounts.

7.5 Settling Velocity of the Particle

When "the retaining dam at the southeast corner" (Shang, *et al.*, 2004, p. 5) broke the water swept the mine tailings up on Kidd Copper. There was an upswelling from the increased velocity of the water scouring the mine tailings and lifting them up into the turbulent water currents. "*Suspension* – The transport of sediment within a fluid body, possible when upward fluid velocities exceed the *settling velocity* of the particles" (Kearey, 2001, p. 261).

"The settling velocity of a particle depends on its density, sphericity, and size. The classic, ... formula for obtaining settling velocities is that of Stokes' and is know as Stokes' law.

$$v = 2/9 \ * (d_1 - d_2/n) \ * gr^2$$

Where v is the velocity of the particle in centimetres per second, r the radius of the particle in centimetres, g is the acceleration of gravity, n is the viscosity of the fluid in poises, an d_1 and d_2 are the densities of the particle and fluid, respectively" (Garrels & Mackenzie, 1971, p. 178). "in general, the sands and silts settle out quickly; the clays remain in suspension and may be carried several hundred kilometers out to sea before settling to the bottom." (Garrels & Mackenzie, 1971, p. 179).

"One of the key factors that determines whether a sediment particle will or will not be transported in turbulent suspension is the relationship between the upward components of flow within the fluid and the terminal speed of fall, or *settling speed*, of the particle. The methods of analyzing the settling speed are based on the simplifying assumption that all the particles are spheres of a single mineral, say quartz, so that the densities of all are the same. The two chief controls on settling speed are (1) density and (2) size (as expressed by diameter of the spheres). Density affects settling according to the first power. Spheres smaller than 0.18 mm settle according to the squares of their diameters" (Friedman, *et.al.*, 1992, p. 240).

"Sedimentation occurs when the settling velocity of a particle exceeds the upward velocity of the air. The settling velocity of a spherical particle is determined by Stokes' law:

$$v_s = \frac{g(\rho_s - \rho_f)\, d^2}{18\eta}$$

Where v_s, is the settling velocity, g is the acceleration of gravity (9.8 m s^{-1}), ρ_s is the density of the particle, ρ_f is the density of the air, d is the diameter of the particle, and η is the dynamic viscosity of air. At sea level, for the standard atmosphere, density = 1.225 kg m^{-3} and viscosity = 1.789 x 10^{-5} Pa s. Settling velocities become significant at diameters greater than 10 µm. For smaller particles, turbulent diffusion is the dominant mechanism and vertical transport is proportional to the concentration gradient. In this case, particles move from regions of high concentration to regions of low concentration. The actual removal of this size particle occurs across a thin laminar air layer that separates the atmosphere from the surface. Thus, the removal process is sensitive to the type of surface" (Eby, 2004, p. 289). Smaller particles as in the case of the mine tailings on Kidd Copper property are subject to turbulent diffusion for the dominant mechanism.

7.6 Field Capacity

"Once the rain or irrigation has ceased, water in the largest soil pores will drain downward quite rapidly in response to the hydraulic gradient (mostly gravity). After one to three days, this rapid downward movement will become negligible as matric forces play a greater role in the movement of the remaining water. The soil then is said to be at its *field capacity*. In this condition, water has moved out of the macropores and air has moved in to take its place" (Brady & Weil, 2002, p. 206).

The Kidd Copper mine tailings was uplifted and redeposited by the force of water, which was the dominant mechanism of uplift. The settling of the mine tailings was within a few hours to a maximum of a few days; therefore the depositional process that occurred on the Kidd Copper impoundment of tailings happened in a very short period of time through very rapid processes.

8.0 Significance of Kidd Copper

The Kidd Copper Site area could be compared to an outside natural observatory that demonstrates stratification forms rapidly. Interpreting the past can indeed be done with the paramenters of the present and true knowledge of depositional processes known.

8.1 Law Defined

"Law. In science, a formal statement of the invariable and regular manner in which natural phenomena occur under given conditions; e.g. the 'Law of superposition'" (Bates & Jackson, 1984, p. 290). "In science, a law is a descriptive principle of nature that holds in all circumstances covered by the wording of the law. There are no loopholes in the laws of nature and any exceptional event that did not comply with the law would require the existing law to be discarded or would have to be described as a miracle" (Isaacs, *et. al.*, 2003, p. 445).

8.2 Theory Defined

"A description of nature that encompasses more than one law but has not achieved the uncontrovertible status of a law is sometimes called a theory" (Isaacs, *et. al.*, 2003, p. 445). For example, the theory of differential rates of cycling has indeterminate conclusions, where "new deposits would reflect the differential rates of cycling of the components, so that the deposits would have a different ratio of rock types from those of the original mass" (Garrels & Mackenzie, 1971, p. 272, 273). Another example of theory is "Charles Darwin's theory of evolution" (Albin, 2004, p. 43).

8.3 Inductive Reasoning

Inductive Reasoning. "The natural-history approach to study geology…The logic is inductive reasoning" (Hsü, 2004, p. vii). "Inductive reasoning, or induction, is reasoning from a specific case or cases to a general rule" (Syque, 2009, p. Inductive Reasoning). "James Hutton in the late 18th century and Charles Lyell in the middle of the last century established the natural-history approach to study geology, and the success of the method is witnessed by the progress of the science over the last 2 centuries" (Hsü, 2004, p. vii).

Stratigraphic Analysis of Kidd Copper Lithologic Units. The approach to the analytics in this report will be based on stratigraphic geology. "Stratigraphy [GEOL] a branch of geology concerned with the form, arrangement, geographic distribution, chronologic succession, classification, correlation, and mutual relationships of rocks strata, especially sedimentary. Also known as stratigraphic geology" (Licker, 2003, p. 355).

Bedded strata are layered units. The accepted rule is that the top is underlain by older stratified units. Superposition defined in the '*Dictionary of Geology & Mineralogy*' published by McGraw-Hill in 2003 is "the order in which sedimentary layers are deposited, the highest being the youngest" (Licker, 2003, p. 361). It states that younger sediments are deposited on top of older, previously deposited sediments, refer to Section 13.9. Therefore, the Law of superposition will be compared to observations *in situ*. The question that will be considered is, are older layers underneath the younger layers? Will a single layer of sediment result when unstratified sediment mixes with water? What will the resultant primary structure be when unstratified sediment is deposited from a catastrophic water event, such as in the case of the flood event on Kidd Copper? How will the sediment distribute itself, in what pattern or discernible order? The Law of superposition will be considered as it states that younger strata are the layers on top.

8.4 Scientific Approach

"Scientific approach: A method for rationally explaining observations about the natural and human world" (Clarke, 2003, p. 324). Scientific method is the process where "scientists gather facts through observations and formulate scientific hypotheses and theories" (Tarbuck & Lutgens, 1999, p.11). "The mathematical science of statistics is the science of drawing valid inferences from samples to populations" (Blatt, Middleton & Murray, 1972, p. 7).

The facts assembled from the data collected on the Kidd Copper property are used in this report to test a number of geologic laws, hypotheses, theories and suppositions. Theory can be tested by the parallel between the Kidd Copper event and a historical flood event recorded in Earth history. The Kidd Copper property provides a natural, real-time, small-scale event that is factually correlated and observed without having to estimate time references.

8.5 Hypothesis Defined

"A hypothesis is a theory or law that retains the suggestion that it may not be universally true" (Isaacs, *et. al.*, 2003, p. 445). "Many scientists believe that science advances only by disproof of its hypotheses" (Gribbin, 2004, p. 6). "Null hypothesis: The state opposite to that suggested in a hypothesis, postulated in the hope of rejecting its form and therefore proving the hypothesis" (Clarke, 2003, p. 320). "If a hypothesis cannot be tested, it is not scientifically useful, no matter how interesting it might seem. Testing usually involves making observations, developing models, and performing experiments" (Tarbuck & Lutgens, 1999, p. 10). Section 14.8-10 gives substantial ideas for different experiments that could be conducted for further verification of the findings reported in this narrative. The tests of reproducibility include point source and sheet source layering in a flume study as well as a re-construction of a spigot dam.

In inferential statistics, hypothesis testing is a process involving the use of data to draw conclusions about the values of a population parameter. Hypothesis tests are conducted by first choosing two disjoint sets of possible parameter values called the null hypothesis (denoted H_0) and the alternative hypothesis (denoted H_1) (Weis, 2008, p. 402).

To test the claim that layers in sediment are a rapid process, the null hypothesis would be $H_0 \neq$ rapid and the alternative hypothesis would be $H_1 =$ rapid.

8.6 Rapid Stratification Hypothesis

Based on the case study of Kidd Copper it can be reasoned that rapid stratification is a primary structure as sediments under water settle out of fluid as stratified layers in a fast rate of deposition. These many layers found in sediment are a one-time process; instead of the previously held theory that these layers form slowly through time in a multi-depositional process.

To support the hypothesis that $H_1 =$ rapid stratification the *in situ* observations in the Kidd Copper property case will include the following: (1) determination of deposit as slow-long or as fast-short period of time; (2) determination of single-uniform layer or multiple layers deposited as primary structure; (3) determination of equal or unequal declivities of adjacent structures; (4) determination of similar or different slope angles in orientation; and (5) determination of symmetrical or asymmetrical profile.

Further support for the hypothesis includes the layering of sediment will not continue laterally in all directions until it thins out as a result of redeposition. Nor will, the adjacent boreholes with the same unit of time for measurement and the same depositional process have the same facies correlation in the rock record.

This scientific report will reject the null hypothesis that stratification is a slow-long process over time and therefore prove the hypothesis.

If the hypothesis of rapid stratification is proven correct then it stands as an axiom to support a 'younger date' for planet Earth because the differences in the adjacent or associated units do not reflect differences in origin but rather different units can reflect the same origin. The hypothesis is that a time-stratigraphic correlation of heterogeneous material can be referenced by the same point in time even when the distribution patterns are dissimilar.

The author asserts that the Kidd Copper redeposited sediment corresponds directly to rapid stratification and that rapid stratification opens the door for a younger earth. This is because the units in the rock record do not need to be correlated along similar patterns thus nullifying the GTS 2004 correlations for time references as the evidence in this report verifies.

8.7 The Multiple Layers are Not Individual Deposits

In the case of Kidd Copper, mine tailings waste was deposited to the site. The mine tailings accumulated over the impoundment area. The rock layers represented the redeposited material when the dam broke. The layers are not individual deposits but instead one deposit. Gary Nichols writes, "the materials are considered in terms of … the places where sediments accumulate into sedimentary rocks and become layers of stratigraphy through time" (2003, p. 3).

To recognize that scarcely any time was involved for the stratification and multiple layers to be present on the Kidd Copper property is important. Rocks did not form layers through time. The multiple layers are not individual deposits. Rather, stratified layers are a rapid process; and a stratified layer is a rapid process. The layers did not have to wait for the process of time to accumulate. It happened quickly.

Science begins with observations and the formulation of testable hypotheses that can be rejected by contrary evidence. The evidence seen through the mine tailings redeposited from a dam break is contrary to some of the established geology laws already mentioned and covered in this report.

The stratification on the Kidd Copper property case revealed heterogeneous deposits were settled out of fluid as stratified layers. Kidd Copper layers had unequal declivities, different slope angles and no symmetrical profile. Instead, of the sediment layers continuing laterally in all directions until thinning out as a result of all redeposition processes there was a random and unoriented distribution pattern.

The observable facts demanded of science support rapid stratification in the multi-layering of strata. The quick multi-layering of the flow from the dam break and the resettling of the mine tailings shows layers are not settled as one deposit on top of another through many events and much time; but rather the evidence supports multi-layered deposits are one event and through rapid stratification. The evidence to support this is that the historical age and the geologic setting of Kidd Copper is well established in documented files from: university reports, geology articles and in aerial photographs, some of which are included in this report, see data support in the References.

The claim that layers in sediment are a rapid process has been tested. The null hypothesis is $H_0 \neq$ rapid and the alternative hypothesis $H_1 =$ rapid. Therefore, the rejection of the null hypothesis leads to the acceptance of the alternative.

A time-stratigraphic correlation of heterogeneous material can be referenced by the same point in time even when the distribution patterns are dissimilar. The Kidd Copper redeposited sediment corresponds directly to rapid stratification and that rapid stratification is evidence of a younger earth will be further explained. Multiple layers have been used to date the age of the Earth. If multiple layers are not indicators of ages, stages, eons or periods of time then how is time to be correlated? Rapid stratification suggests a younger earth. Section 17.0 of this report will primarily focus on the time components to rock dating; and afterwards the rock record will be compared.

8.8 Possible Sources of Error

From major to minor the possible errors in the assertion of a younger deposit can be summarized as follows: (1) Assumed that all variables recorded were accurately reported from the field logs; (2) Assumed that the underlying facts as recorded are true as to the historicity of the Kidd Copper mine site; (3) Assumed that the principles outlined in this report are reflective of current scientific interpretation.

9.0 Geo3Fe3 Geology Assignment

"The site that you will be visiting this year is owned by Crowflight Minerals Inc. and is known as the Kidd Copper Property" (McMaster University, 2004, Geo3Fe3, Assignment #9 Mine Tailings). The boundary conditions were set by the staff of McMaster *Geo3Fe3* for the Mine Tailings assignment, assigned Monday August 30, 2004. Five essential steps were followed:

1) The professors defined the population by giving the University students the opportunity to "walk over the entire site" (McMaster University, August 2004, Assignment #9 Mine Tailings). This was to ensure that every member of the population is equally likely to be chosen to form part of the sample.

2) The students were directed to create "a map that details the different processes that you observe occurring in the different areas" (McMaster University, August 2004, Assignment #9 Mine Tailings).

3) They were to "use a spade at several locations to observe what is occurring in the subsurface" (McMaster University, August 2004, Assignment #9 Mine Tailings).

4) "Finally, you will use the air photos to describe the changes that have occurred at the site over time" (McMaster University, August 2004, Assignment #9 Mine Tailings).

5) To submit a field report of their findings; "An important aspect of this course is to develop the ability to take effective field notes. At the end of the course your field notes will be evaluated" (McMaster University, August 2004, Assignment #3 Field Notebooks).

These field reports are the basis of this analysis and report.

The Geological Staff Team from McMaster University including Dr. Bill Morris, Susan Vajoczki, Dr. Darren Gröcke who requested the digging of holes for the Field camp assignment at Kidd Copper property on August 30, 2004 – boreholes that resulted in the collection of primary research and data for this report. These professors were uninformed as to the development of this study's purpose. This report was written without the staff's or student's prior knowledge, consensus or agreement as to the defence of a younger earth. This report examined the data after the holes were dug and what was done with the accumulated data was the decision of the author. The views represented in this report do not necessarily reflect the views of either the staff of McMaster University or the students who contributed to this work.

9.1 Methodology

Visual inspection was predominantly used for the purpose of assessment of the sediment mass on the Kidd Copper Property. The testing parameters were conducted *in situ* state rather than lab conditions. "Soil survey [GEOL] The systematic examination of soils, their description and classification, mapping of soil types, and the assessment of soils for various … uses" (Licker, 2003, p. 347) was conducted. The geology students conducted original field excavations *in situ* looking "for sedimentary rocks, the major features including dip and strike, the sequence and thickness of beds measured either bed-by-bed or by lithological groupings, the lithology described by using some standard field classification, the major sedimentary structures, including the orientation of those that can be used to reconstruct ancient current directions (paleocurrents) and the presence and kind of fossils" (Blatt, Middleton & Murray, 1972, p. 5). Seismic logs were not undertaken because for the geology students they were not to "use a method that is more elaborate than is warranted by the quality of the data" (Dingman, 2002, p. 125).

9.2 Collection of Data

"The observations made in the field are generally only a selection of the total number that might be made" (Blatt, Middleton & Murray, 1972, p. 6). However, "in order that the measurements should be truly representative of the whole unit under investigation" Blatt, Middleton and Murray suggest "that considerable attention be given to the technique of sampling" (1972, p. 7). "A large number (more than 20 or more than 50 depending on the variability)" (Reineck & Singh, 1973, p. 91). 32 students from McMaster University, studying geology were assigned to study the waste site area on Kidd Copper with the naked eye, taking a large number of samples. 120 boreholes were dug with a shovel to make recordings of field observations for the Geo3Fe3 class, on August 30, 2004.

9.3 Sampling Frequency

"The basic concept in sampling is that of a *random sample*, which consist of a small number of observations chosen to be generally representative of a larger number that constitutes the *population* that is being studied" (Blatt, Middleton & Murray, 1972, p. 7). "The sample must not be chosen haphazardly" (Blatt, Middleton & Murray, 1972, p. 7). "In geology, they may be actual strata (beds) but they may also be any well-defined units… or geographic areas. A random sample is taken within each subgroup and the whole sample is called a *stratified random sample*. The sample is random within each subgroup but *systematic* within the whole population" (Blatt, Middleton & Murray, 1972, p. 7).

"The [sampling] frequency of which a data set is sampled is determined by the number of sampling points per unit distance or unit time, and the sampling frequency is equal to the number of samples (or stations) divided by the record (or traverse) length. For example, if a wave-form is sample 1000 times in one second the sampling frequency is 1kHz (and the Nyquist frequency 500 Hz); if a traverse is 500 m long with 50 stations, the sampling frequency is one per 10 m" (Allaby & Allaby, 2003, p. 475, 476).

9.4 Sampling Station Interval

"The distance between points at which measurements are taken or the time which elapses between measurements; it is equal to the traverse (or record) length divided by the number of stations (or samples). For example, a 250 m ground traverse with 25 stations along it has a sampling interval of 10 m; a wave-form might be sampled every two milliseconds, i.e. with a sampling interval of 2 ms (and a sampling frequency of 500 Hz)" (Allaby & Allaby, 2003, p. 476). In the case of Kidd Copper, a 4 km ground traverse from North to South with 30 boreholes along it has a sampling interval of 133 m.

9.5 Benchmark

The access road to the Kidd Copper property is the bench mark "marker that is firmly implanted in stable ground from which local subsidiary measurements originate and which is used as reference in topographic surveys and tidal readings" (MacDonald, *et al.*, 2003, p. 57). The access road is from the North refer to Figure 4 - Section 1.3. Point 56/4 is used as the marker control point of the intersection of the West-North Transect and the North-South Transect lithographic logs refer to Figure 28 - Section 10.4.

9.6 Field Observations

"Field work forms the basis for the study of sedimentary deposits, both modern and ancient. To make best use of the limited time and resources available for field studies, the geologist must first give some attention to defining what properties he will observe and measure and how he will select his observations and specimens for further study in the laboratory. What are the fundamental properties of sediments? How do we measure them? How can we take a sample that is truly representative of the unit being studied?" (Blatt, Middleton & Murray, 1972, p. 1).

"There are six aspects of sedimentary rocks which it is necessary to consider in the field, and which should be recorded in as much detail as possible. These are: the *lithology*, that is the composition and or mineralogy of the sediment; the *texture*, referring to the features and arrangements of the grains in the sediment, of which the most important aspect to examine in the field is grain-size; the *sedimentary structures*, present on bedding surfaces and within beds, some of which record the *paleaeocurrents* which deposited the rock; the *colour* of the sedimentary rock; the *thickness* and *geometry* of the beds or rock units and of the sedimentary rock mass as a whole;

and the nature, distribution and preservation of *fossils* contained within the sedimentary rocks" (Reineck & Singh, 1973, p. 11).

"Notes and sketches on this [sedimentary rock described in the field] should include the type and spacing of any lamination, the scale and shape of cross lamination, cross bedding, sole structures, lineations, bioturbation, and so on" (Nichols, 2003, p. 35). Point data, including lithological logs from the surface of the waste site are collected from the field observations and reports and discussed later in this report.

The Kidd Copper waste site is a geological observatory for the diagenesis of first layer strata because both the age and the matter content of the original parent rocks are well-established. No guess work is necessary to deduct the initial depositional process. According to Desmond Carroll and Erica Hamilton they would "be examining the product of the first mining cycle" (Carroll, 2004, p. 45; Hamilton, 2004, p. 33). "The product on this are silicates and sulphides, iron sulphides $\rightarrow$ piritite + pyrite when they are exposed to air they rapidly produce sulphuric acid and create a leachate plume" (Carroll, 2004, p. 45).

9.7 McMaster Students Walk Over Assigned Kidd Copper Site Area

Figure 23 - Section 9.7: McMaster Students Walk over the Mine tailings Site in 2004

Credit: Photograph by Andy Weller.

Soil scientist "traverse the landscape along selected transects (straight-line paths), auguring only at enough points to confirm expected soil properties and boundaries" (Brady &Weil, 2002, p. 849). "Landscape [GEOGRAPHY] The distinct association of land forms that can be seen in a single view" (Parker, 1997, p. 198). The 32 students would use the "eyeball"… "Isohyetal method [in other words] "free-hand drawing" (Dingman, 2002, p. 125). Figure 23 - Section 9.7 is a photograph taken by Andy Weller on August 30, 2004 where some McMaster University *Geo3Fe3* students are walking on the Kidd Copper Mine tailings site.

"*Palaeosol* a soil formed on a past landscape, documented for both the *Precambrian* and *Phanerozoic*" (Kearey, 2001, p. 195). "Soil scientists often dig a large hole, called a *soil pit*, usually several meters deep and about a meter wide, to expose soil horizons for study. The vertical section exposing a set of horizons in the wall of such a pit is termed a *soil profile*" (Brady &Weil, 2002, p. 11).

The *Geo3Fe3* students from McMaster University were requested by Dr. Bill Morris to develop a map of the surface area of the Kidd Copper property mine tailings waste site over a period of five hours. "A field of view" (DePree & Axelrod, 2003, p. 295). Start "Time 9:00 am. Weather: Sunny; cloud cover: 85% Temperature = 20°C." (Li, 2004, p. 51). The data was collected and entered into each students' Forestry Supplier's Field Books in 2004. (Dr. Bill Morris and the other McMaster professors are not necessarily in agreement with the findings of this report and their names are solely used for the purposes of requesting the study and for identification purposes; and not an endorsement of the conclusion).

9.8 The Kidd Copper Property is Assayed

The abandoned nickel-copper mine-waste site is at the Kidd Copper Mine, Ontario roughly 2 Km East of Worthington, Ontario, Canada. Topography is "the general configuration of a land surface, including size, relief and elevation" (MacDonald, *et al.*, 2003, p. 432). In Figure 23 - Section 9.7 the topography looks wet and sandy with tufts of grass cover. The grain looks smooth. Hand-shovels were used to unearth the layer of top soil. This area was the primary data collection source and starting place for observations used in the analysis for this report.

9.9 Graphic Log Constructed of the Topograpghy

"The standard method for collecting field data of sedimentary rocks is to construct a graphic log of the sequence" (Reineck & Singh, 1973, p. 13). "Some geologists design comprehensive graphic logs that can be used in the field in order to make sure that a full description of the rocks is recorded" (Blatt, Middleton & Murray, 1972, p. 6). "First, assess the outcrop" (Reineck & Singh, 1973, p. 91). 120 Boreholes were represented by on-site reports and graphic logs from the Kidd Copper property on August 30, 2004.

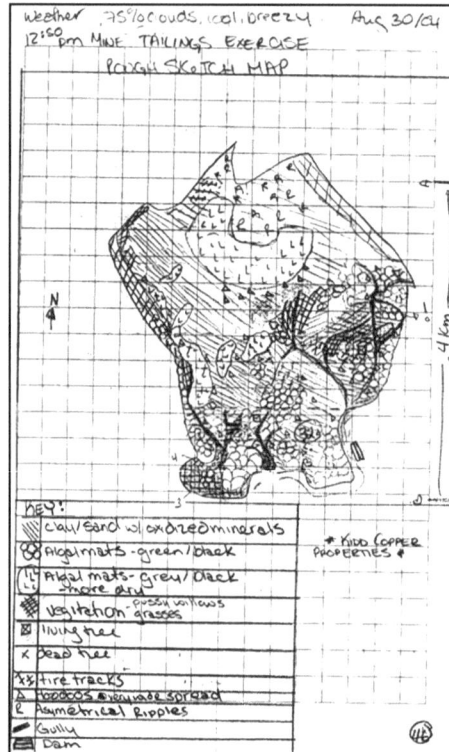

Figure 24 - Section 9.9: Field Sketch from Eden Hynes Student Log, August 30, 2004

The field sketch that was recorded by Eden Hynes on August 30, 2004 is one example that shows the complex surface features found on the Kidd Copper property including: algal mats, clay, dam, gully, hoodoos, sand, trees, and vegetation (Hynes, 2004, p. 46). Refer to Figure 24 - Section 9.9. Underneath each of these topographic features laid the same redeposited mine tailings waste as an accumulation of sediment from the aftermath of the flood. According to "Steno's Law of superposition states that older rocks are found beneath younger rocks. This happens because when sediments settle out of a fluid, the youngest layers will be deposited on top of previous layers" (Albin, 2004, p. 44). Conventional wisdom expects a uniform deposit when the fluids settle because "according to the principle of original horizontality, these layers will initially be set down in a horizontal fashion. Any change from this orientation occurs at another time when the rocks have been tilted by tectonic forces" (Albin, 2004, p. 44). Knowing when the sediment was deposited at Kidd Copper it can be 100% confirmed that tectonic forces were not factors of the various sedimentation layers exposed with stratified orientations; thus nullifying the expectation that tilting only occurs after the sediment initially settles. In the case of Kidd Copper, the mixing of sediment in water caused the diversified orientation in each of the boreholes simultaneously and not at other times.

9.10 Unearth the Layer of Top Soil

Desmond Carroll records the equipment: "shovel, spade, GPS – MAC OA5 046 Garmin GPS 12XL" (2004, p. 43). The first shovel was "*in situ* – (of a rock, soil or fossil) in the situation or position in which it was originally deposited or formed" (MacDonald, *et al.*, 2003, p. 250). The surface deposits on the Kidd Copper property look uniform to the naked eye. Figure 25 - Section 9.10 is a photograph taken by Derrick Li of Loreen Rudd's first shovel to unearth the top layer in the upper zone of transition, breaking the *A horizon* or topsoil to unearth the *B horizon* (MacDonald, *et al.*, 2003, p. 404). More on the soil profile found in Table 29 - Appendix A.

Figure 25 - Section 9.10: Stratified Layers Unearthed in the Upper Zone of Transition

Credit: Photograph by Derrick Li (August 30, 2004).

Reineck & Singh provide information on how to collect and analyze palaeocurrent measurements: (1) take many measurements for accuracy; (2) if similar units collect few measurents; (3) if units are different more measurements (1973, p. 92). "Measurements collected from different sedimentary structures should be kept apart, at least initially. If they are very similar they can be combined. Also, keep separate measurements from different lithofacies at an exposure; they may have been deposited by different types of current, or currents from different directions … The measurements you have taken may not represent the current direction if either the shape or the orientation (or both) of the sedimentary structures has been changed by tectonism. It is important to appreciate that two changes can occur: tilt and deformation. A simple change in the inclination of the plane, or which the sedimentary structure is a part, is described as the tilt. Tilt does not change the shape of a sedimentary structure. Processes which change the shape of a sedimentary structure are described as deformation" (Reineck & Singh, 1973, p. 91, 92).

Tilting or deformation should not be found because tectonic forces were not involved; instead a homogeneous layer with similar distribution and patterns should be found throughout the samples of the boreholes collected when the earth is uncovered.

9.11 Soil Stripes Unearthed After Redeposition on Kidd Copper

For the purpose of this report the data that the McMaster University students collected was in the upper and lower zone of transition. Layering is "a tabular succession of different components in igneous or metamorphic rocks, or the formation of layers in a particular rock; e.g. in plutonic rocks as a result of crystal settling in magma. 'Layering' is preferable to *banding* as it implies three dimensions rather than two" (Bates & Jackson, 1984, p. 292). "Soil stripes [GEOL] Alternating bands of fine and coarse material in a soil structure" (Licker, 2003, p. 347).

Loreen Rudd's first shovel in Figure 25 - Section 9.10 of the mine tailings unearthed the sedimentation in the upper zone of transition revealing an uneven, rust color and white layered appearance of laminae. This hole represents Borehole 64 on the Compilation of Bore Holes on the Kidd Copper Aggregate Base Map; Figure 27 - Section 10.3. The surface deposits look uniform to the naked eye but the underlying layer reveals a stratification of sediment deposits. This section is of an inclined or vertical surface that is uncovered artificially with a shovel (as a strip mine or road cut) through a part of the earth's crust. "A description or scale drawing of the successive rock units or geologic structures showed by the exposed surface" (Licker, 2003, p. 328) is recorded by each Geology student refer to Section 10.2.

9.12 Photographs from Field Observations show Stratified Layers

On August 30, 2004 data acquisition and careful observations and recorded data entries by 32 students enrolled in the *Geo3Fe3* Field Camp. They wrote detailed soil profiles as log entries with notes for clarification. The Geology students used shovels to expose various surface depths of the strata. The area covered was a nickel-copper mine-waste site at Kidd Copper, Ontario. It was approx 4 km square. "Standard distance [was taken] a two-dimensional equivalent of the standard deviation, a normalized distance built from the standard deviations of the eastings and northings for a set of points" (Clarke, 2003, p. 325). GPS coordinates for Borehole 107 "0467286   5137650" (Windus,

2004, p. 36) is one sample of where the mine disposed its waste in the form of mine tailings. Stratification "layers of yellow and grey" was observed to a "depth of 50 cm" (Grenier, 2004, p. 75).

Figure 26 - Section 9.12: Exposed Stratified Layer of Sediment from One Event

The shovel in Figure 26 - Section 9.12 was one of the 120 boreholes unearthed that exposed stratified layers when the earth on the Kidd Copper property mine tailings site was uncovered. This sediment deposit is not uniform in appearance in either the upper or lower zone of transition. This shovel exposed a stratified layer of sediment. This photograph of Borehole 103 was taken by Andy Weller August 30, 2004 showing a depth of four feet down slope bank. Andy's description of the borehole is: "Top two feet yellow clay; Two feet below is gray and looks thick potter's clay with green running through which is copper" (Borehole 103, 2004, 29-43-A). Colored layers are visibly detected.

10.0 Primary Research of 120 Borehole Samples

McMaster University Class Professors, Dr. Bill Morris GSB302, Susan Vajoczki BSB313, and Dr. Darren Gröcke GSB305 extended the discussion of the walk about at Kidd Copper property on August 30, 2004 with the geology students in *Geo3Fe3* Field Camp. These professors assigned Assignment # 9 Due Wednesday, September 1, 2004 that was "use a spade at several locations to observe what is occurring in the subsurface" (McMaster University, August 2004, Mine Tailings).

Thirty-two students traversed the Kidd Copper mine tailings site to collect data on the topography and subsurface, in the upper zone of transition and the lower zone of accumulation, taking notes on the depositional processes they observed. 120 borehole samples were taken (sample size n = 120) which covered 85-90% surface area. Shovels were used to expose various surface depths of the strata and detailed soil profiles are recorded in Section 10.2. For example, the stratification observed at Borehole 115 was seen to a "depth of 75 cm" (Zelek, 2004, p. 31). This is the area where the nickel-copper mine at Kidd Copper, Ontario disposed of its waste in the form of mine tailings. Instrument: Garmin GPS 12 Channel Receiver 12XL, Serial Number 92214873. Magnetic declination of the Whitefish Falls area is 9° West. The Kidd Copper mine tailings waste site area is located at GPS coordinates "0467286   5137650" (Windus, 2004, p. 36).

10.1 Data Acquisition of Soil Profiles

Data acquisition of soil profiles came from an assigned walk-about on Kidd Copper from McMaster University Geo3Fe3 class on August 30, 2004. Field data is pertinent to the analysis of the elucidating conditions, environments and processes of deposition. "Field: The contents of one attribute for one record, as written in a file" (Clarke, 2003, p. 312). Careful observations by 32 students were recorded into each student's Forestry Field Log to provide first-hand observations on the geologic processes.

These students are: Karl Belan, Andrew Benson, RA Brown, Desmond Carroll, Erik Enders, Cetina Farruggia, JJ Gabriel, Justin Grenier, Sarah Gyatt, Erica Hamilton, Nicole Harper, Eden Hynes, Toni Jack, Daria Kilmenko, Jen Lengert, Derrick Li, Karen De Medeiros, Lisa Melymuk, Heather Robertson, Maddy Rosamond, Loreen Rudd, Tripti Saha, Saroj Singh, Stephanie Studdy, Van Pham, Caitlin Vanderkooy, Pete van Hengstum, S Walker, Andy Weller, Zachary Windus, May Xiujuan and Linda Zelek.

Written permission from each student enrolled in the McMaster University's school of Geography and Geology Field Camp (*Geo3Fe3*) was obtained by Loreen Rudd from each of the McMaster University students to photocopy their entries from their Forestry Suppliers Field Books, refer to Appendix H for the 32 permission statements.

On October 14, 2004, original raw data entries and stratigraphic logs with information on location, color and thickness of the formations as well as the type of sediment were photocopied by Loreen Rudd, now Loreen Sherman, who is the author of this report who collected, correlated and compiled the data entries. The first-hand observations of each borehole were taken blindly and without prejudice by the students and without prior knowledge of this research paper or the report's objective, hypothesis or outcome. An abbreviated listing of these entries is included in Section 10.2.

Editor's Note: the use of personal names or brand names in this research paper are for identification purposes only and do not constitute endorsement by the named subjects or McMaster University on the subject matter presented by the author's analysis and views as represented in this report.

10.2 Raw Data of Boreholes Collected August 30, 2004

The compilation of 32 Geology student's field entries provide first-hand observations of exposed sediment supplied from *in situ* site observations after digging their boreholes on the Kidd Copper property on August 30, 2004. This is an abbreviated listing of data and stratigraphic log entries to reference 120 bore holes dug at Kidd Copper. Some GPS coordintates were provided in the boreholes' descriptions to identify the location where that 'boring' was dug.

The Key: Figure 27 - Section 10.3 is a Base Map for the Compilation of Boreholes

Borehole is the number that corresponds to n = 120 on the Base map.
Field Book is the number corresponding to 32 students Forestry Supplier's Field Book.
Page is the student's reference to the borehole contained within their field book.
   *boring = hole dug in ground by shovel

Borehole 1: Karl Belan, Field Book #1, page 36, point 1.

GPS 467190 5137851
-- 5 inches from surface layers of rust coloured
   oxidized tailings
-- 8 inch layer of montmorillonite clay
   with interlayers of oxidized clay
-- Coarse sand beneath that, more thin rust layers
   in sand 12 more inches

-- 2 feet to water table of more clay

Borehole 2: Karl Belan, Field Book #1, page 38, point 2.

GPS 467226 5137734
-- Tiny surface layer of clay 2 cm
-- 30 cm of sand heavily interbedded with oxidized rust coloured layers
-- Clay layer 5 cm
-- Coarse sand layer of 3 cm, interbedded
-- Clay to bottom of 2 foot hole
-- Mini hoodoos on surface around trees caused by Aeolian surface wind

---

Borehole 3: Karl Belan, Field Book #1, page 38, point 3.

-- 20 cm of grey clay with orange layers strands

Borehole 4: Karl Belan, Field Book #1, page 39, point 4.

GPS 467252 51377591

-- 17 cm of very fine grained clay with only slight oxidation layers, brown-orange colour
-- Grey clay layer beneath that layer for 10 cm to bottom of hole

---

Borehole 5: Karl Belan, Field Book #1, page 39, point 5.

GPS 0467124 5137642

-- Grey and orange stained, medium grained sand alternating in thin layers
-- Darker areas indicate higher iron levels
-- Stratified may be ripples

Borehole 6: Karl Belan, Field Book #1, page 42, point 6.

GPS 0467043 5137672

-- Top 25 cm is sand stratified with dark bands of oxygenated material
-- Below 25 cm very fine montmorillonite clay to 2 ft depth

Borehole – no actual bores description only: Andrew Benson, Field Book #2, page 26.

-- Iron oxides leaching to surface through hydrostatic environment
-- Montmorillonite clay abundant under surface
-- Iron oxide bandings found in holes dug
-- Possible Redox reactions
-- Visible lighter band to darker
-- Some green and yellow banding
-- Green possibly CuO yellow possibly a sulphur compound

Borehole 7: R.A. Brown, Field Book #3, page 89, hole 1.

GPS: 0467190 5137851
-- Surface obvious drainage area
   Vegetation on higher points of ground
-- 10 cm sand
-- Oxid
-- 29 cm clay montmorillonite (interlayers of oxidation with clay)
-- Oxid
-- 19 cm sand
-- Montmorillonite clay
-- Grey white

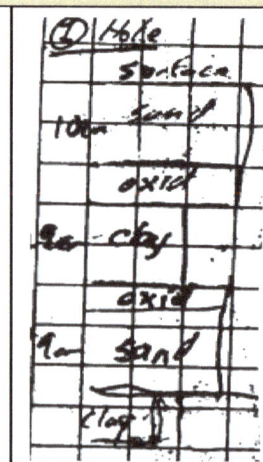

Borehole 8: R.A. Brown, Field Book #3, page 91, hole 4.

GPS: 0467226   5137734
-- Top 2 cm clay
-- 30 cm sand with clay
-- interbedded
-- 54 cm deep hole

Borehole 9: R.A. Brown, Field Book #3, page 92, hole 3.

-- Hole 3
fine sand w (unreadable)
38 cm deep

Borehole 10: R.A. Brown, Field Book #3, page 95, hole 12.

Hole
-- Very fine grain, white/orange clay
-- Some laminations, no definite layers
-- Bottom at gulleys are grey clay
-- Tops are sandy grey-orange

Borehole 11: R.A. Brown, Field Book #3, page 95, hole 13.

-- Hole, sand stratified
-- Possible ripples between laminations
-- Lots of iron oxides

Borehole 12: R.A. Brown, Field Book #3, page 95, hole 14.

GPS 467043   5137672
-- Hole,
-- Stratified clay and sand
-- Large clay layer and dark blue

Loreen Michele Sherman

108

Borehole 13: Desmond Carroll, Field Book #4, page 46, point 2.

-- Section dug, tailings roughly 15 cm, 60 cm of clay material, last
   10 cm is wet moss covered and lots of sand, between sand anoxic
   thin soil layer developing over the liner clay 20cm
-- Sand clay structure
-- Grasses growing, clay is wet again, clay
   in the water table
Found a small stream: 15 cm/75 Bearing 162°
-- Silts, clay interbeds
-- Silt orange
-- Clay grey

Borehole 14: Desmond Carroll, Field Book #4, page 50 point 5.

-- Solid clay
-- Interbedded
   light grey and iron interbeds
-- Dark clay
-- Interbedded dark and light clay
-- Saturated dark clay

Borehole 15: Desmond Carroll, Field Book #4, page 51 point 9.

-- Orange sand →oxidation
       strands grey sand
-- Sand
-- Gray clay
-- Thin band of sand and clay
-- Pyrite band
-- Sand
-- Pyrite band
-- Laminated dark grey
-- Interbedded sand and clay

Borehole 16: Erik Enders, Field Book #5, page 49 point 0.
-- Orange-rust color shows presence of iron
-- Cross-section in most areas shows alternating layers of orange (iron/oxidized) and white (probably Montmorillonite clay)

Borehole w/o reference point description: Cetina Farruggia, Field Book #6, page 41, --.

-- Dig - alternating grey and orange layers
-- Coarse grained sandy
-- Grey clay - fine grained /saturated with water
-- Montmorillonite
-- Low point- chemical change
   Change in color, light to dark- orange
-- Jarosite precipitate
-- Iron sulphide oxidation
-- Ripples
-- More standing water / more oxidation
-- Ferrihydrite
-- Iron oxides and sulphuric acid
-- Hard pan thick layer
-- Active surface

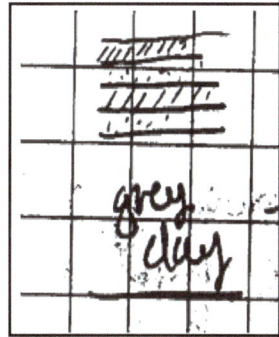

Borehole 17: J.J. Gabriel, Field Book #7, page 48, point 1.

-- Sand highly porous
-- ferrous (orange)
-- Grey – wet
   - porous
-- High concentrations of precipitated salt

Depth 60 cm

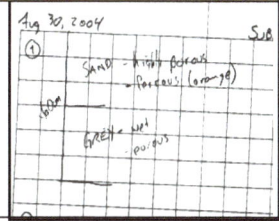

Borehole 18: J.J. Gabriel, Field Book #7, page 48 point 2.

-- Substrate profile
-- Sand
-- 60 cm
-- Clay

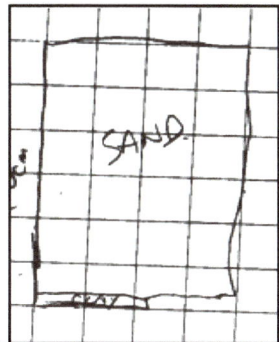

Borehole 19: Justin Grenier, Field Book #8, page 75, site 1.

Dug to sample:
-- Layers of yellow and grey
-- Yellow = Sulphur?
-- Grey = Clay?
-- 50 cm
-- More sand than clay
-- Sand is yellow material

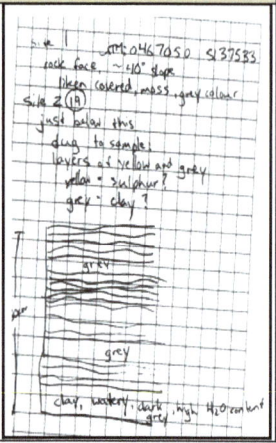

Borehole 20: Justin Grenier, Field Book #8, page 77, site 3.

-- Took sample dug hole 20 cm deep until hit water
-- Sandy / Clay / Sand / Clay / Sand

-- Theorize: clay liner laid at water table, this is how
   tailings could have traveled

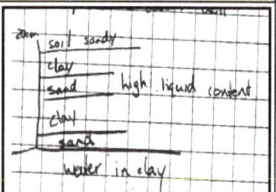

Borehole 21: Justin Grenier, Field Book #8, page 79, site 4.

(Hole dug) ~2.2 m above H_2O table
-- Total depth 30 cm
-- Pyrite layer15 cm down
-- Interbeds of silt (yellow orange)
   clay (grey)
   pyrite layers

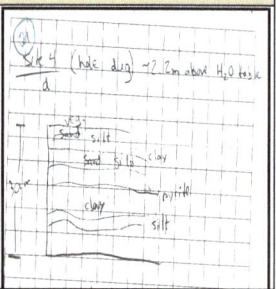

Borehole 22: Justin Grenier, Field Book #8, page 83, site 6.

-- Clay grey with orange banding
-- Dark grey clay
-- Clay with iron banding
-- Interbedded light/dark grey
-- Fine-grain
-- Grey liquefied clay
-- Hole depth 1 m

Borehole 23: Justin Grenier, Field Book #8, page 85, site 9.

(Dug hole) 184 m above sea level
-- Orange sand with grey sand
-- Lenticular beds
-- Surrounding lenticular beds are layers of pyrite
-- Sand→ clay, sand →clay
-- Concentrated iron oxide / clay/ grey /iron oxide /sand
-- Sand
-- Pyrite
Laminated light / dark grey sand/ clay, more sand
Water table
-- Fining upwards sequence

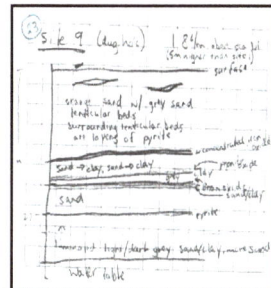

Borehole 24: Sarah Gyatt, Field Book #9, page 59, 60, point 1.

-- Tailings come in, settling occurs, algae stuff grows
-- Salts (precipitation)
-- Dug down 2 feet
-- Liesenganen ¾ of the way was a light-coloured clay with anothengic of orange colour
   and darker wet clay at the bottom
-- Series of redox reaction
-- Presence of iron orange
-- Montmorillonite clay – drill clay

       347 paces N-S

       .6818 m = 509 m

       N- E is a gypsum rich area

Borehole 25: Sarah Gyatt, Field Book #9, page 58 point 2.

-- Copper located about ¾ of the way down -- and heavy clay

Borehole 26: Sarah Gyatt, Field Book #9, page 58 point 3.

-- Yellow

Background: Erica Hamilton, Field Book #10, page 32, 33, point --.

Inco and Falconbridge Navanda controlled site recently got rid of land.

Thought no more big ore bodies here and too expensive to hold.

Then junior mining companies started to work smaller properties.

Crow Flight Resources (go to their website)

Remapped geology of area.

Made new mineral deposit model.

They had to:

1) Clean off covered outcrop of trees.
2) Trenched area.
3) Geophysicists drilled area to inspect covered minerals and then have rock core's logged.

FNX (another junior mining company) hit good spots and found small-high grade deposits.

Assignment => look at the first mining cycle.

(Pyratite & Pyrite => chemically react a lot & quickly.)

Borehole 27: Nicole Harper, Field Book #11, page 56, point A1.

-- 5 inches from surface rust
-- colour oxidated layers
-- % rust coloured
-- 8 inch layer of montmorillonite clay
   with interbedded layers of oxidized clay
-- beneath that more rust layers
-- H_2O table @ 2 ft hit montmorillonite clay

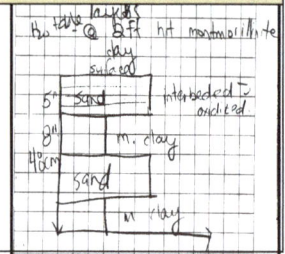

Borehole 28: Nicole Harper, Field Book #11, page 57, point 2.

GPS 0467220 5137734

-- Surface clay
-- clay inter to iron oxide sand
-- 2 feet depth

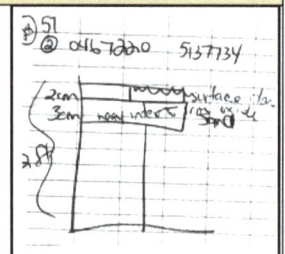

Borehole 29: Nicole Harper, Field Book #11, page 59, point 3.

GPS 0467252   5137591
-- Clay brownish
-- No oxidation present
-- Very fine grain
Depth 17 cm

Borehole 30: Nicole Harper, Field Book #11, page 60, point 4.

GPS 0467124   5137642
-- Sand
-- Stratified with staine (Fe)
   may symbolize with ripples
-- Large Fe deposits
Depth 30 cm

Borehole 31: Nicole Harper, Field Book #11, page 60, point 5.

GPS 0467043   5137672
-- Sand stratified with Fe deposits
-- Montmorillinite

Borehole 32: Eden Hynes, Field Book #12, page 45, 47, details.

-- Miners built damn to seal tailings pond
-- Tailings/waste dumped into area
-- Area under water, algae blooms covered water surface
-- H_2O pumped in SO minerals – sulphides don't oxidate
-- Damn broke => flow pattern
-- Montmillerite clay red
Rill erosion: Flow direction identified b/c small tributaries converge towards main perimeter channels and show us how the water found its way out.
****
-- White/grey orange
-- Iron content, small hoodoos apparent in iron rich areas
=> algae mats likely conscied out
    No removed by running water
-- Sequence of iron rich there whiter body some under identified algae mats
-- Whiter –shiny areasmontmerillimite clay that was once utilized in lubricating drill

Borehole 33: Toni Jack, Field Book #13, page 32, points 1.

-- Ripples on surface of ground
-- Surface is a light sandy color
-- Medium grain (like sand)
-- 1.5 inches down (below surface) a white layer starts => coarse grained with darker grey to white progression about 1 inch thick
-- Ends in a thick dark orange layer also coarse grained

Borehole 34: Toni Jack, Field Book #13, page 33, points 2.

-- Algael mats on surface (green, dark and light)
-- Rain ripples on surface
-- Thinner sandy colored layer = 1 cm which is about half dark orange
-- Thick white layer of clay alternating with orange sediment

Borehole 35: Toni Jack, Field Book #13, page 33-35, points 3.

-- Thin alternating layers of light & dark orange coarse grained sediment
Notes

Ferrihydrite is a crust on the top of areas around points 33 & 34. Many spots are broken up into flakes from disturbance.
A thin, rust colored layer => dry and brittle.
Whitish grey layers of mud montmerillimite clay are formed as standing water evaporates => the heavier coarser grains precipitate first, then the clays => hypothesis: by following alternating layers of orange coarse material and clay you can see when the pipe let out material.

Borehole 36: Toni Jack, Field Book #13, page 40, points 16.

-- Sand (light brown)
-- Very deep (1/2 feet)

Borehole 37: Daria Kilmenko, Field Book #14, page 94, point 16.

-- Iron
-- Orange and white layers
-- Green algae and moss

Borehole 38: Daria Kilmenko, Field Book #14, page 94, point 18.

-- Sand is deposited first because its heavier (coarser) and clay deposited after as its finer

Borehole 39: Jen Lengert, Field Book #15, page 45, point X0.

-- Surface dark with lichen, speckled vegetation
-- 30 cm of moist sand which separates into different
   coloured layers
-- Dark, hard surface
-- Coarse-medium grained sand orange
-- White, fine grained sand
-- Orange layer as above
-- White layers
-- Continues with orange for sample
=> Orange are oxidated layers iron reacting with sulphides

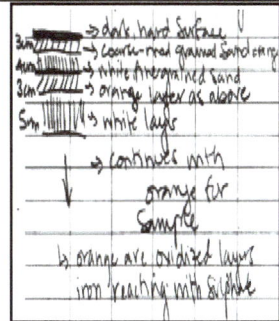

Borehole 40: Jen Lengert, Field Book #15, page 46, point X1.

-- Dark surface
-- White fine-grained sand
-- Orange coarser-grained
-- White sands
-- Orange
-- White
-- Orange with some white
-- Clay, moist consistency

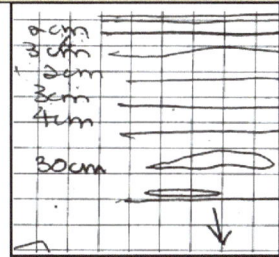

Borehole 41: Jen Lengert, Field Book #15, page 49, point X6.

GPS 0467103   5137623
-- Light and dark surface
-- Less ground vegetation
-- Thicker white sand layers
-- Oxidized layer on rip 3 cm and orange layer
   just above clay
-- Surface
-- 3 cm Orange
-- 20 cm White
-- 20 cm White and orange
-- Clay

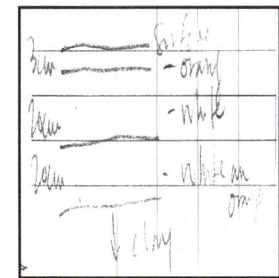

Borehole 42: Derrick Li, Field Book #16, page 54, point W.

-- Iron site is wet
-- Gerry stuff under soil
-- Pebbles found

Borehole 43: Karen De Medeiros, Field Book #17, page 39, holes 1.

GPS 046713, 5137845
-- 2 cm thick dark brown top (clay)
-- Interbedded white, orange and red
   60 cm
-- 2 cm red band
-- Clay all the way down (grey)
-- Top layer
   wave ripples 6 cm
   northern wind
   very fine grain
-- 75 cm deep

Borehole 44: Karen De Medeiros, Field Book #17, page 41, holes 2.

-- Dark brown grey 4 cm
-- 3 cm
-- Dark red chips 20 cm
-- Grey 4 cm
-- 3 cm Red
-- Grey

-- Depth 45 cm

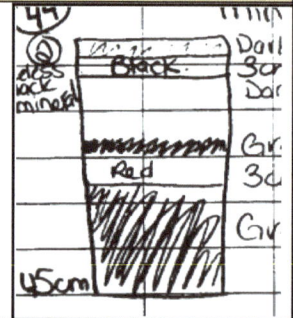

Borehole 45: Karen De Medeiros, Field Book #17, page 42, holes 5.

-- Orange
-- Little red
-- Some grey
-- Depth 30 cm

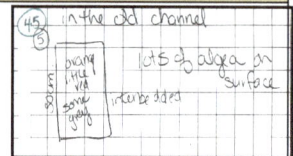

Borehole 46: Karen De Medeiros, Field Book #17, page 43, holes 6.

-- Interbedding orange and
            red with grey
-- Grey
-- 44 cm all the way down

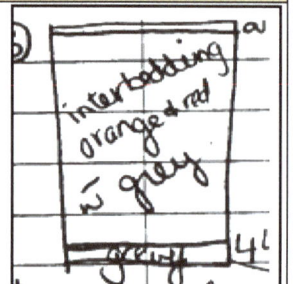

Borehole 47: Lisa Melymuk, Field Book #18, page 42, point 1.

-- Trowel depth
-- 3 cm Orange-white-orange
-- 4 cm Fine white sand
-- 15 cm Orange sand

Borehole 48: Lisa Melymuk, Field Book #18, page 42, point 2.

GPS 0467286   5137650
-- Darker coloured surface
-- O = Orange
-- W = White
-- O
-- W
-- O
-- W
-- W
-- O
-- W
-- O
-- O
-- W
-- Between the white is a layer of grey blue
-- W
-- Depth is 50 cm

Borehole 49: Lisa Melymuk, Field Book #18, page 44, point 5.

GPS 0467194   5137446
-- No vegetation
-- Rust coloured (ferrihydrite) crust on surface
-- Whitish/orange surface
-- Ground crust
-- 30 cm Depth

Borehole 50: Lisa Melymuk, Field Book #18, page 45, point 9.

GPS 0467103   5137623
-- Clay is 40 cm down from surface

Borehole 51: Lisa Melymuk, Field Book #18, page 45, point 10.
-- Ferrihydrite layer within sediment → oxidation front at 15 cm
-- Farther down is clay layer
Indicates recent water presence:
-- Salt deposits from water
-- Ferrihydrite with salt deposits on surface
-- Dead algae → black/brown layer

Borehole 52: Lisa Melymuk, Field Book #18, page 46, point 11.
-- "Aero bar" type bubbles in iron oxide and silica sand layer just below surface (~2 cm down)

Borehole 53: Lisa Melymuk, Field Book #18, page 46, point 15.
-- Oxidation horizon at 60 cm deep
-- Water table at 1 m
-- Oxidation horizon indicates a change in grain size => iron oxide percolates down to top of clay layer, then oxidizes in one spot to create iron hard pan layer

Borehole 54: Heather Robertson, Field Book #19, page 50, 54, point 1.
GPS 0467114   5137842
-- Material brought by pipes as sluri added to H_2O covered area
-- Dam broke and area exposed to air (30 years ago)
-- Channel hard-packed dark grey, smooth, discontinuous
****
-- Darker orange col below
-- Iron oxides
-- 1 m down = black clay
-- Montmorillonite clay- white clay, used to operate drill
-- Very fine grained sand
Soil profile: 75 cm
-- Brown top
-- Interbedded layers dark/ light orange sand
-- White clay 3
-- 5 m strip reddish 3
-- Lighter grey and dark charcoal clay-iron oxides

Borehole 55: Heather Robertson, Field Book #19, page 54, 55, point 2.

Distinct different colours
-- 45 cm
-- Black mineral dumped here (useless)
-- Very dark grey/black
-- Dark brown clay
-- 5 cm White spots throughout
-- 20 cm Interbedded bright red/orange with grey
-- 20 cm Bright grey

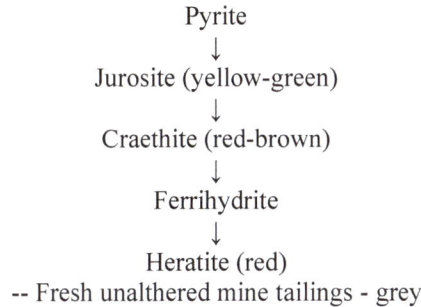

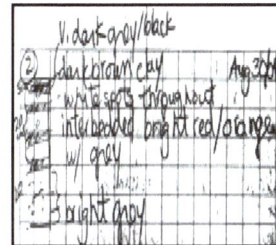

Pyrite
↓
Jurosite (yellow-green)
↓
Craethite (red-brown)
↓
Ferrihydrite
↓
Heratite (red)
-- Fresh unalthered mine tailings - grey

Borehole 56: Heather Robertson, Field Book #19, page 57, point 4.

-- 1 cm Dark black/beige spotted
-- Beige/ bright orange layers
-- Iron waste product oxidized
-- Depth 30 cm

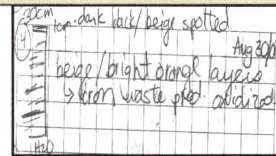

Borehole 57: Heather Robertson, Field Book #19, page 57, point 5.

-- 12 cm Very red

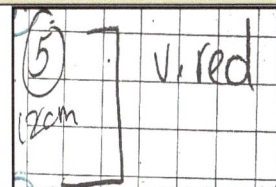

Borehole 58: Heather Robertson, Field Book #19, page 57, point 6.

-- Surface dark grey/black/with moss
-- Light beige/orange and white layers
-- 40 cm

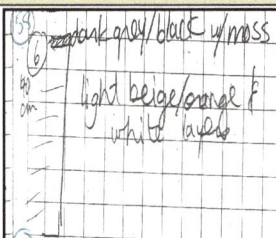

Borehole 59: Heather Robertson, Field Book #19, page 57, point A.

-- Black moss covered surface
-- Light grey/bright orange layers with
  red/brown small layers
-- Light grey
-- 45 cm

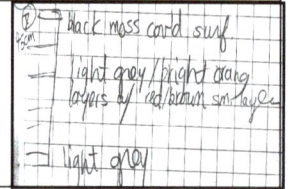

Borehole 60: Maddy Rosamond, Field Book #20, page 73, point 1.

-- Soil Profile taken (sed)
-- Black sand with patterns- mud cracks, moss in cracks
-- Very fine sand
-- Yellow- fine sand
-- Grey, blue clay
-- Black surface- very thin (<5 mm)
-- Total 55 cm depth

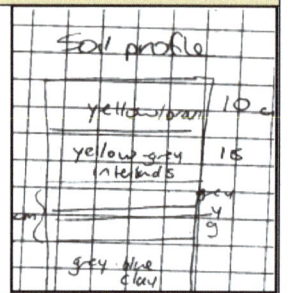

Borehole 61: Maddy Rosamond, Field Book #20, page 75, point 6.

Soil profile on peninsula
-- 20 cm Blue, grey, yellow lam
-- 10 cm Dark blue-grey clay
-- Dark red
-- 30 cm Grey, yellow lam

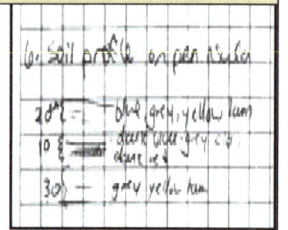

Borehole 62: Maddy Rosamond, Field Book #20, page 75, point 7.

-- 10 Yellow clay
-- Blue Clay

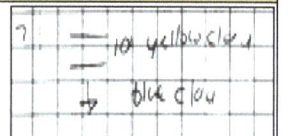

Borehole 63: Maddy Rosamond, Field Book #20, page 75, point 8.

Soil Profile
30 cm Sand

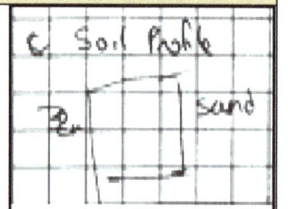

Borehole 64: Loreen Rudd, Field Book #21, page 73, point X1.
-- Depth 1½ feet
-- Rust color and white

Borehole 65: Loreen Rudd, Field Book #21, page 73, point X2.
-- Depth 1½ feet
-- Wet sand
-- Color "iron" orange-brown

Borehole 66: Loreen Rudd, Field Book #21, page 76, point X10.
0 –
-2 – Sand
-4 – Yellow-rust
-6 –
-8 – Grey-white
-10 –
-12 –

Depth of dug-out hole (12 cm)

Borehole 67: Loreen Rudd, Field Book #21, page 76, point X12.
Noticeable colored layers (sand, grey, rust, yellow and white) of mineral deposits distributed randomly to a total depth of 12 cm.

Borehole 68: Loreen Rudd, Field Book #21, page 79, point 23.
Dug hole 1 ½ feet
-- Sand
-- Rust
-- White
-- Dark brown layers

End Time: 16:56 PM

Borehole 69: Tripti Saha, Field Book #22, page 73, point A.
-- See alternating layers of dark brown sand
   and light grey sand when dug up

Borehole 70: Saroj Singh, Field Book #23, page 82, sketch.	
-- Dark Yellow -- Gray -- Weathering processes—once water moves in the gray is deposited -- Montmorillanite clay and the alternating from the sinking of the lighter grains	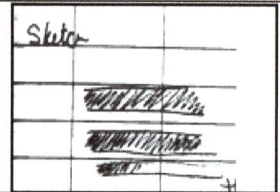

Borehole 71: Saroj Singh, Field Book #23, page 83, Stop1 .	
-- Coarse grained/sandy -- A lot of water in the area -- Laminations about 1 inch thick -- Gray layer Ferrihydrite crust: -- Iron hydrite from the iron sulphide -- the black crusts -- hardpan when it doesn't get disturbed, the black	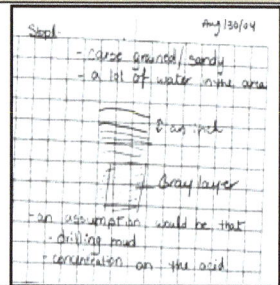

Borehole 72: Saroj Singh, Field Book #23, page 86, a profile.	
-- Gray (much more finer) -- Clay -- Yellow -- Dark brown -- Brown -- Yellow -- Gray Near zone 6 there are copper sulphate deposits	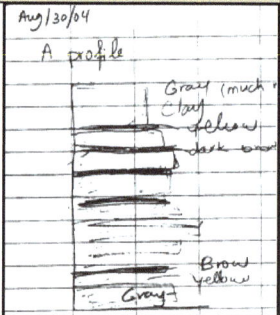

Borehole 73: Stephanie Studdy, Field Book #24, page 83, Soil Profile #1.

-- Grey
-- Red band
-- Grey
-- Red Band
-- Dark Clay
Depth 60 cm

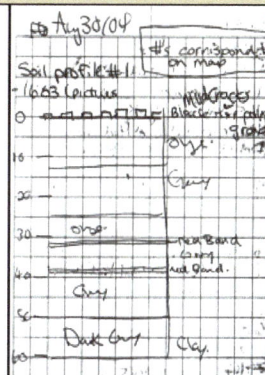

Borehole 74: Stephanie Studdy, Field Book #24, page 83, Soil Profile #2.

Reeds
-- Orange and grey
-- Water table

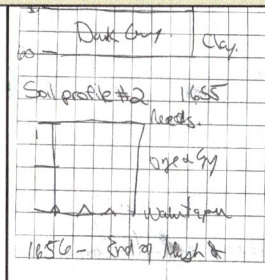

Borehole 75: Stephanie Studdy, Field Book #24, page 84, Soil Profile #3.

Greyish-blue ooz layer

Borehole 76: Stephanie Studdy, Field Book #24, page 87, Soil Profile #4.

Stop 6 Greyish-blue ozz layer

Borehole 77: Van Pham, Field Book #25, page 63, Point 1.

-- Black sand on top with moss growth
-- Successive bands of yellowy-orange and grayish-white
  interbands
-- Depth 60 cm
///////// orange

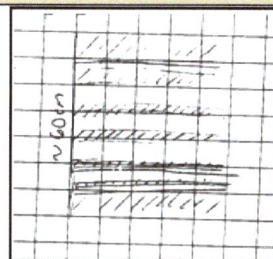

Borehole 78: Van Pham, Field Book #25, page 65, Point 4.

Near top, layer of blue grey oozy clay then bands of orange and grey
-- band of iron below blue grey layer (dark red)
///////// orange

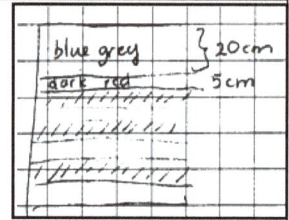

Borehole 79: Caitlin Vanderkooy, Field Book #26, page 114, 115, Point 1.

GPS 467294   5137643
-- Vegetation: small patches low lying weeds, clover grass – great contrast between
   deciduous verses an out solves of area
-- Ground cover: streams of samd (yellowy orange), mainly silt, mudflow layer
-- Under sublayer
-- 1 x orangy yellow sand
-- Cross-section in ground:
-- Smells of sulphuric acid
-- 4 cm cross section
-- Top – mud is 7 cm   Top
   1 x browny grey
   Thin layer orangy yellow   ↓
   Sand 2 cm
   (Reduced sulphides - white layer 4 cm)   Bottom
   1 x mud clay
-- Sharp contact between mud and sand
-- Sand layer (with sulphur elocution)
-- White clay layer (1 cm)
-- Sharp contact
-- Orangy yellow sand (3 cm)

Borehole 80: Caitlin Vanderkooy, Field Book #26, page 115, Point 2.

GPS 469282   5137684
-- 16 cm deep
-- 1 cm organics
-- 15 cm granual clay
-- crumbly clay
-- dark brown

Borehole 81: Caitlin Vanderkooy, Field Book #26, page 115, 116, Point 3.

GPS 467255 5132708
-- Water level 26 cm
-- Organic layer a few mm
-- 26 cm Interbedded of white clay and yellow/orange sand
-- No longer beds, pretty even interbeds
-- 10-17 cm from the bottom there is a greater strand of oxidized sand (yellow/orange)
-- Crispy red sand clumps interbedded in the layer of oxidized iron? Nodules iron platletts (chemical process)
-- 1 cm chunks →5mm thick randomly distributed
-- 5 m from treeline

Borehole 82: Caitlin Vanderkooy, Field Book #26, page 116, 117, Point 5.

GPS 467206 5137768
-- Cross-section 30 cm
Top – 15 cm light brown sand
-- 1 cm thin strips of dark brown sand
-- 3 cm white fine-grain sand
-- 11 cm orangy/yellow sand
-- Sulfate iron
-- 3 m from vegetation

Borehole 83: Caitlin Vanderkooy, Field Book #26, page 117, Point 6.

GPS 467177 5137779
-- 30 cm endice crossbed
-- Organic 1 ½ cm
-- 17 cm banding of yellow
-- 1 cm of dark brown layer
-- 6 cm of salt and pepper
-- More yellow orange sand
-- 6 m from vegetation

Borehole 84: Caitlin Vanderkooy, Field Book #26, page 117, 118, Point 7.

GPS 467153 5137768
-- Reddy brown precipitate crystals – iron oxides chalk like fine grain growing on algae mat
-- In middle of sand bed
-- Cross-bed depth 25 cm deep
-- 2 mm of organic
-- Uneven bands of yellow orange sand and clay
-- 10 cm from top there is a rust colored band
-- Colorless lenses of yellow and white sandy clay
-- 10 m from East line

Borehole 85: Caitlin Vanderkooy, Field Book #26, page 120, Point 10.

GPS 467080   5137638

-- Depth of hole 33 cm
-- 3 mm of organic
-- Interbedded red and white sand, bands of white sands thicker than orangey, red
-- Lense of oxidation going through oxidized iron precipitate
-- 10 m from marsh

Borehole 86: Pete van Hengstum, Field Book #27, page 58, Spot-V.

Spot-V

-- Oxidation lenses outlined is Fe
-- Thick iron oxidized hard pan and fining upwards
   sequence from sand to clay
-- + Montmorillonites
-- + Sand interbeds
-- + Laminated clays (dark to light grey)
-- + Water headed East

Borehole 87: Pete van Hengstum, Field Book #27, page 57, Spot-W.

-- + Liesegang banding of Fe/Ln/Lo/Organisms
-- + Grey, orange, green

-- + Dark gray layer

-- + Orange/lite gray banding

-- + Alternating light gray and dark grey layers

-- + Saturated, non-laminated clays

Depth 50 cm

Borehole 88: Pete van Hengstum, Field Book #27, page 56, Spot-X.

0-10 cm -- + Interbedded clays and sands
 -10 cm -- + Sands orange, clays greys
 -20 cm -- + Low water content
 -50 cm -- + High water content
      Laminated clay, grey
      Clay & no interbeds

Depth 50 cm

Borehole 89: Pete van Hengstum, Field Book #27, page 56, Spot-Y.

-150 cm -- + Layer of vegetation at top
-- + Black, anoxic, organic matter
-- + Highly liquid sand/tailings
-- + Orange, saturated
-- + clay, and sand layer
-130 cm -- + Water table at bottom
15cm in 7 seconds (~2cm/s)

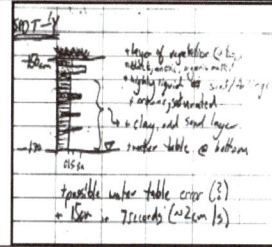

Borehole 90: Pete van Hengstum, Field Book #27, page 57, Spot-Z.

-- + Moss, low lying vegetation
-- + Bushes
-- + Interbedded silts and clays
ΛΛ=> pyrite layer
-- + Grey clay layer
-- + Silty sand
-- + Clay is grey
-- + Silty sand is all orange

Borehole 91: S. Walker, Field Book #28, page 24, Point B.

-- Sand top, grey thick clay (6 cm), orange 6 cm, 3 cm light clay, dark clay
→ Sand top layer varies in this spot < 1 cm to ~ 3 cm. Therefore aeolian.

Borehole 92: S. Walker, Field Book #28, page 24, Point C.

-- Black surface, 1-2 cm orange, 3-2 cm light grey clay, 7 cm orange, 2 cm < grey clay and orange
   down→ black – biological? in dry areas.

Borehole 93: S. Walker, Field Book #28, page 24, Point D.

-- Black surface, 1 cm light brown, 1 cm orange, light grey clay, with few layers small orange.

Borehole 94: S. Walker, Field Book #28, page 24, Point E.

-- Iron oxide platelets at 2 high, only on surface, 2 mm orange, 5 mm light orange, 10 cm light clay,
   orange with clay interbedding (darker orange).

Borehole 95: S. Walker, Field Book #28, page 24, Point F.

-- Orange surface, brown 1 cm, black ~ 2 cm, brown 30 cm, grey clay further down, fluvial.

Borehole 96: S. Walker, Field Book #28, page 26, Point L.

-- Clay 10 cm, orange 1 cm, etc varying layers.

Borehole 97: S. Walker, Field Book #28, page 26, Point M.
-- 15 cm yellow/brown sand ↓ darker clay in river bed.

Borehole 98: S. Walker, Field Book #28, page 26, Point N.
-- 10 cm dirt/mud, 2 cm clay, orange 1 cm, 5 cm layers.

Borehole 99: S. Walker, Field Book #28, page 26, Point O.
-- Moss surface, clay 5 cm, orange 2 cm, clay, orange, etc. → Iron oxide & Gypsum covering area.

Borehole 100: S. Walker, Field Book #28, page 26, Point P.
-- Organic surface, grey 3 cm, 1 cm orange, 3 cm gray, 4 cm brown   → grey orange layering.

Borehole 101: S. Walker, Field Book #28, page 26, Point Q.
-- Thick 1 cm black organic, light gray and orange layering, grey clay after   ~ 20 cm down.

Borehole 102: S. Walker, Field Book #28, page 26, 27, Point R.
-- Light sand, aeolian, small clay and orange sand layers -- 20 cm down there is clay, 5 cm layer *Note: In South East corner there's a dam that broke awhile back, water that entered area cause a lot of erosion and slumping→ sands in area (subsurface) conduct water from the days therefore causing erosion underneath cap layer (vegetation/algae) and slumping and collapse occur later on→ gullies form. -- Grains fine towards dam -- Coarse grains deposited first and fine ones deposited last before dam (clays).

Borehole 103: Andy Weller, Field Book #29, page 43, Point A.
-- 4 Ft down (slope bank) -- Top 2 ft yellow clay -- 2 ft below is gray and looks like thick potter's clay with green   running through it which is copper 

Borehole 104: Andy Weller, Field Book #29, page 45, 46, Point D.

-- Dug down 2 feet
-- Top 1 cm brown, green (algal mat)
-- 18 inches layered brown/gray clay
-- Bottom wet many clay from this point we can follow a line off reddish brown which
    runs into the marsh
-- Montmerillianite layers which are tailings clay to lubricate iron oxide where it seeps
    out and reaches the surface and encounters air.

Borehole 105: Andy Weller, Field Book #29, page 47, Point G.
-- Similar layers of clay but here they are very porous—ask why! Probably gas bubbles.

Borehole 106: Zachary Windus, Field Book #30, page 36, Point 1.

-- Light sand, darker lichen, small grass
-- Fine, sandy white sediment
-- Orange/brown slightly more coarse oxidated
-- Well-layered

Borehole 107: Zachary Windus, Field Book #30, page 36, Point 2.

GPS 0467286   5137650

-- 10 m from tree line
-- More moss covered, surface more compacted
-- Less sand than point 1 [borehole 106]
-- Clay layer, moist, very fine, gray

-- Depth 60 cm

Borehole 108: Zachary Windus, Field Book #30, page 37, Point 3.

-- Clay liquefies but stays solid

-- Very thin layer of muddy sand on top 5cm

Borehole 109: Zachary Windus, Field Book #30, page 37, Point 4.

GPS 0467196   5137517
-- No vegetation, semi-oxidized crust
-- White material prevalent at surface
-- Thin orange layers
-- Darker clay ~ 50 cm down

Borehole 110: Zachary Windus, Field Book #30, page 38, Point 7.

GPS 0467103   5137623

-- Clay 35-40 cm deep

Borehole 111: Zachary Windus, Field Book #30, page 38, Point 8.

GPS 0467089   5137641
- -- Dark clay 35-40 cm down
- -- Ferrihydrite layer, much harder than surrounding layers ~ 15-20 cm down
- -- Iron hardpan
- -- Water infiltrates down to the clay layer, then moves along the gradient
- -- Water oxidizes sediment then precipitates out, leaving hard pan
- -- Connection between layering and geomorphology
- -- Prograding delta causing upward sequence

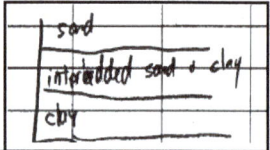

Borehole 112: May Xiujuan, Field Book #31, page 60, Location 1.

- -- Black
- -- Yellow sandy, rusty at spots
- -- Light, crumbly clay
- -- Darker clay

Depth 55 cm

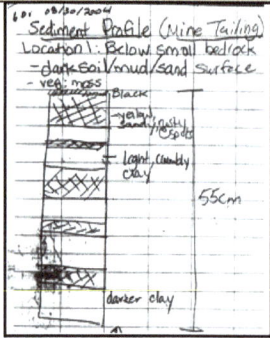

Borehole 113: May Xiujuan, Field Book #31, page 61, 62, Location 2.

- -- Taken in marsh area
- -- 25 cm deep
- *-- Reached water table
- -- Soil profile obscured by water
- -- Clay with orange
- -- Alternating between light grey to bluish grey

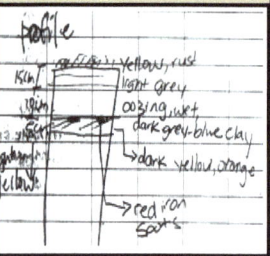

Borehole 114: May Xiujuan, Field Book #31, page 65, Location 5.

- -- Sandy soil
- -- Subsurface mostly liquification, wet clay

Borehole 115: Linda Zelek, Field Book #32, page 31, Location 1.

GPS 0467113   5137845
- -- Brown 2 cm
- -- Darker red orange
- -- Orange
- -- Interbedded orange, darker red and white 60 cm
- -- Dark Gray 10 cm

- -- Depth 75 cm

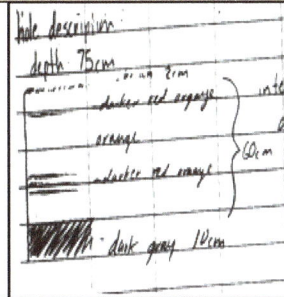

Borehole 116: Linda Zelek, Field Book #32, page 32, Location 2.

GPS 0467005   5137756
- -- Interbedded grey/red
- -- Dark brown clay 4 cm
- -- Dark grey/black
- -- Dark red
- -- Gray colour
- -- Dark red
- -- Gray
- -- Water table (liquid)

- -- Hole description depth 45 cm

Note black mineral that has been dumped in area

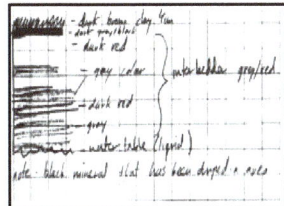

Borehole 117: Linda Zelek, Field Book #32, page 33, Location 3.

GPS 0467015   5137697
- -- Interbedded grey and dark red

- -- Depth 20 cm

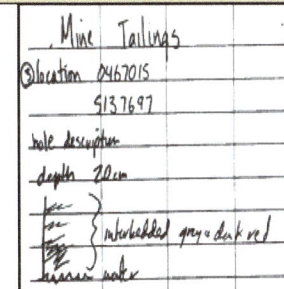

Borehole 118: Linda Zelek, Field Book #32, page 34, Location 4.

GPS 0467075   5137633
- -- Algae on top
- -- Dark red
- -- Grey

- -- Depth 30 cm

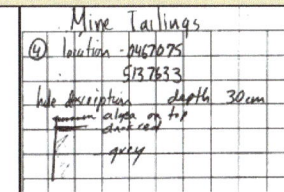

Borehole 119: Linda Zelek, Field Book #32, page 35, Location 6.	
GPS 0467187 5137454 -- Top algae -- Interbedded grey/orange	*(handwritten field notes)*

Borehole 120: Linda Zelek, Field Book #32, page 37, Location 7.	
-- Hole description -- Top layer Algae -- Interbedded grey and orange -- Grey layer -- Depth 40 cm	*(handwritten field notes)*

10.3 Kidd Copper Aggregate Base Map of Borings

"Accurate mapping and description of the rock in the field is the basic starting point of the study of rocks" (Moorhouse, 1959, p. 1). So, this is where the Kidd Copper property analysis will begin. From the walk-about on August 30, 2004, a base map was assembled from 120 soil samples of 32 geology students borings recorded in their field logs in an area encompassing 4 km^2 about 85-90% of the surface ground area covered. Kidd Copper's field data in Section 10.2 is primary research. This collection of raw data provides the base of analysis.

"*Visual hierarchy*: The perceptual organization of cartographic elements such that they appear visually to lie in a set of layers of increasingly importance as they approach the viewer" (Clarke, 2003, p. 328). Figure 27 - Section 10.3 is an aggregate base map of the Kidd Copper major mine tailings deposit exposing 120 boreholes. Kidd Copper's visual hierarchy for the aggregate map of borings is based on the following:

(1) Shown as a detailed sketch map in plan view,

(2) Compilation of the n = 120 boreholes, the attributes circled in points 1-120 are different borehole locations dug on the Kidd Copper property on August 30, 2004. Each soil borings are indicated by dots on the aggregate map to show where the core lithofacies logs were uncovered to analyze.

(3) Where the samples and cross-sections were located with exposed surface depths up to 50 cm.

(4) The dam's washout is located on the South East corner of the Kidd Copper Property.

(5) The higher elevation is on the north boundary of the property. The direction of flow was from north to south, from a higher to a lower elevation, although the slope was minimal, probably <1%.

(6) At Borehole 113 the water table was reached.

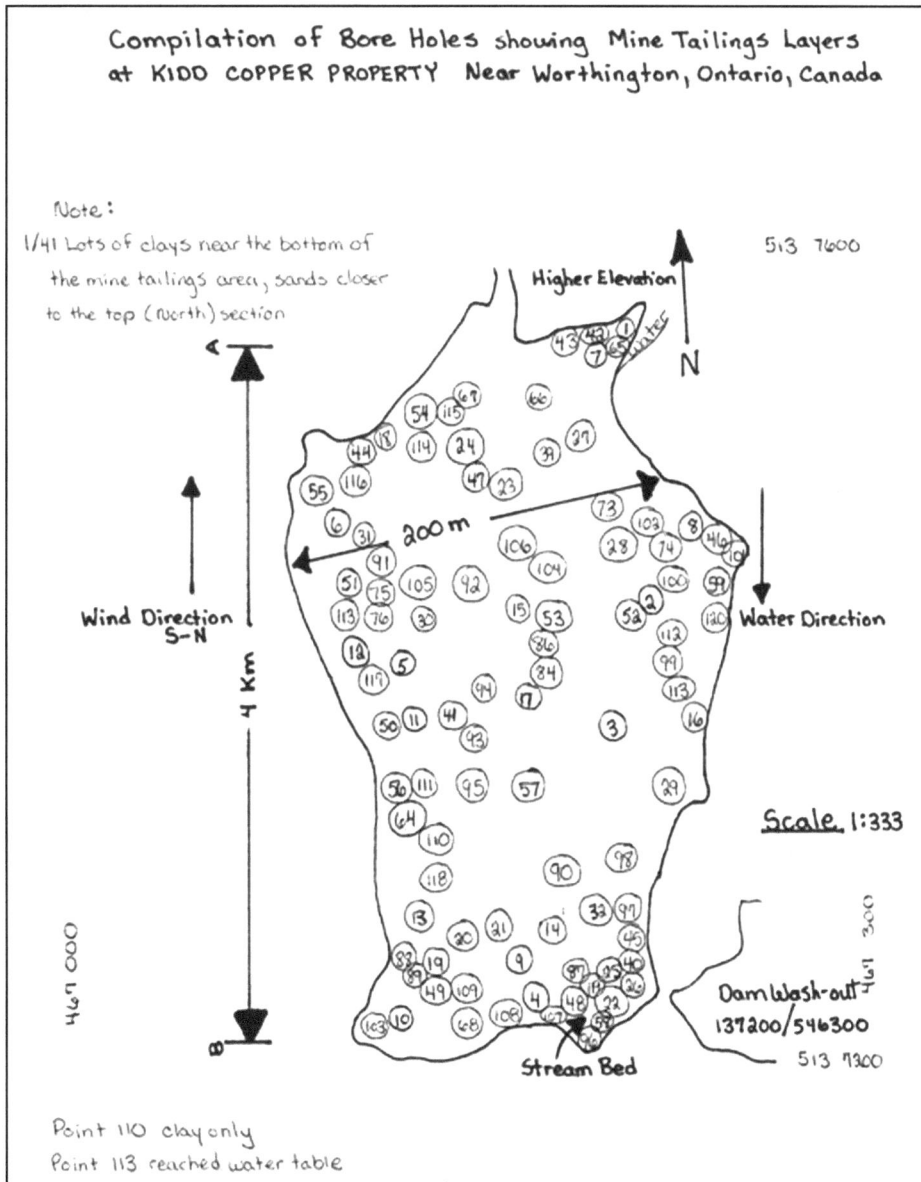

Figure 27 - Section 10.3: Base Map of Kidd Copper Site, a Compilation of Boreholes n = 120

10.4 Correlation of Lithographic Logs Defined

Correlation *n.* [Geology] "a connection of points from well to well in which the data suggest that the points were deposited at the same time (chronostratigraphic) or have similar and related characteristics" (Schlumberger, 2008, Oilfield Glossary).

"*Spatial data*: Data that can be linked to locations in geographic space, usually via features on a map" (Clarke, 2003, p. 325). Kidd Copper data is linked spatially into two major transects: first is the West-North transects and the second is the North-South.

"*Spatial distribution*: The location of features or measurement observed in geographic space" (Clarke, 2003, p. 325). Careful consideration of the spatial distribution of each of the borehole points is noted to analyze the data. The deposition of each borehole point within the spatial distribution of data points is referenced at the same start point in time. Therefore, the expectation is that each point should show similar characteristics in the analysis of the sediment distribution.

Lithology *n.* [Geology] "the macroscopic nature of the mineral content, grain size, texture and color of rocks" (Schlumberger, 2008, Oilfield Glossary).

"Sequence [GEOL] (1) A sequence of geologic events, processes, or rocks, arranged in chronological order. (2) A geographically discrete, major informal rock-stratigraphic unit of greater than group or supergroup rank. Also known as stratigraphic sequence" (Licker, 2003, p. 332). "Such logs can be valuable aids, especially when they are designed with a particular project in mind" (Blatt, Middleton & Murray, 1972, p. 6). "Stratigraphic unit [GEOL] a stratum of rock or a body of strata classified as a unit on the basis of character, property, or attribute" (Licker, 2003, p. 355). The lithographic units were identified in each of the students' logs. This included a West-North Transect and a North-South Transect, where Point 56/4 is the arbitrary intersection. Refer to Figure 28 - Section 10.4.

Two interpretations of these stratigraphic columns are available: (1) diversified strata was laid over time as various sediment deposits; or (2) the diversified strata is the result of a rapid depositional process culminating in rapid straitification. In the case of the Kidd Copper Property, the second interpretation is accepted because all of the sediment layers were deposited at the same time according to site observations.

Figure 28 - Section 10.4 are two geologic cross-sections taken parallel and perpendicular intersecting the plane of a vertical section of the ground on the Kidd Copper property after the deposits from the mine tailings waste site settled from the dam breach. A "geologic cross-section [is] an interpretation of a vertical section through the Earth's surface, most usefully a profile, for which evidence was obtained by geologic and geophysical techniques or from a geologic map" (Allaby & Allaby, 2003, p. 228). Two different lithologic logs are compiled in this report from cross-section A-A' and cross-section B-B' to examine and analyze the mine tailings depositional processes from these simultaneous occurrences and to compare adjacent structures.

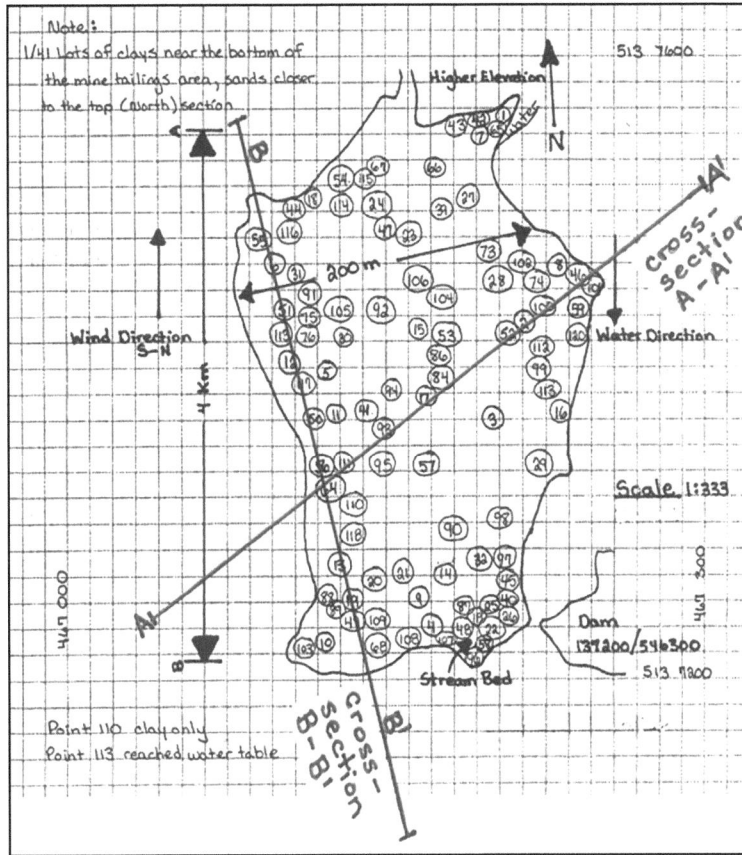

Figure 28 - Section 10.4: Cross-sections West-North Transect and a North-South Transect

10.5 Lithographic Log West-North Transect

"Lowest occurrence" and "first occurrence" are markers for "primary correlation criteria" (Gradstein, *et al*., 2004, p. 278). In the case of the Kidd Copper Property, establishing markers for the time period is difficult because the lowest occurrence of the sediment is identical to the first, no differentiation can be established for time units because all of the lithographic logs marked the same period in ~1970. Point 56/4 was chosen as the 'standard section', located at the intersection of both cross-sections, "the requirement for one of the sections to be adopted as the starting 'standard section,' the stratigraphic thickness measurements of which become the composite units in the composite section" (Gradstein, *et al*., 2004, p. 51).

The correlation of 11 borings in the Lithostratigraphic units of the West-North Transect are examined as to depth in centimetres (boreholes in feet) and the subsequent color codes. Refer to Table 4 - Section 10.5. Color of the soil is a primary indicator of differentiation within each borehole supplied in the Legend for Cross Section A-A' [West – North Transect]. The total depth of Cross Section A-A' [West – North Transect] was ~75 cm or ~2.46 ft.

Table 4 - Section 10.5: A Sequential Lithographic Compilation of Cross-Section A-A'

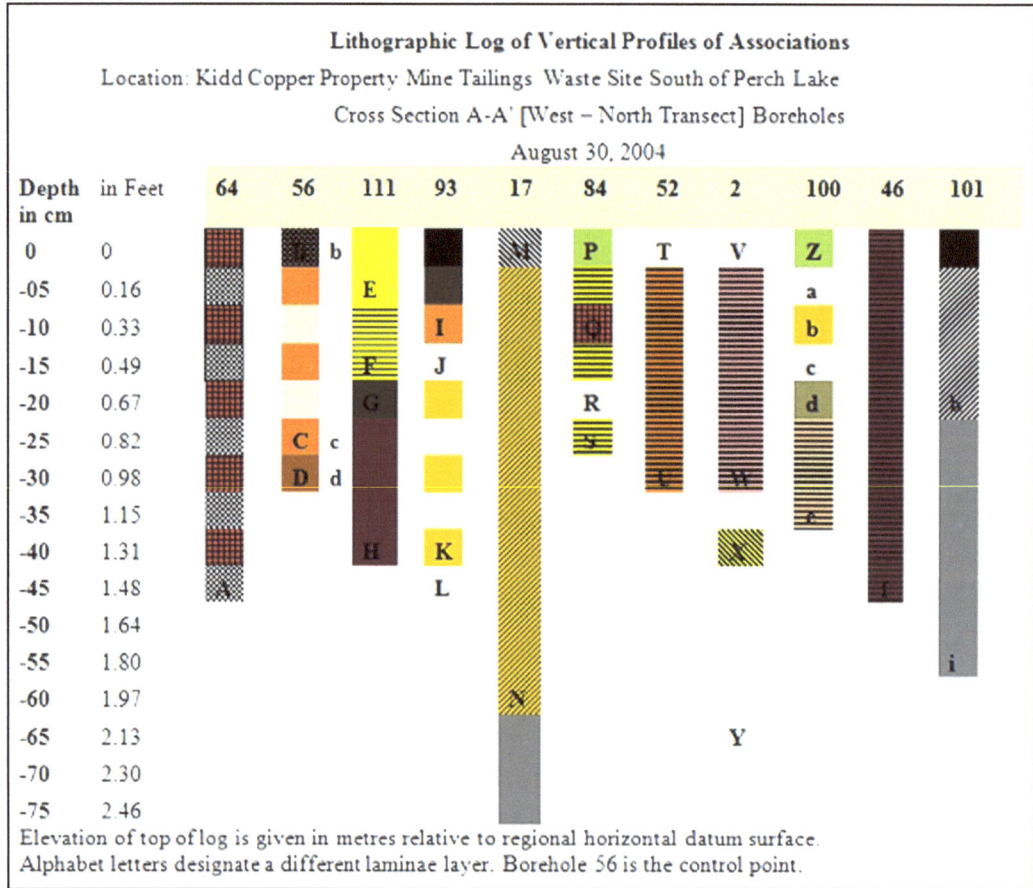

Lithographic Log of Vertical Profiles of Associations
Location: Kidd Copper Property Mine Tailings Waste Site South of Perch Lake
Cross Section A-A' [West – North Transect] Boreholes
August 30, 2004

Elevation of top of log is given in metres relative to regional horizontal datum surface. Alphabet letters designate a different laminae layer. Borehole 56 is the control point.

The following legend to Table 4 - Section 10.5 depicts the multiple color variations & different laminae designations that are marked in the lithographic log of vertical profiles of associations form the Kidd Copper property mine tailings waste site south of Perch Lake, cross-section A-A' [West – North Transect] boreholes.

LEGEND for Cross Section A-A' [West – North Transect]		
MS Word Color Settings	**Symbol**	**In situ Structure Color**
Dark Red – Light Grid		Rust Color
White Background 1– Dark Grid		White
Tan, Background 2, Darker 25% -Dark Trellis		Dark Black/Beige Spotted
Orange Accent 6 Darker 25%		Bright Orange Layers
Tan Background 2 Darker 10%		Beige Layers
Orange Accent, Darker 50%		Iron Waste Product Oxidized
Yellow		Sand
Yellow, Light Horizontal		Interbedded Sand & Clay
Tan, Background 2, Darker 75%		Ferrihydrite Layer
Red Accent 2, Darker 50%		Dark Clay
Tan, Background 2, Darker 25%		Dirt/Mud
White, Background 1, Darker 15%		Clay
Orange		Orange
White, Background 1, Darker 5% - Dark Down Diagonal		Salt
Orange, Lt Up Diagonal		Sand Ferrous, Orange
Wt Background 1, Darker 50%		Grey
Yellow, Dark Horizontal		Yellow-Orange
Olive Green, Accent 3		Organic
Orange, Accent 6, Lighter 80%		Aero Bar Type Bubbles
Orange, Accent 6, Darker 25%, Dark Horizontal		Silica and Sand Layer
Red, Accent 2, Lighter 40%		Sand Interbedded with O_2 Rust
Yellow, Dark Down Diagonal		Coarse Sand Layer Interbedded
Tan, Background 2, Darker 50%		Brown
Orange, Accent 6, Lighter 40%, Dark Horizontal		Grey-Orange Layering
Red Accent 2, Darker 50%, Lt Horizontal		Interbedding Orange and Red w/Grey
Black Text 1		Black Organic
Orange, Accent 6, Dark Up Diagonal		Light Grey & Orange Layering
*The letters A to ZZ are to signify a change from one layer to another.		
*The borehole control Point is 56/4.		

The deposits in Table 4 - Section 10.5 clearly show stratification in all of the vertical log profiles. No similar match of adjacent profiles has been found in the log profiles; yet the sediment was laid at the same point in time, as the aftermath of the flooding of the dam. Instead, a high degree of randomness in the vertical profiles is demonstrated with each color separation. "Collectively, all *features formed during deposition or shortly thereafter* are designated as primary sedimentary structures.

Included are the beds themselves, features on the bedding surfaces, and features within the beds" (Friedman, *et. al.*, 1992, p. 159). "Primary sedimentary structure [GEOL] a sedimentary structure produced during deposition, such as ripple marks and graded bedding" (Licker, 2003, p. 274). The deposited sediment is recorded throughout the field entries of the geology students as layers. Other words used to describe the 11 vertical logs include: band, interbedded, interlayers, interbedding all the way down – 44 cm (borehole 46/6), layer (s), and layering.

A general law of geology is the Law of original continuity. This law states that "a water-laid stratum, at the time it was formed, must continue laterally in all directions until it thins out as a result of nondeposition" (Bates & Jackson, 1984, p. 291; MacDonald, *et al.*, 2003, p. 273). The adjacent boreholes do not conform laterally to the marker as expected.

All the boreholes used for comparison in this report in the lithographic logs did not show lateral distribution of sediment. Each unit is different and does not resemble the marker unit. All lithographic logs did not show a thinning out of the sediment. The Kidd Copper sediment instead showed nonlateral and random distribution in the lithographic logs compiled from the same event. Therefore, based on this evidence The Law of original continuity's conclusion is seriously questioned and can be rejected. In fact, the claim that H_0 at the time it was formed = lateral in all directions can be rejected. Instead, the claim that $H_1 \neq$ is not lateral in all directions can be accepted.

10.6 Lithographic Log North-South Transect

The correlation from 13 borings of the Lithostratigraphic units of the Log North-South Transect are examined as to the depth in centimetres (boreholes in feet) and the subsequent color codes, refer to Table 5 - Section 10.6. Color is a primary indicator of differentiation within each borehole supplied in the Legend for Cross Section B-B' [North – SouthTransect]. The soil depth (m) was an average of ~42 cm or ~1.37 ft.

Other words used to describe the 13 vertical logs include: alternating in thin layers, interbedded, bands, layer(s), and stratified. The total depth of Cross Section B-B' [North – South Transect] was ~60 cm or ~1.97 ft.

"Profile [GEOL] (1) the outline formed by the intersection of the plane of a vertical section and the ground surface. (2) Also known as a topographic profile" (Licker, 2003, p. 275). The Lithologic Log is a North-South transect of the Vertical Profiles of associations on the Kidd Copper property mine tailings waste site South of Perch Lake. This stratigraphic column proves diversified strata were laid as primary sedimentary structures. The deposits clearly show stratification in all of the vertical log profiles.

Table 5 - Section 10.6 is a sequential lithographic compilation of a cross-section B-B' consisting of boreholes to compile a core lithofacies log from 13 samples. The legend shows substantial contrasts in color and identifies many layers.

Table 5 - Section 10.6: A Sequential Lithographic Compilation of Cross-Section B-B'

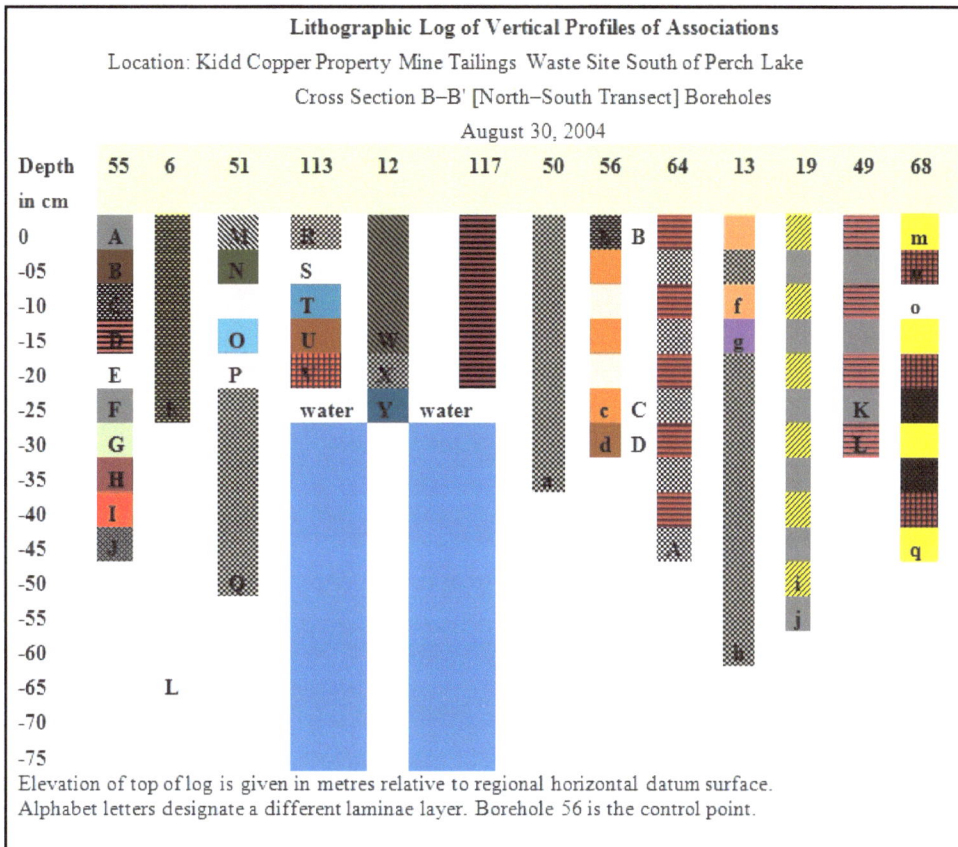

Lithographic Log of Vertical Profiles of Associations
Location: Kidd Copper Property Mine Tailings Waste Site South of Perch Lake
Cross Section B–B' [North–South Transect] Boreholes
August 30, 2004

Depth in cm	55	6	51	113	12	117	50	56	64	13	19	49	68
0	A		M	R				B					m
-05	B		N	S						f			M
-10				T									o
-15	D		O	U	W					g			
-20	E		P	X									
-25	F			water	Y	water		c / C				K	
-30	G							d / D				L	
-35	H						a						
-40	I												
-45	J								A				q
-50			Q								i		
-55											j		
-60										h			
-65		L											
-70													
-75													

Elevation of top of log is given in metres relative to regional horizontal datum surface.
Alphabet letters designate a different laminae layer. Borehole 56 is the control point.

The following legend to Table 5 - Section 10.6 depicts the multiple color variations & different laminae designations that are marked in the lithographic log of vertical profiles of associations form the Kidd Copper property mine tailings waste site south of Perch Lake, cross-section B-B' [North – South Transect] boreholes.

Over 64 different colors were noted and the variation recorded verifies rapid stratification. Multiple layers of sediment can occur in a rapid process; this multiple layering of sediment establishes that rapid stratification can occur as a one-time event.

No similar match on adjacent profiles throughout the cross-sections for color has been found; yet the sediment was laid at the same point in time, as the aftermath of the flooding of the dam. These new sedimentary structures are considered primary sedimentary structures as "all features formed during deposition or shortly thereafter" (Friedman, *et. al.*, 1992, p. 159). The deposited sediment is recorded throughout the field entries of the geology students as layers.

LEGEND for Cross Section B–B' [North–South Transect]		
MS Word Color Settings	**Symbol**	**In situ Structure Color**
Black, Text 1, Lighter 50%		Very Dark Grey/Black
Brown		Dark Brown Clay
Dark Trellis		White Spots
Red, Accent 2, Dark Horizontal		Interbedding Bright Red and Orange w/Grey
White Background 1, Darker 5%		Bright Grey
Olive Green, Accent 3, Lighter 40%		Yellow Green
Red, Accent 2, Darker 25%		Red Brown
Wt Background 1, Darker 50%		Grey
Red		Red
Black Text 1, Lighter 50%, Lt Trellis		Black Mineral
Yellow, Dark Trellis		Sand Stratified with Dark Bands of O_2 Material
Purple, Accent 4, Lighter 80%		Fine Montmorillonite Clay
White, Background 1, Darker 5% - Dark Dn Diagonal		Salt
Olive Green, Accent 3, Darker 50%		Dead Algae – Black/Brown Layer
Aqua, Accent 5		Ferrihydrite Layer
Red, Accent 2, Lighter 80%		Sediment
Tan, Background 2, Darker 25%, Dark Grid		Clay
Blue, Accent 1		Water Table
Orange, Accent 6, Lighter 60%, Dark Grid		Yellow Rust
White Background 1, Darker 5%		Light Grey
Aqua, Accent 5, Darker 25%		Dark Grey – Blue Clay
Orange, Accent 6, Darker 50%		Dark Yellow – Orange
Red, Lt Grid		Red Iron Spots
Tan, Background 2, Dark 50%, Dark Dn Diagonal		Stratified Clay and Sand
Aqua, Accent 2, Darker 50%		Dark Blue
Red, Accent 2, Darker 25%, Dark Horizontal		Interbedded Grey and Dark Red
Tan, Background 2, Darker 25% -Dark Trellis		Dark Black/Beige Spotted
Orange Accent 6 Darker 25%		Bright Orange Layers
Tan Background 2 Darker 10%		Beige Layers
Orange Accent, Darker 50%		Iron Waste Product Oxidized
Red, Accent 2, Light Horizontal		Rust Color and White
Orange, Accent 6		Silt – Orange
Purple, Accent 4		Pyrite
Yellow, 5%, Lt Up Diagonal		Sulphur
Yellow		Sand
Dark Red – Light Grid		Rust Color
White		White
Dark Brown, Dark Trellis		Dark Brown
*The letters A to ZZ are to signify a change from one layer to another.		
*The borehole control Point is 56/4.		

Interpretation of the combined color legends from Cross Section A-A' [West – North Transect] and Cross Section B–B' [North–South Transect] shows no correlation of color between the vertical logs and the adjacent logs, which indicates randomness in the primary sedimentary structures even though the Kidd Copper features formed during deposition or shortly thereafter and are designated primary structures.

The deposition of sediment at ground zero was not deposited homogeneously not was it deposited horizontally as predicted by Sections 6.5 and 6.6. Even though the fine grains were spherical and small they were not evenly distributed upon redeposition. The mine tailings sediment was not distributed evenly; therefore the principle of original horizontality "which says that waterborne sediments settle out and are deposited (laid down) as horizontal layers" (Murck, 2001, p. 49) can be rejected. "Layers of sediment are generally deposited in a horizontal position" (Tarbuck & Lutgens, 1999, p. 191) can be rejected. "*Sedimentary rocks*, especially water-laid strata, are deposited parallel to the surface on which they are deposited and thus horizontally" (MacDonald & Burton, 2003, p. 273) can be rejected.

In fact, the claim that –
H_0 on ground zero = horizontal deposit can now be rejected.
Instead; the claim that $H_1 \neq$ horizontal deposit can be accepted.

11.0 Data Analysis from Field Logs

"Sedimentologists are students of sediments; they describe and analyze sediments and sedimentary rocks" (Hsü, 2004, p. 4). "Their purpose [sedimentologist] is to learn more about the origin of a sediment. Is it a beach deposit, a lime mud laid down on a tidal flat, or an ocean ooze?" (Hsü, 2004, p. 4). As a student of sedimentology, the author of this report studied "sediments and the various depositional environments in which they form … learn how to identify various sediment types and identify clues to interpret their processes and environments of deposition" (Eyles & Boyce, September 2004, Geo3E03 Course Outline). One of the ways for geologists to report their findings is to construct graphic logs, which "give a visual impression of the section, and are a convenient way of making correlations and comparisons between equivalent sections from different areas; repetitions, cycles and general trends may become apparent" (Reineck & Singh, 1973, p. 13). "Many different features of a sedimentary rock can be used as palaeocurrent indicators. Some structures record the direction of movement (azimuth) of the current while others only record the line of movement (trend). Of the sedimentary structures, the most useful are cross-bedding and sole structures (flute and groove clasts); but also structures also give reliable results" (Reineck & Singh, 1973, p. 91).

Field observations from the Kidd Copper site consistently documented: interbedded, layered, interlayers, stratified, sand layers, alternating layers. These descriptions were taken from the written (recorded) observations of thirty-two geology students in-field training examining one hundred and twenty different boreholes. Current, high quality, aerial and sub-surface data was compiled to examine the diagenesis of the site. Three steps of depositional processes were observed in the site area:

(1) The original deposition including the age, parent material and subsequent deposits;

(2) A dam breach casued the upswelling of sediment; and

(3) Redeposition of the sediment after mixing with water.

The collective evidence is overwhelming. The observations support:

    (1) Multiple layers are formed as primary structures;

    (2) Stratification occurred in a very short period of time;

    (3) Randomness of sediment distribution occurs after thorough mixing with water;

    (4) A single event produces stratification and random distribution in adjacent strata.

    (5) An unequal declivity in the adjacent rock units does not mean unrelated events.

The observations on the Kidd Copper property of 120 boreholes after the flooding event strongly supports rapid stratification can be read as a younger geologic event.

11.1 Soil Taxonomy – Distinct Horizons

"Horizons within a soil may vary in thickness and have somewhat irregular boundaries, but generally they parallel the land surface. This alignment is expected since the differentiation of the regolith into distinct horizons is largely the result of influences, such as air, water, solar radiation, and plant material, originating at the soil-atmosphere interface" (Brady &Weil, 2002, p. 11).

"In some soil profiles, the component horizons are very distinct in color, with sharp boundaries that can be seen easily by even novice observers. In other soils, the color changes between horizons may be very gradual, and the boundaries more difficult to locate. However, color is only one of many properties by which one horizon may be distinguished from the horizon above or below it" (Brady &Weil, 2002, p. 13). "Distinguishing between these older nodules or clasts and nodules or clods of the younger soil may be difficult unless they have sharp, ferruginized, or truncated boundaries" (Retallack, 1990, p. 12). "Studies of Quaternary soils and paleosols reveal clearly what a complex thing a soil is and how many factors enter into their formation (Johnson & Watson-Stegner 1987)" (Retallack, 1990, p. 13, 14).

The results of this study on the Kidd Copper property contradict the above statements. The differentiation in the regolith or sediment samples found in the Kidd Copper boreholes was not due to the result of outside influences because the same outside influence applies to all 120 soil samples. The evidence supports a result of random mixing of sediment and the rapid deposition that caused the color changes between the horizons. Against mainstream geologic thought, the logs on the Kidd Copper property had boundaries that were unidentifiable; there was no commonality to correlate. The differentiation was not characteristic of different horizons. One horizon had many colors and fluctuations. Uneven boundaries were prevalent in the soil profiles found on the Kidd Copper Property. Refer to Table 4 - Section 10.5 and Table 5 - Section 10.6. The correlation of these different profiles is based on knowledge of the time factors involved.

"*Soil taxonomy* is based on the properties of soils as they are found today" (Brady &Weil, 2002, p. 79). "Soil taxomony helps in understanding the landscape (Brady &Weil, 2002, p. 117). The facts show the primary sedimentary structures were stratified in a quick process. This is important and should help geologists establish criteria to reject past assumptions about some geological principles and laws. The verification in the rock record that rapid stratification is the result of a fast, one-time occurrence influences the way the rock record is read.

The complexity is that scientists have been trying to correlate sediment along indistinguishable paths. Many soil boundaries are missing. No commonalities exist. The multiple, sharp, ferrrunginized or truncated boundaries over the earth's surface could have been deposited rapidly as in the case of Kidd Copper. This could explain why multiple layers are found in many sediment deposits. Even more, this rapid stratification supports evidence of a younger earth. The layers do not necessarily require long periods of time to form.

11.2 Eye-witness Ground Observations and Testimony

"In the best tradition of the uniformitarianism and the natural-history approach to geology, one should study sediment-gravity flow deposits by reading the observations of actual rockfalls by eyewitnesses" (Hsü, 2004, p. 87). The rationalization is used throughout this report. If one wants to study depositional process then one should study an environmental event to analyze the depositional processes, such is the case of the Kidd Copper study. The UTM coordinates were recorded by the geology students, for example 17T 0467103, 5137623 was one coordinate of a borehole sample that was taken. Primary research was conducted by geologists-in-training. "What makes this site particularly intriguing is the change observed at the property through time" and valuable to this research paper is the additional support of marked papers and field records obtained from the Assignment #9 to the *Geo3Fe3* students at McMaster University in 2004. These observations were made blindly and without prejudice to the outcome of this report.

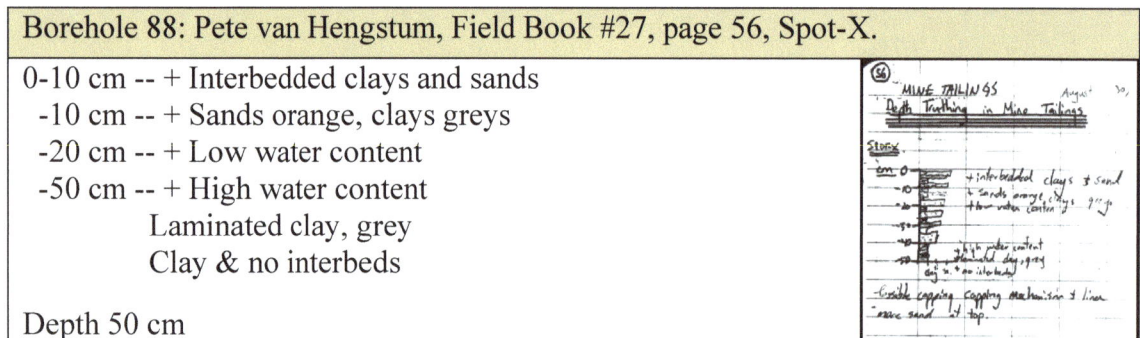

Borehole 88: Pete van Hengstum, Field Book #27, page 56, Spot-X.
0-10 cm -- + Interbedded clays and sands
-10 cm -- + Sands orange, clays greys
-20 cm -- + Low water content
-50 cm -- + High water content
Laminated clay, grey
Clay & no interbeds
Depth 50 cm

Figure 29 - Section 11.2: A Typical Field Observation, Spot X

Soil profiles were taken from the field book data entries of thirty-two different geology students enrolled in the *Geo3Fe3* Field Camp. Refer to Figure 29 - Section 11.2. Borehole 88, Spot X is a typical field observation that clearly describes the alternating layers found on the Kidd Copper property on August 30, 2004. Common descriptions of the site: alternating layer, interbedded, layered, interlayers, stratified and sand layers. Mixed colors: white, orange, blue, grey and red are often recorded. What is evident and observed is rapid stratification; the deposit had a start date in the last few decades and not over a long time period of thousands or even millions of years.

11.3 Quantitative Analysis of Kidd Copper Sediments

Testimonial matrices of descriptors about the sediment on Kidd Copper property include an analysis of the fundamental and derived properties of sediments. These properties include: (1) Composition – kinds of grains and their abundance; (2) Color;

(3) Origin; (4) Size of grains; (5) Shapes of the grains (roundness and sphericity); (6) Packing of the grains; and (7) Orientation of the grains. This is in keeping with collecting information on sediment specimens. "The set must contain all properties necessary to define a rock specimen uniquely; the set must not contain more properties than are necessary for a unique definition" (Blatt, Middleton & Murray, 1972, p. 9).

"Formations and groupings or subdivisions of formations all constitute *rock units*. Rock units, also called *lithostratigraphic units*, are formally defined as bodies of rock identified by their distinctive lithologic and structural features without regard to time boundaries" (Levin, 2003, p. 83). "Such features as texture, grain size, clastic or crystalline, color, composition, thickness, type of bedding, nature of organic remains, and appearance of the unit in surface exposures (or in the lithological record of strata penetrated by wells) are all used to define a rock unit and recognize it in the field" (Levin, 2003, p. 83).

11.4 Matrix of Sediment Types and Descriptors on Kidd Copper

Recall that the satellite known as the Copper Cliff deposit "displays a similar geological setting to the" Kidd Copper site (Shang, *et al.*, 2004, p. 5). "The waste materials at Kidd Copper are derived from the same mineral deposit environment as the more well studied Copper Cliff deposit" (Shang, *et al.*, 2004, p. 2). "Dissolved Fe, Ni and Co … dissolved Fe concentrations" (Coggans, *et al.*, 1991, 464). The description Dr. Bill Morris provided was recorded by Karen De Medeiros on August 30, 2004; she writes, "Pyrite (orange) ↓ Jarosite (yellow green) ↓ Goethite (reddy brown) ↓ Ferrihydrite ↓ Hematitie (Red chips) ↓ Fresh unaltered mine tailings (grey) description from Bill [Morris]" (2004, p. 38). "Silicate, sulphide rocks near here, like pyritite, pyrite (Fe or sulphides) are highly reactive in water, H_2SO_4 produced which dissolves minerals into water, making leachate" (Rosamond, August 2004, p. 67). The minerals on Kidd Copper, like those at Copper Cliff dissolved.

McMaster Professors comment about the composition found on Kidd Copper Property. "Sue's lectures [Susan Vajoczki] mine tailings kept underwater to reduce oxidation, montmorillite clay used in drilling, sent in with tailings" (Rosamond, August 2004, p. 81). "Dr. [Carolyn] Eyles' lecture – some bands not laminations, just due to oxidation (leisingangen)" (Rosamond, August 2004, p. 81).

The random samples were taken from 32 field record entries and compiled in Section 10.2. The identified sediment compositions on the Kidd Copper property are as follows: Forms of Clay are mentioned 79 times; Sand 47 times; various Forms of Iron are mentioned 16 times; Mud is mentioned 4 times; and Silt is mentioned once. The "term *cross-lamination* is used for cross-stratification in which the set height is less than 6 centimetres and the individual cross-layers are less than 1 centimetre thick" (MacDonald, *et al.*, 2003, p. 118).

Table 6 - Section 11.4: The Sediment Composition found on Kidd Copper when Redeposited

Compilation of Sediment Composition					
August 30, 2004, Kidd Copper Property					
Clay 46X	Montmorillonite Clay 14X	Sand 46X	Mud	Hardpan	Iron 4X
Clay material 1X	Clay liquefies but stays solid 1X	Sandy soil 1X	Mud is 7 cm browny grey 1X	Hard pan thick layer 1X	Iron Oxides 4X
Solid Clay 1X	Potter's clay 1X		Mud clay 1X	Iron Hardpan 1X	Iron sulphide oxidization 1X
Thick Clay 1X		Interbeds of Silt 1X	10 cm dirt or mud, 2 cm clay, orange 1 cm, 5 cm clay 1X		Iron oxides and sulphuric acid 1X
Light Clay 1X			Very thin layer of muddy sand on top 5cm 1X		Iron Waste Product Oxidized 1X
Grey Clay 5X					Sand stratified with Fe deposits 1X
Oxidized Clay 1X					Large Fe deposits 1X
Grey liquefied clay 1X					Sequence of iron rich areas 1X
Heavy Clay 1X					Iron Oxide and Silica 1X
Granual Clay 1X					Lighter grey and dark charcoal iron oxides 1X
Crumbly clay, dark Brown 1X					Iron hydrite 1X
Light clay and dark clay 1X					Iron oxide platelets at 2 high, only on surface, 2 mm orange, 3 mm light 1X
Clay, 5 cm layer 1X					
Yellow clay 1X					
Source: Compilation of 32 Geology Students' Kidd Copper Property Field Records Dated 2004-08-30.					

The graphic logs were constructed by observing site findings. Oxidization is mentioned at Kidd Copper 15 times. Ferrihydrite is mentioned 6 times. Pyrite is noted 5 times; Sulfur 4 times: Salts 3 times; Jarosite 1 time and dark red chips 1 time. The chemical interaction of these minerals is beyond the scope of this report. The sediment composition in Table 6 - Section 11.4 records visual observations of the impoundment area. This is a convenient way to make correlations and comparisons between equivalent sections. Each log was analyzed with respect to different areas; repetitions, cycles and general trends that should become apparent. Surprisingly, none were found.

What is important to note in the data distribution is the randomness of the mixing in the mineral composition, refer to Table 7 - Section 11.4. The minerals that were oxidized as well as salts, pyrite, jarosite, ferrihydrite and sulphur compounds on Kidd Copper settled and were distributed randomly in the mine tailings upon redeposition.

Table 7 - Section 11.4: The Mineral Composition found on Kidd Copper after Redeposition

Compilation of Mineral Composition						
August 30, 2004, Kidd Copper Property						
Oxidization 7X	Pyrite 2X	Salts 1X	Jarosite precipitate 1X	Ferrihydrite 3X	Sulphur compound 1X	Dark red chips 1X
Semi-oxidized crust 1X	Pyrite band 1X	High salt 1X		Ferrihydrite crust 1X	Copper sulphate deposits 1X	
Oxygenated material 1X	Pirotite and Pyrite layer 1X	Salt and Pepper 1X		Ferrihydrite Layer much harder than surrounding layers 1X	Reduced Sulphides 1X	
Oxidated Tailings 2X	Pyrite layers 1X			Sand Ferrous 1X	Sulfate 1X	
Lighter grey and dark charcoal iron oxides 1X						
Oxidized sand 1X						
Oxidized organic iron precipitate 1X						
Oxide Sand 1X						
Source: Compilation of 32 Geology Students' Field Records of Kidd Copper Property Dated August 30, 2004.						

11.5 Matrix of Single Color Descriptors Found on Kidd Copper

Table 8 - Section 11.5 is a compilation of the single color descriptors found in the students' reports on Kidd Copper Property. This study provides sufficient depth in the recognition of the single and multi-colored layers found at Kidd Copper to support rapid stratification. In Appendix D-1 a rock colour chart has been provided. The Munsell System could be applied to a future study to classify the colors, where an advanced study on the hue and chroma magnitude of each identified colour could be classified in more detail.

Nine single colors were noted throughout the site area: blue, black, brown, grey, orange, red, rust, sandy (beige), white and yellow. They were mentioned (4X); (4X); (6X); (25X); (18X); (18X); (6X); (5X); (10X); and (4X) respectively. Grey was the dominant color with orange and red followed by white as the predominant hues of the single colors.

Table 8 - Section 11.5: Single Color Descriptors of Kidd Copper Property

Single Color Descriptors									
Compilation from 32 Geology Records on August 30, 2004									
Blue 3X	Black 2X	Brown 1X	Grey 9X	Orange 12X	Red 3X	Rust 5X	Sandy colored 1X	White 5X	Yellow 2X
Dark Blue 1X	Black Clay 1X	Clay brownish 1X	Grey Clay 4X	Orange sand 3X	Dark red chips 1X / Dark red 2X	Copper 1X	Sand (light brown) 1X	Whiter body 1X	Yellow fine-sand 1X
	Black sand with patterns- mud cracks, moss in cracks 1X	Dark Brown Clay 1X	Light Grey 3X	Thick dark orange 1X	Little red 1X		Light sandy color 1X	Whiter and shinier 1X	Yellow clay 1X
		Brown top 1X	Dark Grey Clay 3X	Layer dark orange 1X	Strip reddish 1X		Beige 1X	White sand grains 2X	
		Dark Brown 1X	Dark Clay 3X / Darker Clay 1X	Darker Orange Color 1X	Very Red 1X		Sandy Clay 1X	White material prevalent at surface 1X	
		Brown top clay 1X	Grey liquefied clay 1X	Thin orange layers 1X	Darker Red Orange 1X				
			Grey, blue clay 1X		Crispy red sand clumps 1X				
Source: Compilation of 32 Geology Students' Field Records of Kidd Copper Property Dated August 30, 2004.									

"Law of original continuity [is] a general law of geology: a water-laid stratum, at the time it was formed, must continue laterally in all directions until it thins out as a result of nondeposition or until it abuts against the edge of the original basin of deposition" (Bates & Jackson, 1984, p. 291; MacDonald, *et al.*, 2003, p. 273). Accordingly, the law states that the expected outcome of the adjacent boreholes would laterally conform to the marker. What is seen in one unit should reasonably be seen in the adjacent unit. However, this is not the case in the Kidd Copper single color matrix. The colors did not run laterally into all the directions; instead each direction observed a completely different pattern, that is, that of a stratified and varied color layering, which was continually observed. This indicates that stratification can appear as a one-time layer in the soil profile.

11.6 Matrix of Multi-Color Descriptors

Table 9 - Section 11.6: Multi-Color Descriptions of Kidd Copper Property

Multi-Color Descriptors						
Compilation of 32 Geology Students' Field Records of Kidd Copper Property Dated August 30, 2004						
Brown-Orange Colour 1X	Grey-Red-Grey 1X	Orange – Rust color 1X	Orange and Grey 1X	Dark/Light orange sand 1X	White-Orange clay 1X	Yellow/ Orange sand 3X
Dark brown sand and light grey sand 1X	Grey clay with Orange banding 1X	Orange (iron/oxidized) and white 1X	1-2 cm orange, 3-2 cm light grey clay, 7 cm orange, 2 cm 1X	Light colored clay with an othengic of orange color and darker	White-grey-orange 1X	
Light brown sand 1X	Grey and orange 1X	Orange –then White 2X	Orange, 10 cm light clay, orange with clay interbedding (darker orange) 1X	Color "iron" orange-brown 1X	White and Orange 1X	Yellow and grey 1X
Dark brown sand 1X	Grey- White 1X	Orange and White layers 1X	Interbedded grey/orange 2X	Thick 1 cm black organic, light gray and orange layering, grey clay after ~ 20 cm down 1X	White, orange and red 1X	Yellow-Orange Sand 1X
Dark Brown Layer 1X	Grey clay and orange down 1X	Orange-white-orange 1X	Interbedded grey and dark red 1X	Clay 10 cm, orange 1 cm, etc varying layers 1X	Whitish/Orange 1X	Yellow-Orange 2X
Brown, green (algal mat) 1X	Greyish-blue orange layer 2X	Orange and red with grey 1X	Orange surface, brown 1 cm, black ~ 2 cm, brown 30 cm, grey clay farther Down 1X	Orange-white, Between the white is a layer of blue 1X	White clay and yellow/orange sand 1X	Yellowy-orange and grayish-white interbands 1X
Layered brown/gray clay 1X	Lighter grey and dark charcoal iron oxides 1X	Light & dark orange sediment 1X	Light beige/orange and white layers 1X	Orangy yellow sand 2X	White sand is thicker than orangey red 1X	Colorless lenses of yellow and white 1X
Rust and white 1X	Light grey/bright orange layers with red/brown small layer 1X	Interbedded grey/white 1X	Change in color, light to dark orange 1X	Interbedded of red and white sand 1X		Green and Yellow 1X
Rust colored band 1X	Dark grey/black 1X	Clay 5 cm, orange 2 cm, clay, orange, etc 1X		Interbedded orange, darker red and white 1X		
Sandy grey – orange 1X	Darker grey-to—white progression 1X	Layer of blue grey oozy clay then bands of orange and grey 1X	Organic surface, grey 3 cm, 1 cm orange, 3 cm grey, 4 cm brown → grey orange layering 1X	10 cm dirt or mud, 2 cm clay, orange 1 cm, 5 cm clay 1X	Potter's clay with green running through it which is copper 1X	Black surface, 1 cm light brown, 1 cm orange, light grey clay, with few layers small orange 1X
Source: Compilation of 32 Geology Students' Field Records of Kidd Copper Property Dated August 30, 2004.						

Throughout the Kidd Copper site, multi-colored layering is dominant with a mixture of brown-orange/grey-red/orange-rust/orange-grey/dark with light orange/white-orange/yellow-orange colors distributed throughout the redeposited sediment. Refer to Table 9 - Section 11.6. The orange color was the most prevalent with green being the least visible in the Kidd Copper sediment boreholes. Very little black or green is noted, possibly because these colors are associated with organic materials and surface erosion.

No particular pattern in the layering of the sediment is discernible. Instead, the mixing of colors reflects a variety of patterns and a variety of layers. The diversity of the re-distribution of the sediments as uncovered in the borehole logs was witnessed by all 32 students in their field records. This too is additional evidence that stratification can appear as a one-time layer in the soil profile.

11.7 Origin of the Grain Sediments

The origin of grain sediments was discussed in detail in Sections 1.4 and 2.2, where the characteristics of the depositional setting for the Kidd Copper property mine tailings waste site was noted. An interesting reminder is that the colluvium deposit of slag is homogeneous, the tailings sediment is fine and was dissolved thoroughly in the flow of water released by the dam; however when the logs of the aftermath are examined they clearly reveal a heterogeneous, stratified and multiple layers of deposits.

11.8 Packing of the Grains

"Some sediments (mainly sands of fine grain size) are deposited with a loose, unstable packing (Blatt, Middleton & Murray, 1972, p. 174). "Most aspects of clastic sediments we observe or consider at the outcrop – grain size distribution, grain packing and fabric, stratification, and other primary sedimentary structures" (Harms, *et al.*, 2004, p.6). Eden Hynes writes in her Forestry Supplier's Field Book as she observes the outcrop's surface, "very powdery fine grained texture" (2004, p. 45). Caitlin Vanderkooy adds "chalk like fine grain" (2004, p. 117) to describe the sediment on the Kidd Copper Property.

11.9 Matrix of Grain Size Descriptors

Table 10 - Section 11.9 is a compilation of the sediment grain size found on the Kidd Copper property after the sediment was redeposited from the dam breaking. The thickness of the beds consist of laminae layers: Very thin (<5 mm) and 1 cm chunks →5mm thick, and all are randomly distributed.

The range of grain sizes distributed throughout the Kidd Copper property remains fairly constant between the smallest grains of clay to the larger coarse sand grains. Findings are: Coarse sand 23%; Fine sand 19%; Clay grains 15%: Coarse sand 15%; Very fine silt 8%; Very fine sand 8%; Medium sand 8%; and Fine sand 4%. The average sediment size found on the Kidd Copper property is Very Fine Sand.

Table 10 - Section 11.9: Size of Grains Found on the Kidd Copper Property

Smallest Grains	Size of Grains Compilation from 32 Geology Records on August 30, 2004						Largest Grains
Clay 0.004 mm	Very Fine Silt 0.004- 0.008 mm	Fine Silt 0.008 – 0.015 mm	Coarse Silt 0.03 – 0.06 mm	Very Fine Sand 0.06 – 0.135 mm	Fine Sand 0.135 – 0.25 mm	Medium Sand 0.25 – 0.5 mm	Coarse Sand 0.5 – 1.0 mm
Fine montmorillonite clay 2X	Very fine Grain 2X	Fine- grained 1X	Coarse- grained with darker grey-to— white progression 1X	Very fine sand 2X	Fine sand layers 1x	Medium Grained Sand 1X	Coarse sand 6X
Fine grained clay 2X			Coarse- grained layer 1X		Yellow fine-sand 1X	Medium grain (like sand) 1X	
			Coarse- grained sediment 1X		White fine- grained sand 2X		
			Orange coarser- grained 1X		Fine white sand 1X		
Thickness of Beds [Laminae]							
Very thin (<5 mm) 1X	1 cm chunks →5mm thick randomly distributed 1X						
Source: A Compilation of 32 Geology Students' Field Records of Kidd Copper Property Dated August 30, 2004.							

11.10 Kidd Copper Sediment Grain Size and Shape

Formula for Sedimentary Deposits: Facies – models

$$D = S + T + C$$

Where: D = sedimentary deposit;
S = Structure;
T = Texture; and
C = Composition.

Therefore,
Returning to the outcrop of the Kidd Copper Field samples, it yields the following information:
D = After Flood Redeposited sediment of Kidd Copper Mine tailings waste in 1970s;
S = Undefined layers;
T = Smooth, fine grained; and
C = Clay, silt, sand and minerals.
Shape. "The grains are spherical and of one size" (Coggans, *et al.,* 1991, p. 457).
"Smooth, fine grain" is the description recorded by Karen De Medeiros (2004, p. 41).

11.11 Clayey Fabric Distribution and Orientation Patterns

The findings of the sediment studied at the Kidd Copper property reveals that the orientation of the individual clayey particles is random and each aggregation has an unique preferred orientation. Refer to Figure 30 - Section 11.11 for terminology used to denote clayey fabric distribution and orientation patterns from Practical Sedimentology written by D.W. Lewis (1984, p.67).

Fabric Distribution Patterns

Overall	Individual Particles	Relation to Reference Feature
porphyroskelic (dense groundmass in which grains are set)	random	unrelated
	clustered	normal (perpendicular to . . .)
	banded	
intertexic (grains touching or linked)	radial	parallel
	concentric	inclined (state angle to . . .)
granular (no true matrix)		cutanic (concentrations forming a unit mass adjacent to . . .)

Fabric Orientation Patterns (note magnification used)

Individual Particles

Oriented: Individuals sufficiently oriented that there is roughly continuous birefringence Under Crossed Nicols (UXN); with rotation of the stage, dark extinction lines or bands move across the aggregation of individuals. Can be subdivided into strongly, moderately, or weakly oriented, although attitude of thin section relative to fabric may influence apparent degree of orientation.

Unresolved: At magnification used, appears to be some anisotropism within aggregates that are too small to observe in detail; UXN the mass appears to have random fabric.

Unoriented: At magnification used, fabric is isotropic due to apparent random orientation of individual particles.

Indeterminate: Fabric appears isotropic because of opacity or crystallographic character.

Domains (aggregations of clayey material, each aggregation with a preferred orientation

Asepic fabric: Dominantly anisotropic, with anisotropic domains unoriented with respect to each other.

Sepic fabric: Various recognizable anisotropic domains with various patterns of preferred orientation (see Brewer 1964, for subdivisions).

Undulic fabric: Practically isotropic at low magnifications, and weakly anisotrophic with faint undulose extinction at high magnifications. Domains not distinct.

Isotic fabric: Apparently isotropic matrix at all magnifications.

Crystic fabric: Anisotropic fabric involving recognizable crystals deposited during pedogenesis or diagenesis (see Brewer, 1964, for subdivisions).

Strial fabric: Clayey material as a whole exhibits preferred parallel orientation, giving extinction pattern that is either unidirectional or shows several preferred extinction directions. Common in sedimentary rocks, probably imparted by compaction and other diagenetic processes. May be superimposed on other fabrics, which should be sought when strial fabric is at extinction.

Source: after Brewer, 1964

Figure 30 - Section 11.11: Clayey Fabric Distribution

Douglas Lewis says that the individual particles of clayey fabrics can be distributed in a random, clustered, banded, radial or concentric pattern. For example, "the asepic fabric: [is] Dominantly anisotropic, with anisotropic domains unoriented with respect to each other" (1984, p. 67).

Unorientation is seen in the individual clayey particles, individual logs and the adjacent structures at Kidd Copper. Unoriented particles or random distribution are a distribution pattern found in clayey fabric which verifies that the findings at Kidd Copper are within the constraints of distribution patterns.

The unique orientation at Kidd Copper is not an anomaly, which should be ignored but rather are within the framework of sedimentological distribution. Therefore, the Kidd Copper sediment stands as a reasonable standard to base a diagnostic of sedimentological processes on.

11.12 Matrix of Grain Orientation Descriptors

'No definite layers' is mentioned only one time in the records describing the Kidd Copper sediment as the boreholes were uncovered and observed. The rest of the field entries observed layering descriptions multiple times as the following table illustrates. Borehole 69 is a common comment "see alternating layers of dark brown sand and light grey sand when dug up" (Saha, August 2004, p. 73). "Stratiform [GEOL] (1) descriptive of a layered mineral deposit of either igneous or sedimentary origin. (2) Consisting of parallel bands, layers, or sheets" (Licker, 2003, p. 355).

Table 11 - Section 11.12: Orientation of the Grains found on Kidd Copper after Redeposition

Orientation of the Grains Compilation from 32 Geology Records on August 30, 2004					
Alternating Layers	Alternating layers 2X	Alternating grey and orange layers 1X	Alternating with Orange Sediment 1X	Alternating between light grey to bluish grey 1X	Thin alternating layers of light & dark orange coarse-grained sediment 1X	Alternating layers of dark brown sand and light grey sand 1X
	Sand with Clay 1X	Sand between sand 1X Sand-clay-sand-clay 1X	Clay with orange 1X	Sand clay structure 1X	Dark black/beige spotted 1X	
Band	Rust Colored Band 1X	Sand stratified with dark bands 1X	Iron Oxide Bandings 1X	Lighter band to darker 1X	Thin band of sand and clay 1X	Grey clay with Orange banding 1X
	Green and yellow banding 1X	Pyrite band 1X Banding of Yellow 1X	Clay with iron banding 1X	Successive bands of yellowy-orange and grayish-white interbands 1X	Bands of orange and grey 1X	Bands of iron below blue grey layer 1X

Orientation of the Grains					
Compilation from 32 Geology Records on August 30, 2004					
Interbeds	Interbeds of silt 1X	Interbedded grey/orange 2X	Interbedded 1X Brown top interbedded 1X	Interbedded layers 1X	Interbedded light grey and iron interbeds 1X	Interbedded dark and light clay 1X
	Interbedded orange, darker red and white 1X	Interbedded grey/white 1X	Interbedded grey and dark red 1X	Interbedding orange and red with grey 1X	Interbedded Light-dark grey 1X	Interbedded sand and clay 1X
	Orange, 10 cm light clay, orange with clay interbedding (darker orange) 1X	Interbedded clays and sands 1X	Interlayers 1X	Interbedded of red and white sand 1X	Crispy red sand clumps interbedded in the layer of oxidized iron? Nodules 1X	26 cm Interbedded of white clay and yellow/orange sand 1X
Layers	Layering 1X	Sand Layer 2X	Clay Layer 1X	Thin rust layers 1X	Grey and Orange Stained 1X	Montmorillonite layers which are tailings clay to lubricate iron oxide where it seeps out and reaches the surface and encounters air 1X
	Grey Clay Layer 1X	Fine sand layers 1x	Orange sand with grey sand 1X	Oxidation layers 1X	Colored Oxidated Layers 1X	Oxidated Rust colored layers 1X
	Large clay layer and dark blue	Pirotite and Pyrite layer 1X	Pyrite layers 1X	Layers of yellow and grey 1X (Yellow = sulphur ? Grey = clay? Sand is yellow material).	White layer 2X	Thick white layer of clay 1X
	Coarse-grained layer 1X	Layer dark orange 1X	Layers dark/light orange sand 1X	Beige/Bright Orange Layers 1X	Small layer 2X	Light beige/orange and white layers 1X
	Noticeable colored layers of mineral deposits at different depths to a total depth of 12 cm. 1X	Layer of blue grey oozy clay then bands of orange and grey 1X	Thin layer orangy-yellow 1X	Dark Brown Layer 1X	Clay 10 cm, orange 1 cm, etc varying layers 1X	Light sand, small clay and orange sand layers 1X
	Layered brown/gray clay 1X	Very thin layer of muddy sand on top 5cm 1X	Thin orange layers 1X	Grey Layer 1X	Thick white layer of clay 4X	Anoxic thin soil layer developing over the liner 1X

Orientation of the Grains Compilation from 32 Geology Records on August 30, 2004					
Stratification	Sand stratified 1X	Stratified clay and sand 1X	Orange layers strands 1X	Strands grey sand 1X	15 cm yellow/brown sand ↓ darker clay in river bed 1X	Sand stratified with Fe deposits 1X
	Sequence of iron rich areas 1X	Thin strips of dark brown sand 1X	Colorless lenses of yellow and white 1X	Lense of oxidation going through 1X	Stratified with stain 1X	Separates into different colored layers 1X
Laminations	Laminations 1X	Laminated dark grey 1X	Thin layers 1X	Lenticular beds surrounded with pyrite 1X	Liesegang banding 1X	Lisenganen 1X
	Sand is deposited first because its heavier (coarser) and clay deposited after as its finer 1X	Aero bar type bubbles 1X	Similar layers of clay but here they are very porous— ASK WHY!! – Probably gas bubbles 1X	Ripples between laminations 1X	Prograding delta causing upward sequence 1X	Fining upwards sequence 1X
Source: Compilation of 32 Geology Students' Field Records of Kidd Copper Property Dated August 30, 2004.					

The pattern of the Kidd Copper sediment is unevenly dispersed and follows no natural pattern. Orientation-distributions are without direct lines of intersection. In Table 11 - Section 11.12 the orientation of the grains found on the Kidd Copper property after redeposition represent multiple or heterogeneous fabrics. The distributions of the boreholes bear no relation to each other as individual samples are compared. The impression would leave one to think they are unrelated samples; however they have identical parent and depositional processes, yet bear different outcomes.

Stratification is seen in every entry with noticeable colored layers of mineral deposits at different depths to a total depth of approximately 60 cm. The following remarks were noted in the students' field book entries describing the uncovered boreholes: Clay 10 cm, orange 1 cm, etc varying layers; Separates into different colored layers; 26 cm Interbedded of white clay and yellow/orange sand; Orange, 10 cm light clay, orange with clay interbedding (darker orange); Alternating layers of dark brown sand and light grey sand; Alternating grey and orange layers; Interbedding orange and red with grey; and successive bands of yellowy-orange and grayish-white interbands.

11.13 Were the Layers Formed by Segregation?

One possible explanation for the layering on the Kidd Copper property is that "segregation [GEOL] the formation of a secondary feature within a sediment after deposition due to chemical rearrangement of minor constituents" (Licker, 2003, p. 330). Segregation is "a secondary feature, such as a *nodule* of iron sulphide, that is formed by the chemical rearrangement of minor components within a *sediment* subsequent to its deposition" (MacDonald, *et al.*, 2003, p. 387). "Segregation banding [PETR] a compositional band in gneisses that is the result of segregation of material from an originally homogeneous rock" (Licker, 2003, p. 330). Reasons why this cannot apply: (1) no secondary features were observed within the Kidd Copper sediment; (2) no mechanical rearrangement has occurred to the sediment deposits, the sediment deposits have remained *in situ* since the flood; (3) gneisses were not a part of the chemical composition found at Kidd Copper; and (4) the parent rock was minetailing deposits placed upon goethite as the parent rock.

If researchers did not know the depositional history of the Kidd Copper site, they could falsely conclude that segregation was the answer. Knowing the depositional history is important to understand the depositional processes and make sound conclusions.

11.14 Significance of Boreholes and Matrices

The Kidd Copper site area was left untouched for a period of twenty years and in that time there were several sediment layers of laminae identified. From this we can conclude that the multiple layers are not necessarily multiple events but can be a single event and provide multiple layers.

"If the contacts between facies or facies associations are sharp and/or erosional, there is no way of knowing whether two vertically adjacent facies represent environments that were once laterally adjacent. Indeed, sharp breaks between facies (marked for example by channel scours, or by thin bioturbated horizons implying nondeposition) may signify fundamental changes in depositional environments and the beginning new cycles of sedimentation" (de Raaf, *et al.*, 1965; Walker & James, 2002, p. 6). The assumption from the correlation of facies does not apply in this case. The sharp contrasts do not represent changes in the depositional environment because they were deposited at the same time on the Kidd Copper Property.

11.15 Correlation of Parallel and Perpendicular Cross-sections

"To determine the relationship of geologic events to one another, four basic principles are applied. These are the principles of (1) original horizontality, (2) superposition, (3) lateral continuity and (4) cross-cutting relationships" (Plummer, McGery & Carlson, 2003, p. 177). This report will consider each stratigraphic principle to identify patterns of regularity in the geologic events at Kidd Copper Property.

The time factor for the depositional process of the sediment studied in each log profile was identical in all cases. Each borehole was uplifted and redeposited simultaneously. In theory, the results should be a homogeneous, single layer of sedimentation. Since the depositional process occurred simultaneously, there should be a repetitive pattern. On this occasion, heterogeneous non-comformity patterns distributed in the layers was the distinguished finding. This was confirmed by the irregular patterns and non-existent extremes. "Data extremes: The highest and lowest values of an attribute, found by selecting the first and last records after sorting" (Clarke, 2003, p. 308). Data collected from the 120 boreholes at Kidd Copper indicates the sediment composition, color, grain size and orientation were all randomly mixed. This random mix was not due to data extremes.

11.16 Facies Analysis

"One of the ways of identifying the environment of a facies is by analysing the environments of the facies above and below to come up with the most logical sequence of events" (Selley, 1982, p. 320).

> "Structures preserved on the tops of beds include: those formed by depositional processes (ripple marks, primary current lineations); erosional structures (flutes and scour marks); structures caused by the transportation of an object over the bed (tool marks); and other features such as desiccation and syneresis cracks, sand volcanoes, adhesion ripples and warts, rain prints, and biogenic traces and trails. Structures preserved on the bases of beds (sole marks) include load casts, the casts of flutes, trails and tool marks, and the fill of erosional scours. The external form of sedimentary units (sheet-like, channel-fill, reef or mound, lenticular, etc) is a function of the depositional environment and sometimes a post-depositional compaction" (Allaby & Allaby, 2003, p. 485).

Facies visible on the Kidd Copper property include: ball and pillow; laminations; cross bedding; cross lamination; and cross stratification. Ball and pillow were formed due to post-depositional deformation; whereas, laminations were formed by physical depositional processes. "Internal sedimentary structures include: those formed by physical depositional processes (cross-stratification, flat bedding, lamination, and heterolithic structures); those due to post-depositional deformation (convolute bedding, slump structures, dish and pillar structures, flame structures, ball and pillow structures, etc.); those caused by organic disturbance (bioturbation, trace fossils); or by post-depositional chemical disturbance (enterolithic structures, collapse and solution structures, concretions, etc.)" (Allaby & Allaby, 2003, p. 485).

Facies Analysis. The question to ask is there any repetitive or sequences noted in the lithological logs, and the answer is no. The distribution of sediment was random with no apparent pattern. The multiple layers observed at Kidd Copper were not the result of several depositions made on the property through time; rather the multiple layeres were one deposit which deposited several sediment layers. The sequence boundaries converge with no common pattern in the settling of similar sediments in the same time period throughout the impoundment area on Kidd Copper Property. The spatial distribution in the features observed clearly show no conformable patterns.

The smallest elements – the particles of clay, silt, sand, and minerals, chemical precipitates and other constituents which make up sediments have been carefully studied and observed at Kidd Copper as they relate adjacently, parallel, perpendicular and congruent to each other in the vertical log profiles. The finely-grained particles settled not in a uniform layering but settled as stratified.

11.17 Facies Change

The succession of the facies change and the "abrupt change in the depositional system … as suggested by an end of overlapping of pelagic deposition, increasing facies and thickness differentiation as well as increased input of terrigeneous materials" (Wilmsen & Wood, 2004, p. 223). In this case, the change in thickness is due to an unbalanced, random re-distribution of sediment. At Kidd Copper, no tectonic causes, nor the shifting mid-shelf carbonate depocentre, nor the eustatic sea-level fall need to be invoked to explain the changes in thickness. The facies change can be easily explained by the sediment falling to rest after the upswelling of water across the impoundment area. The facies change is due to a redeposition of sediment by water in a very short period of time, from unstratified to stratified layers.

11.18 Adjacent facies Deposition Changes Gives Clues to Time

"Today, as we travel across swamp, floodplain, and sea, we traverse different environments of deposition. Each of these environments of deposition changes laterally into the adjacent environment of deposition, and each provides its present-day facies that likewise change to adjacent synchronous facies. Geologists record these facies changes on lithofacies and biofacies maps. Because these maps are based on time-rock units, they provide a view of different facies of essentially the same age. If one were able to make such maps of successively different times, it would become apparent that ancient facies have shifted their localities as the seas advanced or retreated or as environmental conditions changed" (Levin, 2003, p. 85).

The geological interpretation provided by Levin is questionable. Levin states that "each of these environments of deposition changes laterally into the adjacent environment of deposition (2003, p. 85). The question is how does one read this with the new information acquired in the Kidd Copper property case? Levin's answer is that "each provides its present-day facies that likewise change to adjacent synchronous facies" (2003, p. 85). The assumption that Levin makes is that the differences are a marked change in an environmental condition "that ancient facies have shifted their localities" (2003, p. 85).   If the environments mentioned by Levin were substituted for the rock units in the profiles recorded earlier in Table 4 - Section 10.5 and Table 5- Section 10.6 and with the factual information gained from the Kidd Copper property case; then it could be inferred that there is no change in the adjacent environments as was predicted by Levin. This is a significant observation because detected in the Kidd Copper strata are numerous changed lateral units with no difference in depositional time. The adjacent synchronous facies are different on Kidd Copper but not because they represent differences in depositional environments or have shifted. Therefore, on the larger scale these synchronous differences could also be reflective of rapid stratification.

Adjacent facies can be various in appearances or altered but this distinguished characteristic is not subject to a time differentiation. The changes may be the result of random processes just as the case of the sequential lithographic logs revealed sharp and distinct contrasts in adjacent profiles based on the Kidd Copper facts. The valid inference gained from the Kidd Copper study is that depositional changes do not give clues to time. Strong evidence supports the claim that any unit that has a change adjacent to synchronous facies could be deposited at the same time, under the same depositional environment, as was the case for the rock units found in the profiles on the Kidd Copper Property.

11.19 Data Summarized from Field Logs

The data from the 32 field logs is consistent across the site, taking into account sides, slopes and interior proximities. Interbedded, layered, interlayers, stratified, sand layers, alternating layers are all descriptors used when documenting the site. All boreholes indicate reference to stratified layers.

These were not laid over a period of time but rather in a short period of time. The layers do not provide a record of a deposition that took place over time with multiple events but records the complete deposition in a one-time occurrence. This is contrary to the expectation that each layer would be a record from the time of deposition of the bottom (oldest) stratum to the time of deposition of the top (youngest) stratum. The deposition of the bottom and top layers occurred in the same event and the stratigraphic record challenges the Law of superposition and the conclusion that different layers represents different time periods.

"Rock strata lack dates, so on the bases of stratigraphy we can only assign relative ages to them" (Murck, 2001, p. 49). The date of the sediment deposit at Kidd Copper is known historically so it is possible to determine the exact age. The conclusion that strata are deposited in chronological order is incorrect based on the new findings from the data collected in this report. The evidence from the findings at Kidd Copper strongly support that each stratum layer does not correlate to a different deposition in the time record as previously understood.

11.20 Significance of the Bouma Sequence and the Kidd Copper Property

The Bouma Sequence classifies sedimentation layers as layers that settled out of suspension and came to rest. Refer to Figure 21 - Section 7.4. The sediment studied in the lithological logs at Kidd Copper settled out of suspension. Evidence of this settling is found in the the Bouma sequence noted on the redeposited units on the Kidd Copper Property, which display flow regimes because the material was deposited from suspension as the turbidity current came to rest. As expected by the Bouma sequence commonalities were found such as: layers were deposited; layers were formed from fine-grained sediment of silt and clay grade; laminated sand characteristics; and with the sand in the top layer as being the finest. All of the divisions in the Bouma sequence that are subject to the influences of the waning flow in the turbidity current are apparent in both Bouma and Kidd Copper. Interesting to note that stratification is a prevalent interpretation in terms of flow regimes. However, Bouma recognizes a pattern that results from this distribution, Kidd Copper does not.

The difference is that Kidd Copper recognizes that these differentiations occurred essentially at the same time because of the historical accuracy of the Kidd Copper records. These divisions were laid down not through a process of time but in a matter of hours. Kidd Copper's deposits were redeposited by turbidity currents in a short period of time. The Kidd Copper sediment was formed into a new accumulation as a result from the deposition of sedimentary material that had been picked up and reworked. Kidd Copper is unique in that it relates the time differentiation to a rapid succession of layering in a small time period.

11.21 Data Interpretation Techniques

"The two subjects [Sedimentology and Stratigraphy] can be considered to form a continuum of scale of observation and interpretation. The study of sedimentary processes and products allows us to interpret the dynamics of depositional environments. The record of these processes in sedimentary rocks allows us to interpret the rocks in terms of environments. To establish lateral and temporal changes in these past environments a chronological framework is required. This time framework is provided by different aspects of stratigraphy and allows us to interpret sedimentary rocks in terms of dynamic evolving environments. The record of tectonic and climatic processes through geological time is held in these rocks along with evidence for the evolution of life on Earth" (Nichols, 2003, p. 1).

"Geologists employ two methods of considering or classifying layers of rock in the field. The first method, *rock stratigraphy*, examines the makeup of rocks, while the second technique, *time stratigraphy*, considers the time at which the layers of rock were deposited. The key unit of measure in rock stratigraphy is called the *formation*. The key unit of measure in time stratigraphy is called a *period*. Formations might indicate previous geologic events – earthquakes, glaciers, volcanoes, or floods" (Albin, 2004, p. 47).

"Time stratigraphy helps the geologist to understand when rock was deposited" (Albin, 2004, p. 47). Asking formative questions such as: How old are they? "How did they form? Why do some show cross-bedding and others graded bedding?" (Blatt, Middleton & Murray, 1972, p. 3).

"*Accuracy*: The validity of data measured with respect to an independent source of higher reliability and precision" (Clarke, 2003, p. 304). "Description should aim at defining the original bed configuration because this emphasis provides the most logical link to interpretation of flow conditions" (Harms, *et al.*, 2004, p. 45).

"In 1965 the property was optioned by Sheridan Geophysics Ltd. And assigned to Kidd Copper Mines Ltd." (Article [2], 2008, *p. nd.*). "The Kidd Copper site is located directly southwest of Sudbury, and the mine actually sits on a mineral intrusion. As a result, the bedrock is rich in goethite" (Shang, *et al.*, 2004, p. 9). "Some time between 1977 and 1983, the retaining dam at the southeast corner of the site gave way" (Shang, *et al.*, 2004, p. 4, 5).

The data collected on the Kidd Copper property is highly accurate because of these known facts about its depositional environment and geologic setting.

12.0 Fast Rates of Deposition Seen Elsewhere

Not all stratigraphic units represent long time intervals as the following examples will show. Many are deposited very rapidlylike: debris flows, turbidity currents, fining upward sequence of sediments, debris flow, mud flow, the Horgen slide, a normal lacustrine mud, Sturzstroms, Grand Banks sediment, mixed-layer mineral, flow layering or flow banding and layered intrusions are examples of other subenvironments that display rapid stratification characteristics. The sedimentation rate depends on the processes involved.

12.1 Many Subenvironments Display Rapid Stratification Characteristics

Geology is a field base science and the observation of rapid stratification is noted elsewhere. In fact, most hillslope processes producing colluvial deposits like the Kidd Copper mine tailings waste site produce stratified sediments. Many processes can deposit stratified sediments very rapidly; for example – volcaniclastic deposits (i.e. ash and tuff layers), glaciofluvial deposits, mass movements, turbidity currents and sturzstroms. "Grain flows rapidly 'freeze' as soon as the kinetic energy of the particles falls below a critical value" (Nichols, 2003, p. 58).

12.2 Debris Flow Characterize both Slow and Fast Sedimentation Rates

Flow is "a type of movement that implies that a descending mass is moving downslope as a viscous fluid" (Plummer, McGery & Carlson, 2003, p. 557). "Pelagic sedimentation rates do vary considerably, but pelagic clays accumulate the most slowly (typically 0.1–0.5 cm ky^{-1}), whereas sedimentation rates for calcareous oozes are typically in the range 0.3–5 cm ky^{-1} and siliceous oozes are the range 0.2–1 cm ky^{-1}" – from fractions of a mm/yr (e.g. pelagic sedimentation, deposition of lacustrine clays) to many tens of metres per year (e.g. deposition of coarse glaciofluvial sediments, debris flow processes etc.) (Rothwell, 2005, p. 77).

Debris Flow. A debris flow is an example of both slow and rapid sedimentation rates. "The general term *debris flow* is used for mass wasting in which motion is taking place throughout the moving mass (flow). The common varieties earthflow, mudflow, and debris avalanche" (Plummer, McGery & Carlson, 2003, p. 208). "*Debris flow* a *sediment gravity* flow process in which particles up to boulder size are supported principally by their buoyancy in, and the *cohesive strength* of, a sediment-water slurry" (Kearey, 2001, p. 68). "Debris flow is a fluid-gravity flow, the fluid is mud which is so viscous that the sediment-water mixture tends to flow viscously" (Hsü, 2004, p. 90). "Debris flow [is] a moving mass of rock fragments, soil, and mud, more than half of the particles been larger than sand size. Soil debris flows may move less than 1 m per year; rapid ones reach 160 km per hour, as in the 1977 Huascaran flow in the Peruvian Andes" (Bates & Jackson, 1984, p.128; MacDonald, *et al.*, 2003, p. 132). The key to note here is that sedimentation rates vary. Rapid stratification is within the parameters of the sedimentation rates.

12.3 Turbidity Currents Rapid Deposition

> "Turbidity currents are individual flow events. They occur over a geologically very short period of time, with most of the deposition occurring in a few hours to a few days. In fact, in the context of geological time turbidities deposit 'instantaneously'" (Nichols, 2003, p. 58).

The science unearthed at the Kidd Copper property supports a very short deposition. What the Kidd Copper study refutes is that "the time taken for the thin layer of suspended sediment to be deposited on the top of the turbidite is many orders of magnitude longer (months to hundreds of years)" (Nichols, 2003, p. 58). The layers of sediment seen on the Kidd Copper property logs support that the magnitude of time was not months or hundreds of years. Instead, the layers of sediment were deposited under a week, likely even within a day or hours within the day. The deposition of the top layers of sediment was deposited as fast as the bottom layers. The dam was not re-built and the mine tailings waste-site remains inactive; therefore the settled deposits remained intact. The thin layers of sediment were not deposited as many orders of magnitude as conventional wisdom suggests. Instead, the subsequent layers noted on the Kidd Copper property was not due to a long process but rather a short process in time.

12.4 Mudflow Fast Rates of Deposition

A mudflow is "a *mass* flow of debris mixed with water with a high proportion of *mud*. Very similar to a *lahar*, but richer in fine-grained material" (Kearey, 2001, p. 176). "A *mudflow* is a flowing mixture of debris and water, usually moving down a channel" (Plummer, McGery & Carlson, 2003, p. 211). Mudflow is "a flow of fine-grained, water-saturated sediment in a stream channel" (MacDonald, *et al.*, 2003, p. 307). "It can be visualized as a stream with the consistency of a thick milkshake. Most of the solid particles in the slurry are clay and silt (hence the muddy appearance), but coarser sediment commonly is part of the mixture" (Plummer, McGery & Carlson, 2003, p. 211). "Typical of semi-arid to arid regions where heavy rains transport earth material collected on hill slopes into pre-existing stream channels" (MacDonald, *et al.*, 2003, p. 307). The point to be made here is that rapid processes are seen elsewhere.

12.5 Landslides Fast Rate of Deposition

"*Landslide (landslip)* – The rapid movement of a mass of soil downslope along a curved or planar *failure* surface, without *deformation* of the *soil structure*" (Kearey, 2001, p.149). Landslide is "a general term for the downward movement, under gravity, of masses of soil and rock material as a result of a variety of processes. Types of landslides include rockfalls, mudflows and slumps" (MacDonald, *et al.*, 2003, p. 270). Again, what is important to note is that rapid processes are seen in other geologic processes.

12.6 Case of Horgen slide

"The Horgen slide was not triggered by an earthquake, but induced by the weight of landfill material. The main mass of the slide consists of lacustrine mud. The slide debris are now found deposited on a subaqueous fan at the foot of the steep slope. The resedimented deposit shows a sequence indicative of the progressive disintegration of the slide material. The base … shows slump folding, with little or no dispersed material. This is overlain by a mixture of large mud fragments with slump folding and soft clasts… This grade upward to the mud-clast facies … Where the disintegration was partial, the original sedimentary structures, such as horizontal lamination or cross-lamination, are partially preserved. Since the sediment particles in both clasts and matrix are sand – or silt-sized, the boundary between the clast and matrix is commonly indistinct" (Hsü, 2004, p. 83).

The Horgen Slide has three commonalities with Kidd Copper. The first is the fact that in both the boundaries between *clast and matrix* is *indistinct*" (Hsü, 2004, p. 83). The second fact is that both hill slope and mudflow events have been recorded observing the rate of sedimentation deposits to occur rapidly. The third is that Kidd Copper and the Horgen Slide are the results of a cataclysmic event. Notice that in both events a progression of layers was observed rather than one layer for one event.

12.7 Primary and Secondary Structures

"Structural features may be primary, i.e. those acquired in the genesis of a rock mass (e.g. horizontal layering), or secondary, i.e. resulting from later deformation of primary structures (e.g. folding or fracturing)" (MacDonald, *et al.*, 2003, p. 417). According to MacDonald the horizontal layering is primary and the folding or fracturing is secondary resulting from later deformation of primary structures. In the case of Kidd Copper, there was no tectonic force involved or other 'later process' to disturb the sediment; instead, the strata on Kidd Copper simply came to rest as a result of the evaporation of the flood waters over the impoundment area. A fixed distinction between primary and secondary structural features was not found in the case of Kidd Copper.

Both primary and secondary structural features can be bent. **Therefore, both horizontal and non-horizontal layers can indicate the start of geological history.**

13.0 Stratigraphic Analysis

The Kidd Copper property offers a way to test stratigraphical principles because the variables of time past and time present are known for the rates of sedimentation, the depositional process is known, the parent material is known, the historical information with regards to the site composition is known.

The data provided by on-site field observations at Kidd Copper mine property shows sediment deposits which challenge these two suppositions by providing contrary results. Instead, observed layering of the mine tailings deposits from Kidd Copper property deposits demonstrates stratification can be young or old on the top and on the bottom layer. Kidd Copper strongly supports the hypothesis of this report that many sediment layers are formed rapidly. This breaks the common rule of thumb where, the top is underlain by older unstratified rocks. Many layers were found as one.

Why is this important? Interpreting the past by reference to the present is a strong fact finder. A precise date for the layering of the deposits is known. Stratigraphy is "the key to understanding almost all Earth processes as stratigraphic analysis provides information about events through geological history" (Nichols, 2003, p. 2).

13.1 New Knowledge and Impact of Kidd Copper Analysis

The idea that the sediment is to be horizontal when it first is deposited underlines the way that mainstream geologists analyze the rock record. The horizontal base of the stratigraphic column is built up as a resulting factor of time. The allowance is that younger rock is deposited on older layers in the stratigraphic column. Kidd Copper's orientation, particle size and fabric distribution were not indicators of old age.

Stratigraphy is "the key to understanding almost all Earth processes as stratigraphic analysis provides information about events through geological history" (Nichols, 2003, p. 2). "The materials are then considered in terms of depositional environments, the places where sediments accumulate into sedimentary rocks and become layers of stratigraphy through time" (Nichols, 2003, p. 3). "Stratigraphic map [GEOL] a map showing the aereal distribution, configuration, or aspect of a stratigraphic unit or surface, such as an isopach map or a lithofacies map" (Licker, 2003, p. 355).

13.2 Steno's Four Principles

"Most rocks occur in layers (called *strata* from the Latin) anywhere from a metre to several tens of metres thick" (Eyles, 2002, p. 68). "The first stratigrapher was Nicolas Steno. He is credited with four principles which are still used to identify the order in which rocks were deposited (their ages relative to each other)" (Eyles, 2002, p. 66).

Steno's four principles are:

1. The principle of original horizontality. "The first fundamental principle of stratigraphy is the *principle of original horizontality*, which says that waterborne sediments settle out and are deposited (laid down) as horizontal layers. The principle of original horizontality is important because it means that whenever we observe water laid strata that are bent, twisted, or no longer horizontal, we know that some tectonic force must have disturbed the strata after they were deposited" (Murck, 2001, p. 49). Furthermore, it "states that water-laid sediments will be deposited as horizontal strata. If plate tectonic forces affect the area, the layers may become distorted by being folded or faulted" (Eyles, 2002, p. 67).

    > "It means that layers of sediment are generally deposited in a horizontal position. Thus, if we observe rock layers that are flat, they have not been disturbed and still have their *original* horizontality. But if they are folded or inclined at a steep angle they must have been moved into that position by crustal disturbances sometime *after* their deposition" (Tarbuck & Lutgens, 1999, p. 191).

    > Law of original horizontality as defined by Dr. James MacDonald, Dr. Christopher Burton, Dorothy Farris Lapidus, Donald R. Coates and Isobel Winstanley state "*sedimentary rocks*, especially water-laid strata, are deposited parallel to the surface on which they are deposited and thus horizontally. Such rocks are assumed to have been horizontal at the start of their geological history, although they may have been folded or tilted subsequently" (MacDonald & Burton, 2003, p. 273).

2. "The second principle of stratigraphy is the principle of stratigraphic superposition, which states that any sedimentary rock layer is younger than the stratum below and older than the stratum above. In other words, sedimentary rock strata are like newspapers laid in a pile day by day, providing a record from the time of deposition of the bottom (oldest) stratum to the time of deposition of the top (youngest) stratum" (Murck, 2001, p. 49). The principle of stratigraphic superposition states that in any undisturbed sequence of strata, any one bed is younger than the bed below it and is older than the bed above it. According to Tarbuck & Lutgens (1999) the underlying assumption in the principle of superposition is that the older rocks are underneath the younger beds (p. 191).

3. The principle of cross-cutting relations states that a feature cross-cutting a layer is younger that the layer. In other words, the "feature that is cut is older than the

feature that crosses it" (Levin, 2003, p. 10). "An igneous rock, fault, or other geologic feature must be younger than any rock across which it cuts" (Allaby & Allaby, 2003, p. 312). Law of crosscutting relationships is "a stratigraphic principle whereby relative ages of rocks can be established; a rock (esp. an igneous rock) is younger than any other rock across which it cuts" (Bates & Jackson, 1984, p. 290).

4. "The principle of fossil succession states that fossils found in lower layers of a sequence of strata are older than those found in higher layers" (Eyles, 2002, p. 67). This report did not develop this point because no fossils were present in the observed sediment found on the Kidd Copper property in the mine tailings.

The Kidd Copper report will assess the first two principles of Steno's because there was no cross-cutting or faunal succession to observe.

13.3 The Principle of Original Horizontality

"The observation that strata are often tilted led Steno to his *principle of original horizontality*. He reasoned that most sedimentary particles settle from fluids under the influence of gravity. The sediment then must have been deposited in layers that were nearly horizontal and parallel to the surface on which they were accumulating. Hence, steeply inclined strata indicate an episode of crustal disturbance after the time of deposition" (Levin, 2003, p. 4).

The principle of original horizontality "means that layers of sediment are generally deposited in a horizontal position. Thus, if we observe rock layers that are flat, they have not been disturbed and still have their *original* horizontality. But if they are folded or inclined at a steep angle they must have been moved into that position by crustal disturbances sometime *after* their deposition" (Tarbuck & Lutgens, 1999, p. 191). "Sedimentary rocks are always deposited as horizontal, or nearly horizontal, strata, although these may be disturbed by later earth movements" (Allaby & Allaby, 2003, p. 312).

"*Original Horizontality*. The state of strata being horizontal or nearly so at the time they were deposited" (Bates & Jackson, 1984, p. 359). "Law of original horizontality [is] a general law of geology: Waterlaid sediments are deposited in strata that are horizontal or nearly horizontal, and parallel or nearly parallel to the earth's surface" (Bates & Jackson, 1984, p. 291). Law of original horizontality – *sedimentary rocks*, especially water-laid strata, are deposited parallel to the surface on which they are deposited and thus horizontally. Such rocks are assumed to have been horizontal at the start of their geological history, although they may have been folded or tilted subsequently" (MacDonald, *et al.*, 2003, p. 273). Therefore, the Law of original horizontality would lead to the conclusion that the first deposition of sediment on ground zero would be deposited horizontally.

"Lamina [GEOL] A thin, clearly differentiated layer of sedimentary rock or sediment, usually less than 1 centimeter thick" found in the Dictionary of Geology & Mineralogy published by Mark D. Licker (2003, p. 177). "Parallel lamination can be produced … from strong currents, referred to as upper plane-bed phase lamination" (Reineck & Singh, 1973, p. 49). Therefore, like the expectations from turbidite deposits the Kidd Copper site redeposited sediment was highly probable to be parallel.

Nearly Horizontal Layers Prediction Based on Conventional Wisdom. On the Kidd Copper redeposited tailings horizontal layers was expected to be found when the earth was uncovered because the Law of original horizontality states that sediments are originally deposited in horizontal layers first. Only *after* deposition, the sedimentary layers may be further tilted or folded. Accordingly, the expectation from the turbidity and the rapid deposit rate identified at the Kidd Copper mine tailings site would be that the redeposited sediment would be found as a horizontally distributed single layer.

13.4 Flood Plains Almost Entirely Horizontal Layers of Fine Sediment

"Some flood plains are constructed almost entirely of horizontal layers of fine-grained sediment, interrupted here and there by coarse-grained channel deposits. Other flood plains are dominated by meanders shifting back and forth over the valley floor and leaving sandy point bar deposits on the inside of curves. Such a river will deposit a characteristic fining-upward sequence of sediments: coarse channel deposits are gradually covered by medium-grained point bar deposits, which in turn are overlain by fine-grained flood deposits" (Plummer, McGery & Carlson, 2003, p. 237-239).

Planar Lamination Predicted Based on Conventional Wisdom. "The initial deposits commonly display planar lamination produced by rapid flow" (Nichols, 2003, p. 120). The expectations for the Kidd Copper mine tailings redeposition would be planar lamination since they were produced by rapid flow.

13.5 The Law of Original Horizontality Tested

"Law of original horizontality states that sediments are originally deposited in horizontal layers. After deposition, sedimentary layers may be further tilted or folded" (Environmental Science 1A03/1B03/1G03, 2002, p. 98). "The principle of *original horizontality* states that beds of sediment deposited in water formed as horizontal or nearly horizontal layers" (Plummer, McGery & Carlson, 2003, p. 177). Plummer, McGery & Carlson state "when the floodwaters recede, these fine-grained sediments are left behind as a horizontal deposit on the flood plain" (2003, p. 237). "Horizontal or nearly horizontal laminae in silt, sand or granules … Many lamination surfaces show parting lineation that is parallel with flow" (Harms, *et al.*, 2004, p. 49). "Most sediments settle in water in horizontal layers with flat upper and lower surfaces" (Busbey, *et. al.*, 1999, p. 50).

According to the principle of original horizontality sediments that are laid down in water will be deposited in horizontal or near horizontal layers. "The principle of original horizontality states that beds of sediment deposited in water formed as horizontal or nearly horizontal layers" (Plummer, McGery & Carlson, 2003, p. 177). Plummer, McGery & Carlson state "when the floodwaters recede, these fine-grained sediments are left behind as a horizontal deposit on the flood plain" (2003, p. 237). "Horizontal or nearly horizontal laminae in silt, sand or granules ... Many lamination surfaces show parting lineation that is parallel with flow" (Harms, et al., 2004, p. 49). "Most sediments settle in water in horizontal layers with flat upper and lower surfaces" (Busbey, *et. al.*, 1999, p. 50).

"Test the principle of original horizontality. Take a tub of water, stir in some mud, and let it settle quietly. Examine the result: the layers of deposited sediment will be horizontal, unless the tub was disturbed during the settling. The principle of original horizontality is important because it means that whenever we observe water-laid strata that are bent, twisted, or no longer horizontal, we know that some tectonic force must have disturbed the strata after they were deposited" according to this geologic law" (Murck, 2001, p. 49).

The principle of original horizontality was tested with the data collected at Kidd Copper. The tub of water was the water that flooded as the dam broke. The mud that was stirred in was the mine tailings. The mixture settled quietly and was not redisturbed. Therefore, the results can be examined to determine what occurs when water-laid strata is observed, as such is the case with the sediment studied *in situ* at Kidd Copper. Refer to Table 12 – Section 13.5 for contrasts between site observation at the Kidd Copper property and The Law of original horizontality. Kidd Copper sediment did not follow this geologic law because the water-laid strata observed from the data collected in this report was bent and twisted. This is opposite to Murck's statement about "water-laid strata that are bent, twisted, or no longer horizontal, we know that some tectonic force must have disturbed the strata after they were deposited" (2001, p. 49).

Table 12 - Section 13.5: Comparing the Law of Original Horizontality with Observed Facts

Law of Original Horizontality Predicts	Actual Kidd Copper Observations
"Waterborne sediments settle out and are deposited (laid down) as horizontal layers" (Murck, 2001, p. 49). Therefore, the predicted findings would be sediments that are laid down in water will be deposited in horizontal or near horizontal layers on Kidd Copper (MacDonald, *et.al.*, 2003, p. 273).	In contrast to the prediction, waterborne sediments settled were deposited randomly without order. The lithographic logs yielded various composites and distribution patterns. The actual observation confirms sediment laid down in water was deposited in variant layers and not in horizontal layers.
Sources: Murck, 2001, p.49; MacDonald, *et al.,* 2003, p. 273. Field observations in this report are from the Kidd Copper Property, August 30, 2004.	

Conclusion: *The Law of original horizontality* explains that sediments like sands, gravels and muds are originally deposited or settled out of water in horizontal layers and will be deposited in horizontal layers. After deposition, sedimentary layers (beds) may be further tilted or folded by tectonic activity but they are horizontal at the time they are formed. However, in the case of Kidd Copper, sediment was not disturbed during the settling process, where the primary sediments settled quietly without disturbance. Contrary to the expected results from the Law of original horizontality, the Kidd Copper strata did not form horizontal layers; instead the stratum that was formed was bent, twisted and stratified layers. The textbook theory and predictions of nearly horizontal layers or planar lamination based on conventional wisdom were not observed when tested with *in situ* site observations at Kidd Copper.

"In science, a law is a descriptive principle of nature that holds in all circumstances covered by the wording of the law. There are no loopholes in the laws of nature and any exceptional event that did not comply with the law would require the existing law to be discarded or would have to be described as a miracle" (Isaacs, *et. al.,* 2003, p. 445).

13.6 Horizontal Lamination is found at Various Levels on First Deposits

Other Supportive Evidence. The flow regime concept was applied by Allen (1963) to the interpretation of fluvial sediments, following an idea of Oomkens and Terwindt (1960). In the interpretation of one particular fining-upward cyclothem, (Allen (1963, p.227) implied that the highest flow regime (plane bed in that example) would be at the base of the channel, with an upward decrease in regime through dunes to ripples at the top of the point bar sequence. Harms and others (1963) also interpreted large – and small-scale trough sets as lower flow regime deposits and horizontal lamination as planebed, upper-flow regime deposits; however, they recognized that horizontal stratification is found at various levels within point bars. "After study of many more sequences, we now know that horizontal stratification is not consistently found on the channel floor, and hence that flow regime interpretations are more complex than implied by Allen. For example, such interpretations must take into account whether the observed deposits formed in active, decaying, or abandoned channels, and also must recognize the frequent changes of stage, or regime, that must have occurred during deposition" (Harms, *et al.*, 2004, p. 77). Harms had recognized that the horizontal stratification is not consistently found. This is congruent with the supposition in this report. Kidd Copper's findings bear that original deposition can have multiple layers as its primary structure in a geologic record contrary to the point of origin held by The Law of original horizontality that horizontal layers are laid first.

13.7 The Principle of Stratigraphic Superposition

"Hutton's work was complemented by the observations of the English surveyor William Smith, who noticed that sedimentary layers of rock, also called beds or strata, occur in the same sequence throughout England. He concluded that the strata were deposited in chronological order, with the lower strata being the older ones" (Discovery Books, 1999, p. 32). This scientific report examines if Hutton's conclusion is accurate. Are strata deposited in chronological order with the lower strata being the oldest?

"The second principle of stratigraphy is the *principle of stratigraphic superposition*, which states that any sedimentary rock layer is younger than the stratum below and older than the stratum above. In other words, sedimentary rock strata are like newspapers laid in a pile day by day, providing a record from the time of deposition of the bottom (oldest) stratum to the time of deposition of the top (youngest) stratum. Newspapers, of course, have dates on them, so it is possible to determine the exact age of a given newspaper within the pile. Rock strata lack dates, so on the bases of stratigraphy we can only assign relative ages to them" (Murck, 2001, p. 49).

13.8 All Geologic Chronology is based on The Law of Superposition

"*Law of Superposition*. A general law upon which all geologic chronology is based: In any sequence of sedimentary strata (or of extrusive igneous rocks) that has not been overturned, the youngest stratum is at the top and the oldest at the base; each bed is younger than the bed beneath, but older that then bed above it. The law was first clearly stated by Steno in 1669" (Bates & Jackson, 1984, p. 291, 292).

"In 1667 Steno developed the principle of superposition: 'in a sequence of layered rocks, any layer is older than the layer next above it'" (Nichols, 2003, p. 2). Stratigraphy is "the key to understanding almost all Earth processes as stratigraphic analysis provides information about events through geological history" (Nichols, 2003, p. 2).

> "The smallest elements – the particles of sand, pebbles, clay minerals, pieces of shell, algal filaments, chemical precipitates and other constituents which make up sediments – are considered first, together with the processes which move and deposit them. The materials are then considered in terms of depositional environments, the places where sediments accumulate into sedimentary rocks and become layers of stratigraphy through time" (Nichols, 2003, p. 3).

The Law of superposition states that sediments that have not been disturbed will have the youngest sediments on top and the oldest sediments on the bottom. So, reasoning according to the Law of superposition is that the youngest sediment on a geologic structure will be on top. The Law of superpostion says that layers are the result of time. Superposition reads the many layers as different depositions of sediment.

Figure 31 - Section 13.8 is an "Analog, a representation where a feature or object is represented by another tangible medium. For example, a section of the earth can be represented in analog by paper map, or atoms can be represented by ping-pong balls" (Clarke, 2003, p. 305). The sections of the soil layers are described as layers of cake.

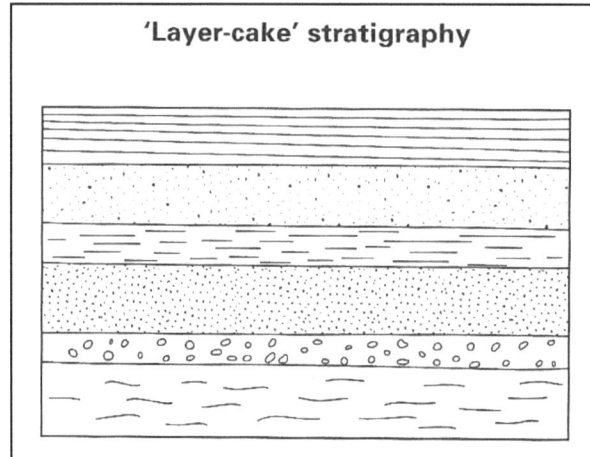

Figure 31 - Section 13.8: 'Layer-cake' Stratigraphy (Nichols, 2003, p. 232)

13.9 The Law of Superposition Defined

"*The Law of superposition*, which states that in an underformed sequence of sedimentary rocks or lava flows, each layer is older than the one above it and younger than the one below it" (Tarbuck & Lutgens,1999, p. 6). "Law of superposition of strata (principle of superposition) strata are deposited sequentially, so that in an undisturbed sedimentary succession each layer of rock is younger than the layer beneath it" (Allaby & Allaby, 2003, p. 312). "Superposition [GEOL] (1) The order in which sedimentary layers are deposited, the highest being the youngest. (2) The process by which the layering occurs" (Licker, 2003, p. 361). Superposition is "a principle or law stating that within a sequence of undisturbed sedimentary rocks, the oldest layers are on the bottom, the youngest on the top" (Plummer, McGery & Carlson, 2003, p. 565).

"Law of superposition [GEOL] the law that strata underlying other strata must be the older if there has been neither overthrust nor inversion" (Licker, 2003, p. 180). "Subsequent earth movements may overturn and invert this sequence" (Allaby & Allaby, 2003, p. 312). Law of superposition "in any sequence of sedimentary strata that is not strongly folded or tilted, the youngest *stratum* is at the top and the oldest at the bottom" (MacDonald, *et al.*, 2003, p. 273). "Law of superposition states that younger sediments are deposited on top of older previously deposited sediments" (Environmental Science 1A03/1B03/1G03, 2002, p. 98).

"The principle of *superposition* states that within a sequence of undisturbed sedimentary or volcanic rocks, the layers get younger going from bottom to top. Obviously, if sedimentary rock is formed by sediment settling onto the sea floor, then the first (or bottom) layer must be there before the next layer can be deposited on top of it. The principle of superposition also applies to layers formed by multiple lava flows, where one lava flow is superposed on a previously solidified flow" (Plummer, McGery &

Carlson, 2003, p. 177). "Provided the rocks are the right way up... the beds higher in the stratigraphic sequence of deposits will be younger than the lower beds" (Nichols, 2003, p. 232).

"The principle of *superposition* states that within a sequence of undisturbed sedimentary or volcanic rocks, the layers get younger going from bottom to top. Obviously, if sedimentary rock is formed by sediment settling onto the sea floor, then the first (or bottom) layer must be there before the next layer can be deposited on top of it" (Plummer, McGery & Carlson, 2003, p. 177). The observable facts revealed at the Kidd Copper site contradict this principle. The observations are evidenced by visible, sediment in stratified layers. No long-time estimate is required. The facts support a short-time ~1970 for the deposits and unearthed in 2004.

The outcome of the comparisons in Kidd Copper's vertical logs recognizes no no uniformity in the correlation of lithographic logs over the impoundment area at Kidd Copper, see Table 4 - Section 10.5 and Table 5 - Section 10.6. These logs are disjointed, random composites of sediment that have no uniform or similar correlation pattern in their spatial distribution. The fact rests that on Kidd Copper a mixture of sediment in multiple layers formed a primary structure. The fact remains that the process that moved them was a cataclysmic flood event. Each of these deposits are related but in the vertical logs they do not show distinguishable patterns to read as related.

Conclusions can be drawn from Kidd Copper data: (1) the youngest beds of sediment do not have to be the top layer; (2) the oldest layer does not have to be the bottom layer; (3) layers can be formed at the same time; (4) the distribution of sharp breaks in the adjacent layers does not necessarily reflect separate depositional periods; (5) distinct horizons do not necessarily mark different depositional periods; (6) flood-event such as Kidd Copper can overturn one homogeneous layer and redeposit many layers; (7) the mixing of sediment can provide either a homogeneous or a heterogeneous sediment layer as a primary structure; (8) sediment deposited in water does not only form only as horizontal layers; and (9) marking time through correlation of layers is not reliable.

The Kidd Copper case study is a strong foundation to establish as a base line to purport geological theories into supported observations. These observations are the baseline of real science.

13.10 Igneous Rock Expresses Layers

Two examples are provided to show the reader that layers are common throughout the rock cycle and visible not only in sedimentary rocks but in igneous rocks, too. The first example, *flow layering* or *flow banding* is "a feature of an *igneous rock* expressed by layers of contrasting colour, texture or sometimes mineralogical composition, formed by *magma* or *lava flow*. *Flow layers* vary in thickness from less than one millimetre up to many centimetres" (MacDonald, *et al.*, 2003, p. 186). "Flow layers may be planar, folded into smooth open forms or intricately contorted. *Planar layers* originate with *laminar flow* within the moving magma; *folded* or *contorted layers* indicate a transition from totally laminar to nonlaminar or *turbulent flow* as the moving magma responds to local changes, such as obstruction" (MacDonald, *et al.*, 2003, p. 186).

The second example, "*Layered intrusion.* – A major igneous *intrusion* within which the *magma* has crystallized as a series of layers with contrasting mineralogy and chemical composition" (MacDonald, *et al*, 2003, p. 274).

Could it be that there are no differences in the depositional processes within these rock classifications – sedimentary – metamorphic – and igneous? Could it be that the same rock processes exist between them? Igneous rock, such as magma and ash tufts are representative of rapid stratification. True, the rock type is of a different compositional material but could the same force and process that forms igneous rock, manifest itself on sedimentary rocks? The possibility exists.

13.11 Banded Structures

Band is "an informal stratigraphical term for a *stratum* or *lamina* distinguishable from adjacent layers by colour or lithological difference" (MacDonald, *et al.*, 2003, p.49). "Band [GEOL] A thin layer or stratum of rock that is noticeable because its color is different from the colors of adjacent layers" (Licker, 2003, p. 34). "Banded [PETR] pertaining to the appearance of rocks that have thin and nearly parallel bands of different textures, colors, and minerals" (Licker, 2003, p. 34). "Banded – (of a vein, sediment or other deposit) showing alternating layers differing in texture or colour, and possibly in mineral composition, for example *banded iron formation*" (MacDonald, *et al.*, 2003, p. 49). "Banding [PETR] (1) the series of layers occurring in a banded structure. (2) In sedimentary rocks, the thin bedding of alternate layers of different materials" (Licker, 2003, p. 34). "Banded ore [GEOL] ore made up of layered bands composed either of the same minerals that differ from band to band in color or textures or proportion, or of different minerals" (Licker, 2003, p. 34). Banded ore is "ore that consists of bands or, strictly speaking, layers of different minerals or the same minerals differing in texture or colour" (MacDonald, *et al.*, 2003, p. 50).

The point to consider here is that banding or layering is common throughout the rock record and points to the possibility that they were not formed separately but at the same time. If within one rock multiple layers exist, if within the rock record of the mantle multiple rock layers exist, why could they not have been formed quickly?

13.12 The Law of Original Horizontality and Superposition Illustrated

"The term *scale* in rock outcrops means two things: the size of the exposed section, and the distance from which it is observed. The example shown here is the side of a mountain (Mt. Carpegna in the Appennines, Italy) and represents a large-scale outcrop, indeed one of the largest possible. Several hundred meters of superposed, parallel strata are visible in the picture, and many of them can be traced laterally for more that 1 km (we say they are *continuous* within the outcrop limits). The *style* of stratification is thus clearly expressed … the intersection of *bedding planes* with the topographic surface, mark *individual beds* or bundles of beds (*bed sets*): geologists can speak of *levels* or *horizons*. … Although even a very large outcrop covers only a part of the area of a subsurface section, it gives you an idea of what would appear if you could dig a trench with a gigantic shovel in the fill of a sedimentary basin" (Lucchi, 1995, p. 19).

"The principles of original horizontality and of superposition … are well illustrated by this picture [refer to Figure 32 - Section 13.12 of Plate 1] … The formation to which this outcrop belongs is named after another locality of the Apennines (Monte Morello, near Florence)" (Lucchi, 1995, p. 19).

This point demonstrates that mainstream geologists interpret that each layer represents one deposition of sediment. Appendix J lists some of the divisions in Geology and the Earth Sciences. According to Ricci Lucchi, the beds grew regularly through a slow period of time; and the layer-cake bedding is interpreted as a growth model which reflects the outcome of the principles of original horizontality and or superposition.

Interpretation of Figure 32 – Section 13.12 Plate 1, "the pile of strata grew regularly, one bed after the other, in the most classical way you can imagine sedimentation to occur. There are, anyway, other *growth models* for bedding and sedimentary bodies: this one is identified as *vertical accretion, aggradation*, or, more colloquially, *layer cake*" (Lucchi, 1995, p. 19).

The problem with dating sediment that has already existed before anyone can remember how it was deposited is that unknown parameters must be assumed. This missing information makes it difficult to accurately determine with precision the original or subsequent depositional factors. For example, if the Kidd Copper property was unknown to time or depositional processes one could come to the erroneous conclusion that Kidd Copper too was thousands or even millions of years old based on its stratified layers. However, this is not the case. So what if other beds or layers in the Geologic record were laid in a quick process in time too? What if the rock outcrops that are extensive have been misread?

Plate 1
Plane-parallel bedding: panoramic view

Before reviewing the various types of sedimentary structures, beds and bedding are introduced as they show up at various scales, both on the Earth's surface (outcrops) and subsurface (instrumental records of acoustic waves, or seismic sections).

The term *scale* in rock outcrops means two things: the size of the exposed section, and the distance from which it is observed. The example shown here is the side of a mountain (Mt. Carpegna in the Apennines, Italy) and represents a large-scale outcrop, indeed one of the largest possible. Several hundred meters of superposed, parallel strata are visible in the picture, and many of them can be traced laterally for more than 1 km (we say they are *continuous* within the outcrop limits). The *style* of stratification is thus clearly expressed, whereas the resolution of the image is not sufficient to tell whether the lines you see, which represent the intersection of *bedding planes* with the topographic surface, mark *individual beds* or bundles of beds (*bed sets*): geologists can speak of *levels* or *horizons*. This problem of resolution is the same as you have in seismic sections (see plate 2), which means that the scale is comparable. Although even a very large outcrop covers only a part of the area of a subsurface section, it gives you an idea of what would appear if you could dig a trench with a gigantic shovel in the fill of a sedimentary basin.

The principles of original horizontality and of superposition, mentioned in the introduction, are well illustrated by this picture. The pile of strata grew regularly, one bed after the other, in the most classical way you can imagine sedimentation to occur. There are, anyway, other *growth models* for bedding and sedimentary bodies: this one is identified as *vertical accretion, aggradation,* or, more colloquially, *layer-cake*.

The formation to which this outcrop belongs is named after another locality of the Apennines (Monte Morello, near Florence); this means that this section is not the *type section,* which serves to define, according to stratigraphic rules, a lithostratigraphic unit such as a formation. In Alpine chains, geologists still use also names for categories of formations, e.g., *flysch* and *molasse.* These terms refer to assumed tectonic and sedimentary conditions of deposition in different development stages of basins associated with fold-thrust belts (*orogenic basins*). The M. Morello Formation is a "flysch" of Paleocene age, made mostly of carbonate turbidites that filled a subsiding abyssal plain and were later detached from their oceanic substratum by tectonic compressions that built the Apennines chain.

Figure 32 - Section 13.12: Plate 1 Plane-Parallel Bedding: Panoramic View

13.13 Application of the Law of Superposition

Can we expect that multiple layers of deposits of various sediments to signify a short depositional time period? Answer given by current geological thinking is "no."

"The spectacular Grand Canyon has been created by the Colorado River cutting down through horizontal layers of rock, laid one upon the other. The deeper the layers, the older they are, and the rocks at the very bottom of the Canyon are two billion years old" (Gribbin, 2004, p. 151).

"Applying the Law of superposition to the beds exposed in the upper portion of the Grand Canyon, we can easily place the layers in their proper order. Among those that are pictured, the sedimentary rocks in the Supai Group are the oldest, followed in order by the Hermit Shale, Coconino Sandstone, Toroweap Formation, and Kaibab Limestone" (Tarbuck & Lutgens, 1999, p. 191). The underlying assumption in this principle is that the older rocks are underneath the younger beds; "applying the Law of superposition to these layers exposed in the upper portion of the Grand Canyon, the Supai Group is oldest and the Kaibab Limestone is youngest" (Tarbuck & Lutgens, 1999, p. 191). Figure 33 - Section 13.13 is the Upper Portion of the Grand Canyon (Tarbuck & Lutgens, 1999, p. 191). The oldest layer, the Supai Group on the bottom in the photograph of the Upper Portion of the Grand Canyon. "The distinctive property of most sedimentary rocks is their *stratification*, which refers to the layers or bed as each one is deposited on top of the earlier ones. Just as the bottom book in a pile must have been the first one put down, so the lowest bed was the first one deposited, and each successive bed was formed at a later time" (Pearl, 1955, p. 25, 26). Kidd Copper provides evidence that the layers did not form at a later time as conventional wisdom suggested. Instead, the observation from the research at Kidd Copper is that each layer is from the same depositional period and not from a different one. This raises the possibility that other rock units may also be concurrent with the layer above or below it.

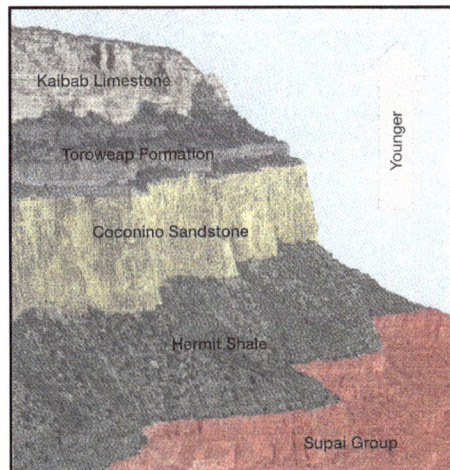

Photo by E.J. Tarbuck
Figure 33 - Section 13.13: Grand Canyon Exposed Beds – Upper Portion

"The geology of the Grand Canyon… can be analyzed in four parts: (1) horizontal layers of rock; (2) inclined layers; (3) rock underlying the inclined layers (plutonic and metamorphic rock); and (4) the canyon itself, carved into these rocks" (Plummer,

McGery & Carlson, 2003, p.176). Therefore, the Grand Canyon is given as evidence that a long depositional process must be required in Earth's history based on the assumption that each layer represents a different depositional period.

Figure 34 - Section 13.13 is a cross section of the Grand Canyon that "dates back more than two million years" (Coupe, 2004 , p. 97). This picture is one that mainstream geologists paint. Who knows how accurate the dates are? Who was there to verify their accuracy? Who can provide historical assurances that their calculations are correct? The picture that is painted is portrayed to be accurate but how accurate is it when no observations, no testimonials, no raw data and no valid methods have been found to substantiate the claims found in the conventional principles, laws and theories?

Instead, accurate depositional samples on reliable testimonial evidence are given to support the possibility that the rock record can be read another way. The layers do not need to be read as multiple time units. Indeed, there is evidence to open the doorway to a 'younger earth'.

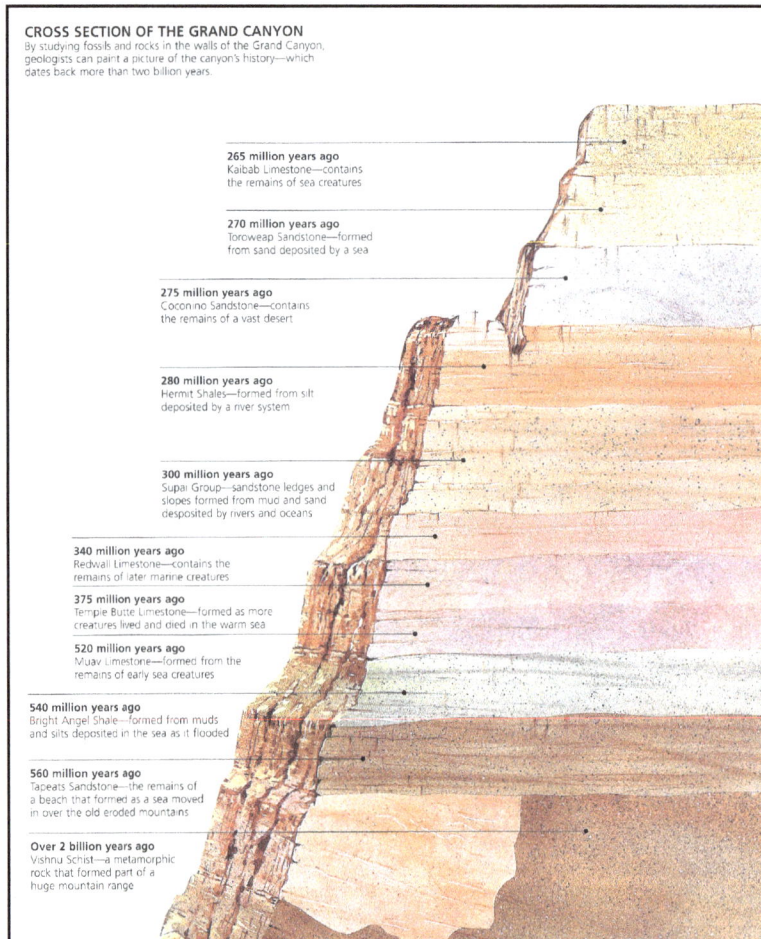

CROSS SECTION OF THE GRAND CANYON
By studying fossils and rocks in the walls of the Grand Canyon, geologists can paint a picture of the canyon's history—which dates back more than two billion years.

265 million years ago
Kaibab Limestone—contains the remains of sea creatures

270 million years ago
Toroweap Sandstone—formed from sand deposited by a sea

275 million years ago
Coconino Sandstone—contains the remains of a vast desert

280 million years ago
Hermit Shales—formed from silt deposited by a river system

300 million years ago
Supai Group—sandstone ledges and slopes formed from mud and sand desposited by rivers and oceans

340 million years ago
Redwall Limestone—contains the remains of later marine creatures

375 million years ago
Temple Butte Limestone—formed as more creatures lived and died in the warm sea

520 million years ago
Muav Limestone—formed from the remains of early sea creatures

540 million years ago
Bright Angel Shale—formed from muds and silts deposited in the sea as it flooded

560 million years ago
Tapeats Sandstone—the remains of a beach that formed as a sea moved in over the old eroded mountains

Over 2 billion years ago
Vishnu Schist—a metamorphic rock that formed part of a huge mountain range

Figure 34 - Section 13.13 Geologists Paint a Picture of Grand Canyon's History

13.14 The Principle of Cross-cutting Relationships on Rocks

"Law of cross-cutting relationships states that if a rock unit intrudes or cross-cuts another sedimentary or rock layer, the disturbed unit is always older than the intruding body of rock" (Environmental Science 1A03/1B03/1G03, 2002, p. 98).

"The principle of *inclusion* states that fragments included in a host rock are older than the host rock" (Plummer, McGery & Carlson, 2003, p. 182).

Igneous. Law of Cross-cutting Relationships "an *igneous rock*, or any other geological feature, is younger than any rock it transects" (MacDonald, *et al.*, 2003, p. 273). "For example, an igneous intrusive is certainly younger than the rocks it injects; younger sedimentary rocks lie above older rocks unless there has been a major disturbance of the rock sequence" (Garrels & Mackenzie, 1971, p. 63).

Sedimentary Rocks. "The principle of *cross-cutting relationships* states that a disrupted pattern is older than the cause of disruption" (Plummer, McGery & Carlson, 2003, p. 178). See Figure 31 - Section 13.8, "a layer cake (the pattern) has to be baked (established) before it can be sliced (the disruption)" (Plummer, McGery & Carlson, 2003, p. 178). "Sedimentary rocks are laid down in layers called strata or beds. The layers look like a giant 'multidecker' sandwich. When the Earth's crust moves, the 'sandwich' may be bent and squashed. The layers of rock are broken and folded, bringing rocks from deep down up to the Earth's surface" (Williams, 1993, p. 24).

The main concept here is that the folding, twisting or bending in either igneous or sedimentary rocks is from a secondary process *after* the first layer has been disrupted.

13.15 The Principle of Fossil Succession

"Correlation by fossils: certain fossils existed during distinct periods of geologic time. The presence of these fossils in a sedimentary layer allows a relative age for that sedimentary layer to be determined. Likewise, similar fossils present in different stratigraphic sections may be used to correlate sedimentary layers" (Environmental Science 1A03/1B03/1G03, 2002, p. 98).

Law of faunal assemblages is where "similar assemblages of fossil fauna or flora indicate similar geological ages for the rocks in which they are embedded" (MacDonald, *et al.*, 2003, p. 273). Faunal assemblages were not recorded at the Kidd Copper property so are beyond the scope of this report.

However, many 'monotonous' sandstone-shale formations that had previously been neglected by most geologists because they contained no fossils or recognizable marker horizons" (Blatt, Middleton & Murray, 1972, p. 5). The methodology in this report is based on original excavations *in situ*. This report establishes lithographic criteria as foundational markers to identify its younger earth concept where in contrast conventional geology uses faunal assemblages.

14.0 Uniformitarianism – Fundamental Principle

"Uniformitarianism is a fundamental principle of modern geology" (Tarbuck & Lutgens, 1999, p. 4). "Hutton argued that the forces of erosion, deposition and volcanism observable at the present day can be used to explain ancient conditions on Earth '*uniformitarianism*'" (Eyles, 2002, p. 14). "In order to understand the past we have to know how planet Earth works today; we assume similar geological processes happened in the past. This fundamental principle is called *uniformitarianism*, the foundation of much of what geologists do" (Eyles, 2002, p. 3, 4).

14.1 Fathers of Modern Geology

The father of modern geology is the English scientist Sir Charles Lyell (1797 – 1875) as shown in Figure 35 - Section 14.1. His main contribution to the young science of geology was the introduction of the concept of 'uniformitarianism'"the geological principle formulated by James Hutton in 1795 and publicized by Charles Lyell in 1830 that geological processes occurring today have occurred similarly in the past, often articulated as, 'The present is the key to the past" (Schlumberger, 2008, Oilfield Glossary). This geologic principle proposed "that past events occurred at the same rate as they do today" (Busbey, *et. al.*, 1999, p. 19).

Uniformitarianism as defined by the Dictionary of Geological Terms 3rd edition prepared by the American Geological Institute on page 546 states "the fundamental principle that geological processes and natural laws now operating to modify the earth's crust have acted in much the same manner and with essentially the same intensity throughout geologic time, and that the past geologic events can be explained by forces observable today; the classical concept that the "present is the key to the past" (Garrels & Mackenzie, 1971, p.65).

Uniformitarianism is "the principle of James Hutton (1726-97), a leading figure of the Scottish Enlightenment and widely regarded as the founder of modern geology. In his *Theory of the Earth*, Hutton postulates that the laws of nature now prevailing have always prevailed and that, accordingly, the results of processes now active resemble the results of like processes of the past" (MacDonald, *et al.*, 2003, p. 442).

James Hutton "advocated the theory that contemporary geological processes are the same as those that have always prevailed" (Isaacs, *et. al.*, 2003, p. 385).

Figure 35 - Section 14.1: A Father of Modern Geology, Sir Charles Lyell

Uniformitarianism "simply states that the *physical, chemical and biological laws that operate today have also operated in the geologic past*. This means that the forces and processes that we observe presently shaping our planet have been at work for a very long time. Thus, to understand ancient rocks, we must first understand present-day processes and their results" (Tarbuck & Lutgens, 1999, p. 4). "The acceptance of uniformitarianism meant the acceptance of a very lengthy history for Earth" (Tarbuck & Lutgens, 1999, p. 5). "The accumulations of deep units of sedimentary rock were formed by the same process of sediment deposition, and at the same rate, as those that can be observed in the modern world. Since most geological processes were shown to be very slow, and the rock units that were formed by these processes were often very often very large, Lyell concluded that the Earth was extremely old. He was the first person to talk of its age in millions of years" (Busbey, *et. al.*, 1999, p. 19).

"The idea that present-day geologic processes, such as the transport of sediment by streams, can provide scientific insights into ancient geologic processes is a fundamental principle of geology. This principle is called *uniformitarianism*, and it was first stated by Hutton. The principle of uniformitarianism says that the processes operating in Earth systems today have operated in a similar manner throughout much of geologic time; it is another way of saying that 'the present is the key to the past.' We can examine any rock, however old, and compare its characteristics with those of similar rocks that are forming today. We can then infer that the ancient rock very likely formed in a similar environment, through similar processes, and on a similar time scale" (Murck, 2001, p. 48).

14.2 Actualism – Principle of Uniformity

"Uniformitarianism [GEOL] classically, the concept that the present is the key to the past; the principle that contemporary geologic processes have occurred in the same regular manner and with essentially the same intensity throughout geologic time, and that events of the geologic past can be explained by phenomena observable today. Also known as actualism; principle of uniformity" (Licker, 2003, p. 384).

"Uniformitarianism (sometimes called actualism or gradualism)" (Fox, 2000, p. 7). "Actualism is the principle that natural laws governing both past and present processes on Earth have been the same" (Levin, 2003, p. 7). "*Uniformitarianism (actualism) –* The present viewed as the key to the past. Essentially the converse of *Catastrophism*" (Kearey, 2001, p. 281).

It can be argued the doctrine of uniformitarianism implies that all changes are at a uniform rate. This concept is contrary to what is known about minor local catastrophes as established by the evidence in this report on the Kidd Copper property case. Later on, *the principle of uniformitarianism* was revised to mean that it "does not imply the invariance through time of conditions now prevailing; what is meant is that changes or modifications of the Earth's crust are produced only in accordance with the laws of physics and chemistry" (MacDonald, *et al.*, 2003, p. 442).

> "Term uniformitarianism is a bit unfortunate, because it suggests that changes take place at a uniform rate. Hutton recognized that sudden, violent events, such as major, short-lived volcanic eruption, also influence Earth's history. In some countries *actualism* is used in place of uniformitarianism. The term actualism comes closer to conveying Hutton's principle that the same processes and natural laws that operated in the past are those we can actually observe or infer from observation as operating at present. It is based on the assumption, central to the sciences, that physical laws are independent of time and location. Under present usage, uniformitarianism has the same meaning as actualism for most geologists" (Plummer, McGery & Carlson, 2003, p. 176).

"The fact that the flood of Genesis 6-9 occurred destroys the theory of uniformitarianism and its stepchild (organic evolution)" (Fox, 2000, p. 8). Noting that significant problems exist with the fundamental principle of geology, uniformitarianism can be refuted. The principle of uniformity can be refuted, remember "there are no loopholes in the laws of nature" (Isaacs, *et. al.,* 2003, p. 445).

14.3 Chemical Uniformitarianism

"Many of the major minerals of the earth's crust are represented fairly well by the five-component system $Na_2O-K_2O-Al_2O_3-SiO_2-H_2O$" (Garrels & Mackenzie, 1971, p. 370). "Ion exchangers, such as the clay minerals, are rapidly responsive to their aqueous environment. In some waters, the exchange positions of montmorillonite-group minerals may be occupied by K^+, whereas in others the K^+ may be displaced by Na^+. "For such an exchange, the Na form of the mineral can be considered as one component and the K form as another" (Garrels & Mackenzie, 1971, p. 363).

Conventional Wisdom Predicts Predictable Proportions in Units Cycled. "The cycing of sediments through time should lead to differences in the proportions of rock types of a given age as observed today and also those proportions at the time of deposition" (Garrels & Mackenzie, 1971, p. 275, 276). "The sediments deposited at any given time in the earth's history are the same as those deposited at any other time" (Garrels & Mackenzie, 1971, p. 276). First, the different proportions of rock types at Kidd Copper are observed in Table 6 - Section 11.4 of the sediment composition found after redeposition. These sediments do not show an even distribution or a proportional distribution in the transect samples of sediment; therefore this prediction can be rejected.

14.4 Law of Nature

"Law of nature. A generalization of science, representing an intrinsic orderliness of natural phenomena or their necessary conformity to reason" (Bates & Jackson, 1984, p. 291).

> "If the term uniformitarianism is to be used in geology (or any science), one must clearly understand what is uniform. The answer [according to Harold L. Levin] is that the physical and chemical laws that govern nature are uniform. Hence, the history of the Earth may be deciphered in terms of present observations on the assumption that natural laws are invariant with time. These so-called natural laws are merely the accumulation of all our observational and experimental knowledge. They permit us to predict the conditions under which water becomes ice, the behavior of a volcanic gas when it is expelled at the Earth's surface, or the effect of gravity on a grain of sand settling to the ocean floor. Uniform natural laws govern geologic processes such as weathering, erosion, transport of sediment by streams, movement of glaciers, and movement of water into wells" (2003, p. 7).

The scope of this report will further discuss the accumulation of the present observations made on the Kidd Copper Property. These observed geologic processes at the Kidd Copper mine tailings site will help to unveil the past. The analytical approach is to take the knowledge recorded from *in situ*, testimonial evidence and apply the resulting evidence to accept or reject theory. Conventional wisdom is challenged.

14.5 Present-day Processes

"An understanding of present-day sedimentary environments provides the key to interpreting sedimentary rocks" (Nichols, 2003, p. 328). "Sedimentary rocks include sandstones, limestones and mudstones" (Selley, March 1999, *p.nd*). "By comparison of present-day processes and their products with the characteristics of sedimentary rocks, the physical, chemical and biological conditions under which the sediment formed can be determined" (Nichols, 2003, p. 4).

"The condensation and differentiation of the earth were probably completed in their major aspects within 1 billion years, and for the past 4 billion years surface conditions apparently have not differed drastically from today's, except for the progressive increase of oxygen in the atmosphere" (Garrels & Mackenzie, 1971, p. 31).

> "James Hutton, a Scotsman who is regarded as the father of modern geology, realized that geologic features could be explained through present-day processes. He recognized that our mountains are not permanent, but have been carved into their present shapes and will be worn down by the slow agents of erosion now working on them. He realized that the great thicknesses of sedimentary rock we find on the continents are products of sediment removed from land and deposited as mud and sand in seas. The time required for these processes to take place had to be incredibly long. Hutton upset conventional thinking (the world was believed to be less than 6,000 years old) by writing in 1788, 'We find no sign of a beginning – no prospect for an end'" (Plummer, McGery & Carlson, 2003, p. 176).

Deducing information without a beginning or an end is difficult for the earth scientist trying to establish geological history. The significance of the observations collected is relevant to modern day geology because the interpretation of the sediment processes can be defined with a beginning and end-point. Correlating strata on Kidd Copper property is easily interpreted because guesswork is eliminated to the context. The history and geologic processes are known. A track record of the phenomena found at the site can be traced because of the records collected. No guessing is involved as to when the process started or the end date because the historical records of the Kidd Copper property are known. The present-day transformation and redeposition of sediment on the Kidd Copper property occurred in a relative short period of time.

A time-stratigraphic correlation of heterogeneous material can be referenced by the same point in time even when the distribution patterns are dissimilar.

14.6 Reproducibility

How can one check the accuracy of conventional wisdom, textbook or any theory? The answer is what support comes from observations and reproducibility. If contrary evidence accumulates then the hypothesis can be rejected. Science is supported on evidence that is seen.

Science is based on historical evidence and not the interpretation of theoretical applications. For example, in Section 19.11, relying on Plate 59 is not an accurate base to determine past or unknown depositional processes as the time reference is not verified with eye-witness testimony as is the case of Kidd Copper where each borehole was known to be historically uplifted and redeposited simultaneously. Each borehole underwent the identical depositional process within the same time frame.

Correlation of Sedimentary Strata. The common factors between all the lithographic logs at Kidd Copper includes: (1) same parent material; (2) same cataclysmic event; and (3) the same element of time. The stratification at Kidd Copper is random without a common correlating factor among the adjacent sediment structures as to color descriptors. This random distribution is noted in impact physics as well where "random distribution on sand was noted after impact" (Daily Planet, October 2009, Discovery). In the physics' case an experiment was conducted blasting sand with a 'bullet' propelled through it. The sand dart went through the sand and the distribution pattern was analyzed. This randomness of sediment is a curious phenomenom in impact physics and seen in the sediment on the Kidd Copper Property.

Uncomformable Pattern. All of the profiles at Kidd Copper were heterogeneous displaying non-comformity pattern distribution in the layers. No repetitive pattern to the adjacent sedimentary strata was seen. The layering was unequal in distribution. This was a distinguished finding. This was verified by the irregularity of patterns and non-existent extremes. Each borehole showed different layers to the adjacent strata preventing traditional correlation. All of the distinctive layers characterize one sediment deposit. Each layer does not provide a separate record of a different deposition but rather is a record of the complete deposition of a one-time occurrence. Therefore, without knowing the rapid deposition of events leading to the layers it would be easy to conclude that several depositions were made on the property through time; rather than the known fact that one deposition deposited several sediment layers in no conformable pattern. There is sufficient evidence for a heterogeneous layering of sediment and documentation to show a one-time depositional process.

Interpretation of the combined color legends from Cross Section A-A' [West – North Transect] and Cross Section B–B' [North–South Transect] shows no correlation of color between the vertical logs and the adjacent logs, which indicates a random sediment structure can be a primary sedimentary structure. The Kidd Copper features formed during deposition or shortly thereafter and are designated as primary sedimentary structures. This validates stratified sediment is a primary structural deposition.

No Correlated Pattern. "Other strata, such as layers of tephra, are deposited essentially instantaneously" (Friedman, *et. al.*, 1992, p. 175). The random deposition of the particles at Kidd Copper is consistent with the theory that stratified layers are a quick process through time and that seeing many layers does not mean they are of different depositional processes because other formations in geology show this characteristic, such as the tephra layers Friedman mentions. Strong evidence supports the claim in this report that any unit that has a change adjacent to synchronous facies could be deposited at the same time, under the same depositional environment, as is the case for the units found in the profiles on the Kidd Copper Property.

Scientific Question: Could the classification of other non-consistent spatial patterns be misclassified? And, if so, where in the rock record would these mistakes exist?

Sediment Rates Vary. Rates of sedimentation are not constant to the sediment layering as Gradstein, Ogg & Smith assumed (2004, p. 117). Neither are the rates proportional. Rapid processes are seen in other geologic processes, such as: turbidity currents, mudflows, debris flows, and landslides. The build up of these landforms was not the result of slow processes through time. Like these landforms the stratified layers on the Kidd Copper property are not the result of slow processes through time. If stratified layers are not a measurement of multiple time units as in the Kidd Copper layers, then what about elsewhere on Earth?

The valid inferences from the Kidd Copper case include: (1) younger beds of sediment do not have to be on the top layer; (2) the oldest layer does not have to be on the bottom layer; (3) layers can be formed at the same time; (4) the distribution of sharp breaks in the adjacent layers do not reflect separate depositional periods; (5) distinct horizons do not mark different depositional periods; (6) flood-event overturned at least one layer and redeposited many layers; (7) mixing of sediment can provide either a homogeneous or a heterogeneous sediment layer as a primary structure; and (8) marking time by correlating layers is not accurate.

In fact, the claim that H_1= rapid stratification of all layers

$H_0 \neq$ lower strata are not the oldest; $H_0 \neq$ top layer is not the youngest layer of strata. Again, the null hypothesis can be rejected and the alternative accepted.

The valid inferences in this report offer support to the hypothesis and the claim that layers in sediment are a rapid process. The hypothesis was tested in the case of Kidd Copper. The hypothesis was tested based on geologic principles identified throughout this report and conventional wisdom has been challenged. There is evidence to support the claim that multi layering of strata cannot be interpreted as individual time events. The evidence in the land base at Kidd Copper identifies stratification visibly apparent on the first deposition where the sediment was clearly layered from the first few cm of land surface down to 75 cm.

Another way to test a hypothesis is to determine if a reproducible result will occur and if it does, then it strengthens the position of the assertion, which then leads to forming strong conclusions. If a pattern is repeated from the same factors then it can be concluded that if '*xxx*' occurs then '*yyy*' results. Refer to Table 24 - Section 19.24, which considers some geologic assertions found throughout this report by detailing the responses found from the Kidd Copper property data.

14.7 A Fundamental Assumption

"A fundamental assumption made in interpreting the processes of sedimentation from the character of sedimentary rocks is that the laws which govern physical and chemical processes have not changed through time" (Nichols, 2003, p. 4). With this supposition the information gained from Kidd Copper property will help unveil the past.

Function of Age. "Because of the assumption of equal probability of exposure to erosion of all equal masses, the same source materials of the total mass would always be available to erosion, so that the result of deposition of a constant ratio of rock types, derived from a different ratio of source rock types, results in changes in the proportions of rock types as a function of age" (Garrels & Mackenzie, 1971, p. 273). Accordingly, it was noted that no major erosional changes occurred. The Kidd Copper site area is virtually remained untouched; therefore the exogenetic processes play a relatively small factor in the reshaping of its landform. Jiali Shang, Bill Morris and Philip Howarth noted "in 1977, the major mine-tailing deposit … displays almost the same distribution as the present day" (2004, p. 4).

14.8 Models for Future Replication

Models assume four major functions: (1) act as a *norm*, for purposes of comparison; (2) act as a *framework* for future observations; (3) act as a *predictor* in new geological situations; and (4) "act as an integrated basis for *interpretation* for the system that it represents" (Walker & James, 2002, p. 7). "Models are constantly being refined as more information becomes available, and as the significance of the various constituent parts of the model becomes better understood" (Walker & James, 2002, p. 6). A model is "a working hypothesis or precise simulation, by means of description, statistical data, or analogy, of a phenomenon or process that cannot be observed directly or that is difficult to observe directly" (Bates & Jackson, 1984, p. 331).

The tests and experiments provided in this report are to allow avenues for future ways to test the concepts outlined in this report and to substantiate or disprove the findings herein. In contrast, many geologic time zones fail to pass the test of reproducibility simply because no one lived in that time zone and no observations or testimonials could be provided to substantiate what happened. Not so, in the case of Kidd Copper. Observations and testimonials offer validation to the findings. The author provides additional ways to provide further validation with the following suggestions.

14.9 Test of Reproducibility

"Analysis: The stage in science when measurements are sorted, tested, and examined visually for patterns and predictability" (Clarke, 2003, p. 305) is the part where reproducibility can validate the theorem or hypothesis. The layering of sediment deposits on Kidd Copper is reproducible in many depositional environments. Other rock formations that have multiple layers and stratification have been included in the main part of this report to show the reader how sensible the findings at the Kidd Copper property are; such as found in Sections 13.10 and Section 13.11.

More experimental future reproducibility projects are suggested in Appendix F to test the findings in this report but for the space and scope of this narrative the author who is not an expert engineer will leave it to the reader and others to replicate the suggestions for these experiments and to double-verify that the findings in this report are accurate.

14.10 Experimental Design of a Future Flume Study

"The large acid requirement for the formation of sedimentary rocks is illustrated by asking whether the process could be duplicated in the laboratory" (Garrels & Mackenzie, 1971, p. 245). Here is one suggestion for a later reproducible experiment. The flume study with two methods is outlined as follows:

Date: Future Date 20xx

Title of Experiment: Water-filled Sand Box

General Points: "It is possible to set up a laboratory scale model of bed-form migration, using dimensionless depth, velocity, and grain size as modeling parameters, so that a flow on a workably large flume scale can represent without distortion all the phenomena of flow and sediment transport in a much larger flow" (Harms, *et al.*, 2004, p.36).

Purpose: A laboratory experiment that can be done to falsify/verify the results is simply to build a glass-framed enclosed box which can hold water and then add uniform-grains of colored sand which will be mixed by adding water to the enclosed water-filled sand box and blowing the water through the sand.

Example: Water-filled sand box.

Method: "Unidirectional flows" (Harms, *et al.*, 2004, p. 5).

A) *Point Source*: The inflow of water is concentrated on the middle sediment layer. This should cause equal pressure and subsequently cause an equal distribution of the mixed sediments as they are thoroughly mixed and re-settle in the second containment storage area. How do these colored layers redeposit after disturbance?

In Table 13 - Section 14.10 eight different colored sediment layers are placed into containment 1. An access hole is provided to allow the sediment and water to move from the initial deposit to the final deposit area. The arrow diagram represents the strong inflow of water that will cause the upliftiing and the turbulent flow, which will thoroughly mix all eight layers of sediment. The mixture will be pushed through the access hole and be redeposited in containment 2.

The inflow of water is highly pressurized causing a turbulent flow and a thorough mixing of the sediments. The pressure remains constant on one particular sediment source midway in the sedimentation. The pressure from the water will push the mixture through the access hole. This should create a similar distribution of sediment on both sides as the sediment is redeposited.

Table 13 - Section 14.10: Point Source Layering

Point Source Layering		
Containment 1: Initial Deposit	Access Hole	Containment 2: Final Deposit
A Green fine-grained sediment	Depth1	Is the top half the same as the bottom half?
B Blue fine-grained sediment		
C Red fine-grained sediment		Patterned or Unpatterned Distribution?
D Yellow fine-grained sediment	Depth2	
E Purple fine-grained sediment		Random or uniform patterns noted in each depth level?
F Orange fine-grained sediment	Depth3	
G Grey fine-grained sediment		
H Brown fine-grained sediment		

Is the top half the same as the bottom half when the sedimentation is redeposited? It would be expected from conventional geology that it should be. However, according to the findings on the Kidd Copper property this will not happen and instead an unpatterned distribution will occur; thus validating the findings on the Kidd Copper Property, which concurs that an unequal distribution occurs from the same source point.

Alternative suggestion is to extend the containmnent 2 to see if changes occur.

B) *Sheet Source*: How do these colored layers redeposit after disturbance?

In Table 14 - Section 14.10 eight different colored sediment layers are placed into containment 1. An access hole is provided to allow the sediment and water to move from the initial deposit to the final deposit area. The arrow diagrams represent the strong inflows of water that will cause the uplifting and the turbulent flow, which will thoroughly mix all eight layers of sediment.

The mixture will be pushed through the access hole and be redeposited in containment 2.

Table 14 - Section 14.10: Sheet Source Layering

Sheet Source Layering		
Containment 1: Initial Deposit	**Access Hole**	**Containment 2: Final Deposit**
A Green fine-grained sediment	Depth1	Patterned or
B Blue fine-grained sediment		Unpatterned?
C Red fine-grained sediment		
D Yellow fine-grained sediment	Depth2	Distribution? **?**
E Purple fine-grained sediment		Proportional?
F Orange fine-grained sediment	Depth3	
G Grey fine-grained sediment		Heterogeneous?
H Brown fine-grained sediment		Homogeneous?

The inflow of the sheet of water is highly pressurized causing a turbulent flow and a thorough mixing of the sediments. The pressure remains constant. The pressure from the water will push the mixture through the access hole. Is the sediment redeposited in a homogeneous or heterogeneous distribution pattern? It would be expected that it will be homogeneous because all of the fine sediment is thoroughly mixed. However, according to the findings on the Kidd Copper property this will not happen and instead an unpatterned distribution will occur; thus validating the findings on the Kidd Copper Property, which concurs that a heterogeneous distribution patterb occurs from mixing of sediment.

Thought Question: "Geologists... are more concerned with what bed geometries can indicate about what the generating flows were like" (Harms, *et al.*, 2004, p. 5). So what do they indicate?

Data and Observations: "Observational knowledge of bed configurations" (Harms, *et al.*, 2004, p. 5).

Graphing Exercise: "The summarization of experimental data … is accomplished in a series of diagrams based on the variables of velocity, flow depth, and grain size …this is a simple yet correct way to present these data for geologic application. Through these diagrams, bed forms can be related to flow conditions" (Harms, *et al.*, 2004, p. 1).

Procedure: "Current velocities in most of the important clastic environments are commonly great enough to transport the bed sediment, and transportational bed geometries are generated in a wide range of grain sizes from silts to gravels … the nature of the geometries depends strongly, albeit complexly, on strength of the generating flow" (Harms, *et al.*, 2004, p. 5).

"Three-dimensional plot of flow depth, flow velocity, and sediment size. Two kinds of section through this three-dimensional diagram are most useful for visualizing relationships among bed phases: depth-velocity diagrams for a given sediment size, and size-velocity diagrams for a given flow depth. The former arise naturally from flume work, because typically a larger number of runs at various depths and velocities are made for a given sand size. Size-velocity diagrams are most easily constructed once a series of such depth-velocity diagrams is available" (Harms, *et al.*, 2004, p. 15).

Experimental Method: "The challenge for the future lies in imaginative use of primary sedimentary structures, as well as the host of other chemical and biological characteristics, to identify significant differences within these general environmental settings" (Harms, *et al.*, 2004, p. 2).

Formula: "When fluids like air and water flow over beds of loose sediment at velocities great enough to transport some of the grains, a great number of different kinds of bed geometries can result, depending on the nature of the flow and the nature of the sediment" (Harms, *et al.*, 2004, p. 5).

Calculations: A combination of engineering: fluid dynamics, sediment transport mechanics, sedimentary physics and "flow-visualization and measurement techniques" (Harms, *et al.*, 2004, p. 7).

Special Techniques: "Hydrodynamic problems behind bed configurations" (Harms, *et al.*, 2004, p. 5).

Discussion: "Engineers have studied the movement of sand in large experimental channels (flumes) and produced structures that seem very like the ones that geologists see preserved in the field" (Blatt, Middleton & Murray, 1972, p. 4).

Test: Does the sand fall in layered colors in the boxes? Does the sand settle in a uniform layer? What is the distribution? What are the proportions?

Interpretation: "In making hydraulic interpretations from primary sedimentary structures, the underlying (and usually implicit) assumption is that a given flow over a given sediment bed will produce a definite bed configuration, and therefore a definite internal geometry of stratification, and that for any other realization of the same flow conditions the form and internal stratification of the bed, while different in detail, would show the same kinds of significant features and would be characterized by nearly identical values of any given statistical measure or morphology" (Harms, *et al.*, 2004, p.5, 6).

Further Comments: "Studies of modern and ancient sedimentary sequences are used to illustrate and interpret environments of deposition. Fluvial sediments, perhaps the most extensively studied and best understood, are reviewed to show how experimentally derived generalizations are applied or qualified to interpret natural environments" (Harms, *et al.*, 2004, p. 1).

Results: What is typically seen in these sand boxes is a layering of colored sand rather than a uniform settling. Expected outcome: random, unordered distribution and multiple layers of sediment.

Conclusion: "These bed forms are in turn related to characteristics of stratification, the commonly preserved record of flow" (Harms, *et al*, 2004, p. 1).

Sources: Harms, J.C., Southard, J.B., Spearing, D.R., and Walker, R.G. (2004). Depositional Environments as Interpreted from Primary Sedimentary Structures and Stratification Sequences. *ProQuest Company*. Michigan.

14.11 Spigot Dam Re-Construct Model

The re-construct model details information on how to model a spigot dam for a future experiment to replicate the findings in this report. The author is not an engineer; and so leaves the dam modelling to the engineers to work out the exact measurements and materials. Henceforth, the following information is just to get the initial parameters.

Date: Future Date 20xx

Title of Experiment: Spigot Dam Re-Construct Engineering Project

General Points: "The great majority of tailing dams are built on the concept of a single spigot, some use the concept of multiple parallel spigots. The tailings dam must not be on an ore deposit. The volume of the tailings containment must be calculated to account for disposal volumes, runoff of snow or rain, and the pumping out of reclaim water (Abulnaga, 2002, p. 12-5-1). "Flow depths in natural environments are commonly much greater than in flumes" (Harms, 2004, *et al.*, p. 16).

Purpose: "A set of mathematical conditions to be satisfied, in the solution of a differential equation, at the edges or physical boundaries (including fluid boundaries) of the region in which the solution is sought. The nature of these conditions usually is determined by the physical nature of the problem" (DePree & Axelrod, 2003, p. 94). In the case of the re-construction of a spigot dam on the Kidd Copper property several factors could be studied: (1) rate of velocity of water in proportion to the upward movement and capture of impoundment settlement; (2) the time involved to resettle and; (3) the findings of the resettled sediment in spatial distribution.

Method: "The guidelines for constructing a tailings dam have been established by the International Commission on Large Dams (1982)" (Abulnaga, 2002, p. 12).

Copper Cliff satellite site has similar geological setting to kidd copper, so that: "The elevated tailings impoundments are a regional ground-water recharge zone where precipitation infiltrates downward through the tailings and migrates radially away from the impoundments. The elevated water table in the CD and M impoundments is maintained, in part, by continued pumping of water into ponds within the impoundment. The water table slopes gently to the southeast, but hydraulic gradients increase sharply near the tailings dam" (Coggans, *et al.*, 1991, p. 450).

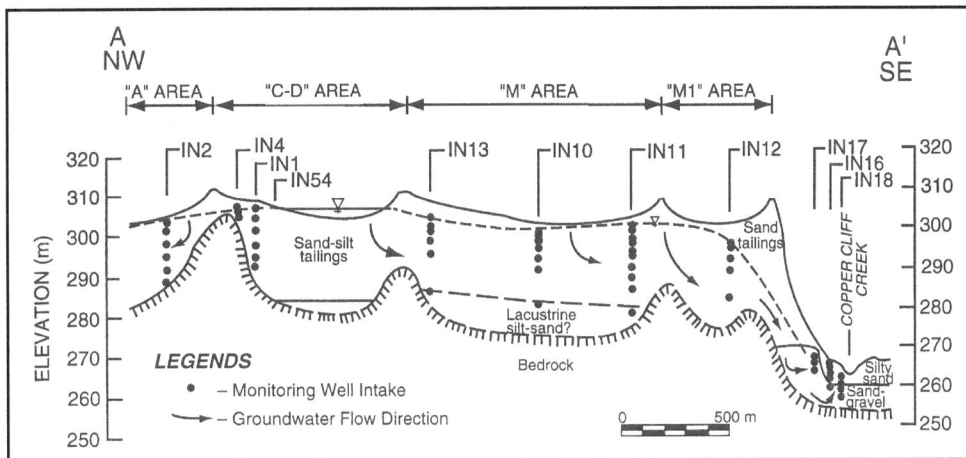

Figure 36 - Section 14.11: Cross section along A-A'

Figure 36 - Section 14.11 "Cross section along A-A' showing stratigraphy, ground-water flow direction and piezometer network" (Coggans, *et al.*, 1991, p. 449).

And "describes the hydrogeochemistry of the tailings water along a cross section of tailings deposited between 1940 and 1960, paralleling one of the major ground-water flowpaths within the impoundments (cross section A-A')" (Coggans, *et al.*, 1991, p. 449). The results show "the water table at the piezometer nests along cross section A-A' (Figure 36 - Section 14.11) varied between 0.7 and 5.1 m below ground surface and the vadose zone increased in thickness near the southeast edge of the impoundments" (Coggans. e*t al.*, 1991, p. 450)

Thought Question: What is the settling speed of the sediment upon redeposition?

Derived Equation for settling speed: where c = settling speed (cm/s).

$$\Psi = \log 2\ c,$$

Data Display, Observations and Manipulation: Rate of deposition can be collected. Traction current can be observed.

Graphing Exercise: "As in physics, chemistry, and biology, it is common also in soil science to study changes with time as the independent variable, or *x* axis in a graphical representation" (Retallack, 1990, p. 261). Resource estimation can be considered by further study of these variables: Lake behind dam; channels; flow through; gulleys; and wash out area as noted in some of the *Geo3Fe3* log entries.

Procedure to Build a Tailings Dam: "One method of constructing the walls of a dam is to use the coarse material in the tailings. The fines or slimes are allowed to sink to the bottom of the tailings pond or to form beaches between the water and the dam" (Abulnaga, 2002, p. 12-5-1). "A bulldozer relocates and then compacts the material by rolling over it. The banks are gradually built this way" (Abulnaga, 2002, p. 12-5-1).

Copper Cliff Satellite Site Experimental Method:

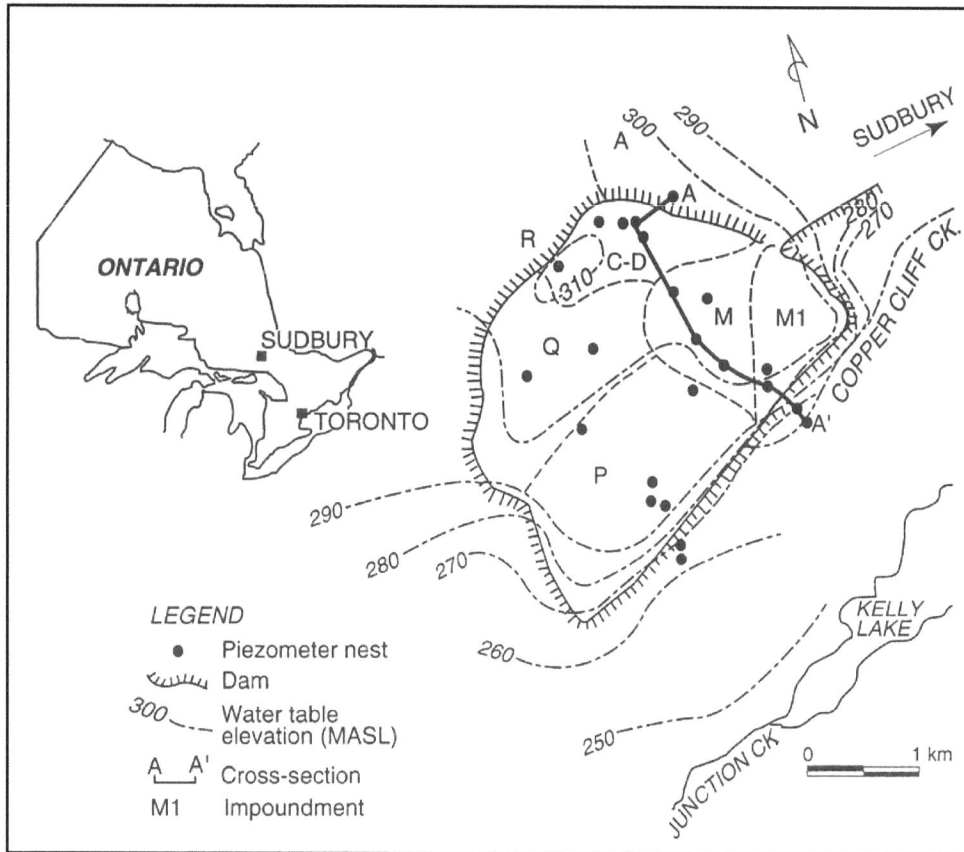

Figure 37 - Section 14.11: Location of INCO Ltd. Copper Cliff Tailings Impoundments

Figure 37 - Section 14.11 shows the location of the INCO Ltd. Copper Cliff tailings impoundments and water table configuration" Coggans, *et al.,* 1991, p. 448). "Samples were collected at 10 cm increments, to a maximum depth of 1.4 m. (Coggans, *et al.*, 1991, p. 448).

Inputs: The inputs for the tailings deposition consist of: a topographic surface of the basin, design information in the form of spigot points, dam centerlines and drainage basin from the waste facility. The elevation is known in the area of the waste tailings as 850 ft and the elevation of Perch Lake is 900 ft according to the topographic map in Figure 4 - Section 1.3.

The flow path of the dam is the same as the stream flow, which is north to south.

Tailings slope is ~2° or flat.

Tailings will flow around dodgelegs and islands.

Model Outputs: (1) "The type of information required (e.g., peak flows, flow volumes, ground-water heads, soil water contents, evapotranspiration rates); (2) The required accuracy and precision of the output; (3) The locations for which the output is required; (4) The time intervals for which the output is required" (Dingman, 2002, p. 27).

Formula: Variables [for the dam] (Hsü, 2004, p. 30) are as follows:

- area A: L^2
- volume V: L^3
- velocity μ: $L\,t^{-1}$
- acceleration a: $L\,t^{-2}$
- force F: $M\,L\,t^{-2}$
- energy E: $sM\,L^2\,t^{-2}$
- power P: $M\,L^2\,t^{-3}$
- pressure, Stress ρ, σ: $M\,L^{-1}\,t^{-2}$
- viscosity η: $M\,L^{-1}\,t^{-1}$

Calculations: For further future modelling the volume of the dam is calculated along the following perimeters so that a mathematical model can be constructed, as to "estimate flood flows" (Dingman, 2002, p. 4). For example, Dingman explains how to contour the watershed from the dam break (p. 11). Delineation of the watershed – tracing the divide (Dingman, 2002, p. 10)

"The properties of water dictate how it responds to the forces that drive the hydrologic cycle" (Dingman, 2002, p. 7). "If *q* represents an instantaneous water-movement rate and *x* the instantaneous value(s) of the variable(s) that control that rate, we can often use physical principles to derive the functional relation between them: $q = f(x)$" (Dingman, 2002, p. 5, 6).

Variables: In Appendix F a list of variables can be found to reconstruct dam and flow patterns.

Calculations of Dam Capacity: Water holding; storm-water capacity; drainage; flow rate and run off could be considered, if the dam was to be broken and flooded.

The volume of the dam is calculated along the following perimeters to construct a mathematical model.

$$V = l * w * h$$

Mathematical equation:

V = Volume
l = length
w = width
h = height

Dam's length to be calculated =

Dam's width to be calculated =

Bulk Density of broken rock = 1.87 (tonnes/cubic meter) specific gravity. Cubic meter is a measurement of volume or capacity.

"Sediment Transport Equation: One of a number of equations designed to predict the *sediment yield* of a given flow from its *shear velocity* or *shear stress*, the *grain size distribution*, the *hydraulic radius* of the channel and the characteristics of the fluid" (Kearey, 2001, p. 238).

"Manning Equation: A formula for estimating stream velocity (*v*) from *channel roughness,* slope (*s*) and *hydraulic radius* I: $v = kRsn$, where k = constant (1 in SI units), n = Manning roughness coefficient of the stream bed" (Kearey, 2001, p.163). Refer to Appendix G for values.

Discharge: "The *discharge* of a stream is the volume of water that flows past a given point in a unit of time. It is found by multiplying the cross-sectional area of a stream by its velocity (or width * depth * velocity). Discharge can be reported in cubic feet per second (cfs), which is standard in the United States, or in cubic meters per second (m^3/sec)" (Plummer, McGery & Carlson, 2003, p. 229).

Discharge (cfs) = average stream width (ft) * average depth (ft) *average velocity (ft/sec)

For example, "a stream 100 feet wide and 15 feet deep flowing at 4 miles per hour (6 ft/sec) has a discharge of 9,000 cubic feet per second (cfs). In streams in humid climates, discharge increases downstream for two reasons: (1) water flows out of the ground into the river through the stream bed; and (2) small tributary streams flow into a larger stream along its length, adding water to the stream as it travels. To handle the increased discharge, these streams increase in width and depth downstream" (Plummer, McGery & Carlson, 2003, p. 229).

Special Techniques: "Impoundment – The process of forming a lake or pond by a dam, dike or other barrier" (Bates & Jackson, 1984, p. 255). The tailings impoundment perimeter distance will be compared with the dam's perimeter so that the reflection of the total surface area will effectively construct an upstream dam. Height of Dam = 1-5 ft above mine tailings. This is an important calculation because the Height of Tailings = production/tonnes by meter. Rates of sedimentation deposits can be compared.

Discussion: The diagenesis of the Kidd Copper Area could be further studied. Diagenesis is "the sum of all changes, physical, chemical and biological, to which a sediment is subjected after deposition but excluding *weathering* or *metamorphism*. Some diagenetic changes (*Halmyrolysis*) occur at the water-sediment interface; however, most diagenetic activity occurs after burial. Unstable forms of minerals are converted to stable forms and new (*authigenic*) minerals deposited, often as concretions. Examples of diagenetic activity include the modification of *clay minerals* to *illite* or *chlorite*, the recrystallization of *argonite* to *calcite*, and the formation of *glauconite*" (MacDonald, *et al.*, 2003, p. 138). "*Diagenesis* – A term for the gradual and successive chemical-physical changes that take place in sediments previous to or during their consolidation. Diagenesis may also include the numerous processes of lithification but is a useful term only when particularly applied to the more or less contemporaneous chemical alteration of sediments" (DePree & Axelrod, 2003, p. 224).

Test: Pore water – is between grains in sediment. "*Pore fluid (pore water)* – A solution occupying the pore spaces in soil or rock, whose composition reflects its origin, subsequent mixing and *diagenetic* interaction with the host sediment. Plays an important role in diagenesis" (Kearey, 2001, p. 209). How?

Interpretation: Data mining is "revisiting existing data to explore for new relationships using new and more powerful tools for analysis and display" (Clarke, 2003, p. 308). The suggestion would be to revisit the actual site area and do a time-analysis comparative study. The topography and sub-surface geology could be studied further as to the effects of time.

Further Comments: Future models could establish: (1) "Residence time is also called *turn-over time*, because it is a measure of the time it takes to completely replace the substance in the reservoir" (Dingman, 2002, p. 24); (2) Peak flow rate; and (3) Response to an actual event.

Results: To verify the Rate of deposition = grain size and outflow/ Time and meters.

Versus: Many years to accumulate sediment = muds 1cm.

Conclusion: The Kidd Copper sediment analysis on August 30, 2004 showed that sediment accumulated quickly and that stratification of the layers is a rapid process. Primary structures can have multiple layers. Time is not a factor in the laying down of sediment. The verification that multiple layers can be deposited in a short time period is seen in other geologic processes. So, certain structures in the rock record that are commonly accorded long times may instead be designated as quite short since no one was around million years ago to support the old age theory. The accepted old age of the planet Earth can be questioned. This validation would support the new theory that the earth's rock record supports a 'younger age' for Planet Earth.

14.12 Acceptance Testing

"Man is constantly adding to his knowledge of the world, but to do any good it must be shared – by the people – Alistair Cooke (b.1908)" (Busbey, *et. al.*, 1999, p.13). What will the effect of the awareness that rapid stratification changes the interpretation of geologic time and re-calibrates the Geologic Time Scale? Many different branches of science will be altered to a significant point with these findings. Refer to Appendix J. A detailed model could be –reconstructed detailing the findings and reproducing rapid stratification as outlined earlier in this Section. Future research could take several directions with depositional rates being looked at and examined on other catastrophic site areas. "A rigorous theory of intelligent design will be a useful tool for the advancement of science in an area that has been moribund for decades" (Behe, 2003, p. 231).

Acceptance testing will be the next step in the further projection of rapid stratification as a fundamental factor that substantiates a younger Earth. Like Dingman said about acceptance testing, either "acceptance or rejection of the model" (2002, p. 30).

15.0 The Stratigraphic Record and Correlation

"Since depositional environments of different sediment types are unique, a vertical sequence of sediment provides evidence into environmental changes through time. This concept is referred to as stratigraphy and mapping of the vertical column of sediment is known as stratigraphic logging. The relative age of rock units in a stratigraphic section can be determined from the following four concepts: the Law of original horizontality; the Law of superposition; the Law of cross-cutting relationships; and correlation by fossils" (Environmental Science 1A03/1B03/1G03, 2002, p. 98).

"A rock facies is a body of rock with specified characteristics which reflect the conditions under which it was formed (Reading & Levell 1996). Describing the facies of a sediment involves documenting all the characteristics of its lithology, texture, sedimentary structures and fossil content which can aid in determining the processes of formation. If sufficient facies information is available, an interpretation of the depositional environment can be made" (Nichols, 2003, p. 5). The Kidd Copper property provides much information, which can be ascertained as fact based upon eye-witness testimony and observations which eliminates the guesswork. Methodology is important in this report because it compares conventional wisdom and theory with observations.

15.1 Stratigraphic Analysis of the Rock Record

In the case of the Kidd Copper property "sedimentological and stratigraphic analysis of the rock record using the material exposed at the surface, [and uncovered]… in the subsurface, provides us with the means to reconstruct the history of the surface of the Earth as far as the data allow" (Nichols, 2003, p. 328).

> "Geologists read sedimentary sequences – the order in which the rock layers were laid down. Their interpretation depends on two key facts: Sedimentary rock is always laid down horizontally and if the rock layers have been undisturbed, the oldest rock will lie on the bottom. So it follows, for example, that if the rock layers are not horizontal, then the rock must have changed position over time. It also follows that the more bent and fractured and tilted the rock layers, the more change that has taken place – and, therefore, the older they are. Sometimes the layers are buckled over so far that, all these millions of years later, geologists have to figure out which way is up, an exercise in deduction called younging" (Discovery Books, 1999, p. 40).

The Kidd Copper property provides a geological assessment that can be observed and the account of its history profiled against the stratigraphic rock record to determine if these assumptions are correct.

Will the rock layers at Kidd Copper mine tailings waste site be horizontal *in situ*? Conventional wisdom and textbook theory would support this outcome.

15.2 Law of the Correlation of Facies

Facies is "the aspect, appearance, and characteristics of a rock unit, usually reflecting the conditions of its origin; esp. as differentiating it from adjacent or associated units" (Bates & Jackson, 1984, p. 176). Facies is defined as "the sum of the characteristics of a sedimentary unit" (Middleton 1973; Nichols, 2003, p. 62). "These characteristics include the dimensions, sedimentary structures, grain sizes and types, colour and biogenic content of the sedimentary rock" (Nichols, 2003, p. 62). Refer to Figure 38 - Section 15.2.

The facies model assumes "four main functions: (1) it must act as a *norm*, for purposes of comparison; (2) it must act as a *framework* and *guide* for future observations; (3) it must act as a *predictor* in new geological situations; (4) it must act as an integrated basis for *interpretation* for the system that it represents" (Walker & James, 2002, p. 7).

"The relationship between depositional systems in space, and the resulting stratigraphic successions developed through time was first emphasized by Johannes Walther (1894, in Middleton, 1973) in his *Law of the Correlation of Facies*. Walther stated that 'it is a basic statement of far-reaching significance that only those facies and facies areas can be superimposed primarily which can be observed beside each other at the present time. Careful application of the law suggests that in a vertical succession, a *gradational* transition from one facies to another implies that the two facies represent environments that were once adjacent laterally. If the contacts between facies or facies associations are sharp and/or erosional, there is no way of knowing whether two vertically adjacent facies represent environments that were once laterally adjacent" (Walker & James, 2002, p. 6).

"Indeed, sharp breaks between facies (marked for example by channel scours, or by thin bioturbated horizons implying nondeposition) may signify fundamental changes in depositional environment and the beginning new cycles of sedimentation (de Raaf, *et al.*, 1965). These sharp breaks, or *bounding discontinuities*, are now used to separate stratigraphic sequences and allostratigraphic units" (Walker & James, 2002, p. 6). The concept of vertical facies sequence has been respected since Johannes Walther expounded his Law of Facies in 1894, where comfortable vertical succession of facies, allowed no major break as products of environments which were originally laterally adjacent. The geologic community interprets changes as breaks in time and allowances for different depositions of sediment to accumulate.

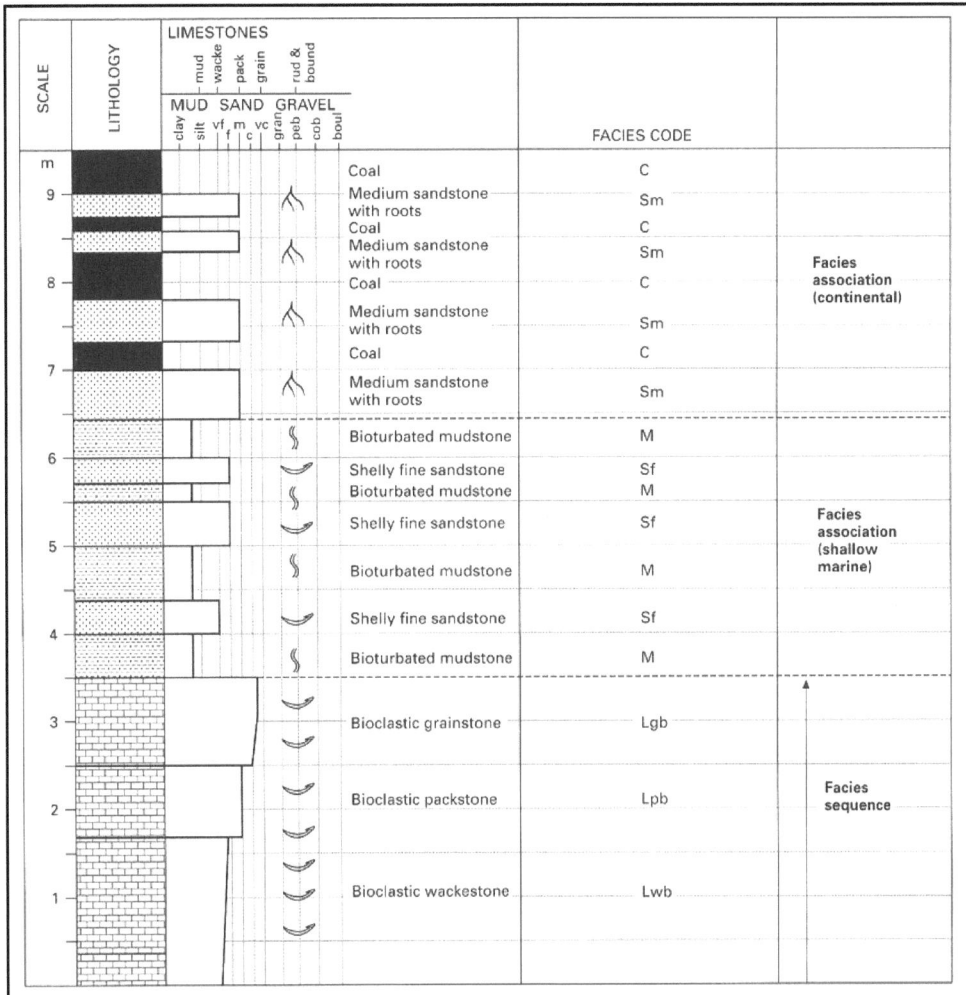

Figure 38 - Section 15.2: Facies Associations, Facies Sequences & Facies Codes (Nichols, 2003, p. 65)

To arrive at this conclusion one has to assume that continuous, harmonious layering signifies a one-time event. Any disruption or association that is different in appearance means that a different depositional occurrence happened. The reason to support this conclusion according to Walker & James is that "there is no way of knowing whether two vertically adjacent facies represent environments that were once laterally adjacent" (2002, p. 6). The rocks were laid down without modern man being able to testify about what happened. Therefore, the science in man asks for an explanation and uses his powers of reasoning to formulate a plausible explanation. However, man's reasoning may be incorrect because of insufficient information or inaccurate assumptions.

What are required are observable facts. Here the question rises does a continuous, harmonious layering signifiy a one-time event only? Walker & James in 2002 made this conclusion based on the assumption that one could not verify whether two vertically adjacent facies represent environments that were once laterally adjacent because the depositional start date for the process was unknown. Today, new information is available. The observations from the Kidd Copper property are taken from what is known and well established. Interpreting the Kidd Copper lithographic logs as a one-time event means that the sharp breaks and different patterns do not reflect different stages of deposition. This report establishes that disharmonious layers represent a one-time event based on site observations from the Kidd Copper property and that these changes do not mean changes in cycles. The interpretation from Kidd Copper strongly supports stratified layers do not represent the beginning of new sedimentary cycles because the Kidd Copper stratified layers were known to be from the same event.

15.3 Correlation of Lithostratigraphic Units

Lithostratigraphy is (1) "the element of stratigraphy that deals with the lithology of strata, their organization into units based on lithologic character, and their correlation" (Bates & Jackson, 1984, p. 299); and (2) Lithostratigraphy *n.* [Geology] "the study and correlation of strata to elucidate Earth history on the basis of their lithology, or the nature of the well log response, mineral content, grain size, texture and color of rocks" (Schlumberger, 2008, Oilfield Glossary).

Lithostratigraphic unit is "a body of rock that consists dominantly of a certain lithologic type or combination of types, or has other unifying lithologic features" (Bates & Jackson, 1984, p. 299).

"*Correlation* usually means determining time equivalency of rock units. Rock units may be correlated within a region, a continent, and even between continents" (Plummer, McGery & Carlson, 2003, p. 183). "Correlation by similarity of rock types is more reliable if a very unusual sequence of rocks is involved. If you find in one area a layer of green shale on top of a red sandstone that, in turn, overlies basalt of a former lava flow, and then find the same sequence in another area, you probably would be correct in concluding that the two sequences formed at essentially the same time" (Plummer, McGery & Carlson, 2003, p. 184). "Correlation between two regions can be made by assuming that similar rock types in two regions formed at the same time" (Plummer, McGery & Carlson, 2003, p. 184). See Figure 72 - Appendix M.

Extracting from the above quotations, correlation is based on how similar a rock unit is to another even if the rock formations are continents apart. The basis of reasoning is the conviction that "formations are defined in terms of their lithological characteristics and not their ages, the correlation of lithostratigraphic units in the context of a geological map does not provide anything more than general information about the temporal relationships between different parts of the area" (Nichols, 2003, p. 262). In other words, if the rock unit has similar characteristics to another then they are said to be correlated.

One way that correlation is done in the field is to "trace physically the course of a rock unit ... to correlate rocks between two different places (Plummer, McGery & Carlson, 2003, p. 183). *Physical continuity* generally works for a short distance because the units are not continuously exposed between regions (Plummer, McGery & Carlson, 2003, p. 184).

So, what does one look for? One searches for a "formation [which is] a fundamental lithostratigraphic unit used in the local classification of strata. Formations are classified by the distinctive physical and chemical features of their rocks that distinguish them from other formations although formations do not have to be lithologically homogeneous. The boundaries of formations may be *sharp*, *gradational* or *interdigitating*" (MacDonald, *et al.*, 2003, p. 192).

15.4 Correlation of Sedimentary Strata

"Correlation – determination of the equivalence of sedimentary strata – can also be accomplished by comparing the physical attributes of stratigraphic units in two, or more, different regions" (Garrels & Mackenzie, 1971, p. 66).

Table 15 - Section 15.4 is a list of a few of the architectural elements found on the Kidd Copper Property. There is much diversification in the architectural elements; and clearly the primary structure in each vertical log showed a heterogeneous configuration at their original deposition. A one time event led to many layers being deposited at the same time. These log profiles support that a primary structure can be a heterogeneous layer. The evidence from the architectural elements support that the stratification is random, without a common correlating factor, among the adjacent sediment structures, as to color descriptors.

Three common factors between all the lithographic logs:

(1) Same parent material;

(2) Same cataclysmic event; and

(3) The same element of time for deposition and redeposition.

The results found in Table 4 - Section 10.5 and Table 5 - Section 10.6 display inclined and cross-laminae but what is surprising is that no two lithographic logs are the same. According to conventional wisdom and the way correlation works, Kidd Copper would not be correlated because the physical attributes are dissimilar. However, this is not the case, because the redeposited sediment has much in common in each of the Kiddd Copper log profiles in relation to history, process and time indicators. Therefore, Kidd Copper can be correlated but unexpectedly without similar physical attributes like color.

Table 15 - Section 15.4: Kidd Copper Architectural Elements

Borehole 13: Desmond Carroll, Field Book #4, page 46, point 2.

-- Section dug, tailings roughly 15 cm, 60 cm of clay material, last
  10 cm is wet moss covered and lots of sand, between sand anoxic
  thin soil layer developing over the liner clay 20cm
-- Sand clay structure
-- Grasses growing, clay is wet again, clay
  in the water table
Found a small stream: 15 cm/75 Bearing 162°
-- Silts, clay interbeds
-- Silt orange
        -- Clay grey

Borehole 17: J.J. Gabriel, Field Book #7, page 48, point 1.

-- Sand highly porous
-- ferrous (orange)
-- Grey – wet
  - porous
-- High concentrations of precipitated salt

Depth 60 cm

Borehole 19: Justin Grenier, Field Book #8, page 75, site 1.

Dug to sample:
-- Layers of yellow and grey
-- Yellow = Sulphur?
-- Grey = Clay?
-- 50 cm
-- More sand than clay
-- Sand is yellow material

Borehole 49: Lisa Melymuk, Field Book #18, page 44, point 5.

GPS 0467194   5137446
-- No vegetation
-- Rust coloured (ferrihydrite) crust on surface
-- Whitish/orange surface
-- Ground crust
-- 30 cm Depth

Borehole 68: Loreen Rudd, Field Book #21, page 79, point 23.

Dug hole 1 ½ feet
-- Sand
-- Rust
-- White
-- Dark brown layers

      End Time: 16:56 PM

Borehole 114: May Xiujuan, Field Book #31, page 65, Location 5.

-- Sandy soil
-- Subsurface mostly liquification, wet clay

Borehole 117: Linda Zelek, Field Book #32, page 33, Location 3.

GPS 0467015   5137697
-- Interbedded grey and dark red

-- Depth 20 cm

Source: Abridge of 32 Geology Students Field Record of Kidd Copper Property dated August 30, 2004

"*The basic unit of sampling*, suggested by Otto (1938), is the sedimentation unit. It was defined as that part of a deposit that was laid down under essentially uniform chemical and physical conditions. For example, a bed of sand might constitute a sedimentation unit. "If it was later covered by another bed of sand, it would be important not to take a sand specimen that included parts of both beds. Such a sample could not provide a meaningful estimate of some property of the sand, such as grain size, because it would include sand from two different populations" (Blatt, Middleton & Murray, 1972,

p. 7). The samples taken from Kidd Copper were not taken from different populations, in fact, the samples were from the same bed; thus the architectural elements should be easy to correlate. Knowing this one would expect from conventional wisdom that the architectural elements from the same bed would share many similarities.

"A stratigraphic log includes the thickness of formations (distinctive lithological units that can be traced through the subsurface) seen in the section, the contacts between formations, the type of rock/sediment, the grain size and any other visible properties that may be contained within the rock units such as fossils and sedimentary strucutures" (Environmental Science 1A03/1B03/1G03, 2002, p. 98). Moreover Table 15 - Section 15.4 compares the various units on the Kidd Copper property within a close proximity range. The architectural elements in the stratigraphic units were noticeably different in color and composition. The architectural elements were not identical as expected from mixing the same elements together. The most common correlation was the fact that mixtures of color existed within each layer. Even the surface texture of each profile was different. The units on the Kidd Copper stratigraphic logs hold few, if any, commonalities and would be interpreted and possibly indexed as separate units in time without knowing that they came from the same sediment deposit.

Herein lays a perplexing fact about the basic unit of sampling. If the sample is taken from two different beds it cannot be properly correlated because it would lack meaning and so different findings are to be discarded; perhaps because they have no relevancy or relationship. This interpretation based on conventional wisdom is incomplete because different architectural units were found on the same bed in the case of Kidd Copper. The units marked as Boreholes 13, 17, 19, 49, 68, 114 and 117 in Table 15 - Section 15.4 are all samples of the architectural elements found in vertical logs on August 30, 2004. Each of these log descriptions describes the same sediment layer deposited with the same sequence of events from the impoundment on the Kidd Copper Property. One descriptor was 'dark brown layers', another 'ferrihydrite', another 'high precipitated salt', another 'interbedded grey and dark red', another 'moss covered', another 'sand', another 'silt', and another 'no vegetation'. It sounds like they were from different regions although these vertical logs were all taken from the Kidd Copper impoundment surface and sub-surface area. On every borehole the adjacent environments were the same although the architectural elements are varied.

One cannot discard information gathered because of differences in the vertical logs. Instead, new evidence shows these lithographic tables provide sufficient evidence to support that heterogeneous layering of sediment is a one-time depositional process. This, in turn validates rapid stratification as a primary structural depositional process. Therefore, many architectural elements can represent the same population and can be correlated together.

What is significant about this is that different physical attributes do not represent varied depositional environments. This evidence strongly refutes that layers are the result of the accumulation of sediment over time. The resultant conclusion is that differences rather than similar characteristics were found in each of the Kidd Copper vertical logs description of the architectural elements; albeit such a close proximity to each other and an unexpected finding according to previously held convictions under conventional wisdom from mainstream geologists. "In 1667 Steno developed the principle of superposition: 'in a sequence of layered rocks, any layer is older than the layer next above it'" (Nichols, 2003, p. 2). Today, this principle can be rejected because the layered 'rocks' were not sequential and each layer did not represent a different time or deposition so that all layers were of the same age in the Kidd Copper case.

15.5 Walther's Facies Law

The Law of the Correlation of Facies is another name for Walther Facies Law. "The lateral distribution of different kinds of soils across an ancient landscape can be reconstructed from a vertical section by using a common kind of geological inference often called Walther's Facies Law. This states simply that different kinds of sediment deposited side by side in nature will be preserved on top of one another in a sedimentary sequence" (Retallack, 1990, p. 119). Facies change is "a lateral or vertical change in the characteristics of sediments or rocks. It reflects a change in the depositional environment" (MacDonald, *et al.*, 2003, p. 172).

"Features of recent sediments and ancient sedimentary rocks can be combined and condensed into idealizations or models that characterize particular sedimentary environments" (Walker & James, 2002, p. 1). First example, the general features of Aeolian facies deposition include: wind-blown sand along marine shorelines. "*Aeolian stratification* – The *bedding* and *lamination* formed as wind-blown sediment, usually of *sand grade*, accumulates on a dry surface" (Kearey, 2001, p. 4).

Second example, "Lutites are rocks composed of particles smaller than those of sandstone, commonly in the range of a few microns, and represent finer debris carried by the erosional agents of running water, ice, and wind. The more exclusive term *shale*, which denotes lamination or fissility in addition to grains smaller than those of sandstones, is also applied to these fine-grained rocks. The clay minerals are dominant constituents of lutites but fine-grained quartz and feldspar are common" (Garrels & Mackenzie, 1971, p. 39). Therefore, a soil horizon is "a layer of a soil that can be distinguished from adjacent layers by physical properties (structure, texture, color) or chemical composition" (MacDonald, *et al.*, 2003, p. 404). Table 16 - Section 15.4 "summarizes some key relationships among the sedimentary rocks. Sandstones and shales are initially deposited as particles of sand and mud. With burial and time, cementation and a certain amount of mineral reconstitution take place, and the coherent material is regarded as rock. When post-depositional changes become drastic, the rock is said to be metamorphosed, and new names are given to the altered products. Sandstone is altered to quartzite, limestone to marble, shale to slate, and so on" (Garrels &

Mackenzie, 1971, p. 40). Post-depositional changes were drastic on Kidd Copper but metamorphism was not accepted as the cause because of the rapid process.

Table 16 - Section 15.4: Some Characteristics of Sedimentary Rocks

Rock: Sandstone and Lutite Compared						
Rock	**Some sediment equivalents**	**Some metamorphic equivalents**	**Origin**	**Grain Size**	**Major minerals**	**Wt % of sedimentary rock mass**
Sandstone	Sand	Quartzite	Clastic particles	1/16 – 2mm	Quartz Feldspar Rock fragments	15-20
Lutite	Silt/mud/clay	Slate Schist	Clastic particles	<1/16 mm	Clay minerals Quartz Feldspar	70-80
Source: (Garrels & Mackenzie, 1971, p. 40).						

15.6 Law of Original Continuity

Law of original continuity is "a general law of geology: a water-laid stratum, at the time it was formed, must continue laterally in all directions until it thins out as a result of nondeposition or until it abuts against the edge of the original basin of deposition" (Bates & Jackson, 1984, p. 291; MacDonald, *et al.*, 2003, p. 273). Accordingly, the law states that the expected outcome of the adjacent boreholes would be laterally conformed to the marker. What is seen in the one unit should be seen next to it. This is important because the Kidd Copper deposition occurs at the same time with all the boreholes used for comparison in this report; therefore the adjacent logs are predicted to reveal the same units for deposition as found with the marker unit. However, this is not the case in either of the lithographic logs compiled from the Kidd Copper property where the borehole control marker point 56/4 is noticeably different than any of the other profiles. Refer to the data compiled in Section 10.0.

15.7 Lateral Continuity

"A principal of *lateral continuity* has been proposed, which states that layered sediments extend in a lateral fashion in all directions from any single point. The importance of this concept is that it allows one to predict the position of stratigraphic layers far from any given locality. For instance, if you discover a particular composition of layers, that composition will extend for some distance, laterally, unless it is interrupted by another natural obstacle" (Albin, 2004, p. 44, 45).

"The principle of *lateral continuity* states that an original sedimentary layer extends laterally until it tapers or thins at its edges. This is what we expect at the edges of a depositional environment, or where one type of sediment interfingers laterally with another type of sediment as environments change" (Plummer, McGery & Carlson, 2003, p. 178).

> "Whenever one observes the exposed cross-section of strata in a cliff or valley wall, one should recognize that the strata, as originally deposited, should continue laterally for a distance that can be determined by field work and drilling. If lateral continuity is not observed and the lack of continuity is not related to one of the reasons given above, then the cause may be displacement of strata by faulting or erosional loss of strata. When geologists stand on a sandstone ledge at one side of a canyon, it is the principle of original lateral continuity that leads them to seek out the same ledge of sandstone on the far canyon wall and then to realize that the two exposures were once continuous" (Levin, 2003, p. 4, 5).

"The *principle of original lateral continuity* was the third of Steno's stratigraphic axioms. It pertains to the fact that, as originally deposited, strata extend in all directions until they terminate by thinning at the margin of the basin, end abruptly against some former barrier to deposition, or grade laterally into a different kind of sediment" (Levin, 2003, p. 4). This is not the case at Kidd Copper, where the abruptions found in the lithographic logs and sediment profiles are disjointed as the settling process was abrupt and rapid leaving a heterogeneous distribution as sediment cover; instead of the thinning suggested by Steno. The third of Steno's stratigraphic axioms is not supported.

The Prediction of Lateral Continuity is Not Supported. In the case of the architectural elements found on the Kidd Copper property lateral continuity is not found. The reason for this is that the Law of superposition is inaccurate, which provides the reason as to why the axiom that supports lateral continuity is inaccurate. Neither the law of superposition is validated by the observations and facts found on the Kidd Copper Property; nor is the axiom that lateral continuity exists between all exposed cross-sections, or recognized strata, as originally deposited. The irregular patterns are contrary to the expected outcomes from the Law of superposition that holds layers are the result of time and different depositional processes or periods. Plummer, McGery & Carlson write, "The principle of *superposition* states that within a sequence of undisturbed sedimentary or volcanic rocks, the layers get younger going from bottom to top" (2003, p. 177). The Kidd Copper strata was not disturbed but had layers; these layers could not be read as younger on top because each layer was formed at the same date.

The adjacent units observed at Kidd Copper do not extend laterally in all directions from the borehole control marker point 56/4 even though conventional wisdom predicted they should. All the exposures on Kidd Copper in this report are continuous from the impound site even though they do not look like they are. The conclusion is that sediment layers do not continue laterally in all directions until thinning out as a result of all redeposition processes such as the flooding event at Kidd Copper substantiates.

The Law of superposition and its axiom, the principle of lateral continuity, remain unsupported when examined and compared to the facts observed in the case of the Kidd Copper Property.

15.8 Law of Equal Declivities

"Law of equal declivities. Where homogeneous rocks are maturely dissected by consequent streams, all hillside slopes of the valleys cut by the streams tend to develop at the same slope angle, thereby producing symmetrical profiles of ridges, spurs and valleys" (Bates & Jackson, 1984, p. 290). This is the rationale behind the suggested reproducible flume study in Table 13 - Section 14.10 on point source layering to determine; are sediments always, sometimes or never distributed equally?

Food for thought is that since the same process on the same sediment occurred on Kidd Copper the outcome can be tested. Did the sediment lay uniform or in a random order? Did the sediment at Kidd Copper distribute itself equally; after all should not an equal distribution be observed from the sediment processes like the Law of equal declivities suggests? If the process and the object are the same should not the end results be the same? Is it not rational to conclude that equal matter and equal processes provide equal results? Is it not rational that sediment from the same slope when lifted up and deposited land in the same manner?

Already established from the Kidd Copper lithographic logs, the Law of superposition and its axiom, the principle of lateral continuity, remains unsupported by the facts examined. The stratification on the Kidd Copper property case revealed heterogeneous deposits were settled out of fluid as stratified layers. Moreover, these layers had unequal declivities, different slope angles and no symmetrical profile. Again, conventional wisdom is challenged with the evidence provided in this case.

15.9 Same Sequence is Visible in Rock Outcrops

Conventional Wisdom Predicts Same Sequence Means Continuity. "Another thing geologists know to be true – no matter where they are – is that if the same sequence is visible in the rock outcrops in different places, it is likely that these beds are or once were continuous. Weathering and erosion may have removed the intervening rock" (Discovery Books, 1999, p. 40). As previously discussed, in the case of the 120 boreholes uncovered on the Kidd Copper property the same sequence should be transparent because the same depositional process occurred through the site area. However, this is not seen in the observations because of multiple differentiations rather than the continuous flow of strata to yield the same outcome. In fact, the different outcomes made visible were a part of the continuous bed of sediment uncovered on the impoundment for evaluation, contrary to what mainstream geologists hold true that beds are only continuous if similar.

15.10 Unconformities

"An *unconformity* is a surface (or contact) that represents a *gap in the geologic record*, with the rock unit immediately above the contact being considerably younger than the rock beneath" (Plummer, McGery & Carlson, 2003, p. 182). "Breaks in the rock record are termed unconformities. An unconformity represents a long period during which deposition ceased, erosion removed previously formed rocks, and then deposition resumed. In each case uplift and erosion are followed by subsidence and renewed sedimentation" (Tarbuck & Lutgens, 1999, p. 193). "Most unconformities are buried erosion surfaces" (Plummer, McGery & Carlson, 2003, p. 182). "Erosional breaks, or unconformities, between sedimentary sequences also aid in correlation" (Garrels & Mackenzie, 1971, p. 66).

> "The sequence may contain a gap in time that geologists call an unconformity, a place where a younger layer of rock lies directly on top of a much older layer of rock. The intervening 'middle-aged' layers that you would expect to see are absent. The unconformity represents the boundary or contact between these two rock layers. The missing layers have been eroded away, and overlain a few million years later with fresh layers of sediment" (Discovery Books, 1999, p. 42).

"Unconformities are classified into three types – disconformities, angular unconformities, and nonconformities – with each type having important implications for the geologic history of the area in which it occurs. In a *disconformity*, the contact representing missing rock strata separates beds that are parallel to one another. Probably what has happened is that older rocks were eroded away parallel to the bedding plane; renewed deposition later buried the erosion surface" (Plummer, McGery & Carlson, 2003, p. 182).

"An *angular unconformity* is a contact in which younger strata overlie an erosion surface on tilted or folded layered rock. It implies the following sequence of events, from oldest to youngest: (1) deposition and lithification of sedimentary rock (or solidification of successive lava flows if the rock is volcanic); (2) uplift accompanied by folding or tilting of the layers; (3) erosion; (4) renewed deposition (usually preceded by subsidence) on top of the erosion surface" (Plummer, McGery & Carlson, 2003, p. 182, 183). Angular unconformity – easily seen with the naked eye – "sediments rest on top of rock that has been deformed before being eroded" (Discovery Books, 1999, p. 42).

"A *nonconformity* is a contact in which an erosion surface on plutonic or metamorphic rock has been covered by younger sedimentary or volcanic rock. A nonconformity generally indicates deep or long continued erosion before subsequent burial, because metamorphic or plutonic rocks form at considerable depths in Earth's crust" (Plummer, McGery & Carlson, 2003, p. 183).

"Paraconformity marks a break in the sedimentary sequence where the fossil record, or faunal succession, is the only clue that something is missing" (Discovery Books, 1999, p. 42). The term *discontinuity* is "any interruption in sedimentation that may represent a time interval in which sedimentation has ceased and/or erosion has occurred" (MacDonald, *et al.*, 2003, p. 143).

The problem with unconformities is that any break in the sequence can be subjected to an arbitrary assertion as to what the missing information was. Who was there in the Eras/Epochs/Periods or other time designations to confirm if the assumed information in accurate? True, something is missing but what?

16.0 Rock Dating Methods and Principles

"Steno applied a very simple rule that has come to be the most basic principle of relative dating – the Law of superposition. The law simply states that in an underformed sequence of sedimentary rocks, each bed is older than the one above and younger than the one below" (Tarbuck & Lutgens, 1999, p. 191). "The *principle of superposition* states that in any sequence of undisturbed strata, the oldest layer is at the bottom, and successively higher layers are successively younger" (Levin, 2003, p. 3). Relative dating is "the chronological arrangement of fossils, rocks or events with respect to the geological time scale, without reference to their *absolute ages*" (MacDonald, *et al.*, 2003, p. 367). The principle of Relative Dating is problematic because in a heterogeneous layer such as examined at Kidd Copper, the bottom layer of sediment in each of the log profiles was neither the oldest sediment nor the youngest because each sediment layer in the Kidd Copper profiles had the same start date. Refer to the raw data found in Section 10.2.

16.1 Geologic Time Units Defined

Geochronological unit is "an uninterrupted interval of time in geological history during which a corresponding chronozone was formed, i.e. a division of time delineated by the rock record. In order of decreasing magnitude, geological time units are the *eon, era, period, epoch, age* and *chron*" (MacDonald, *et al.*, 2003, p. 203). See Appendix M.

Geochronology is "the measurement of time intervals of the geological past, accomplished in terms of absolute age-dating by a suitable radiometric procedure, e.g. the potassium-argon method, or relative age-dating by means of sequences of rock strata or varved deposit analysis.

16.2 Relative Age Dating

"Relative dating means that rocks are placed in their proper *sequence of formation*. Relative dating cannot tell us how long ago something took place, only that it follows one event and preceded another" (Tarbuck & Lutgens, 1999, p. 191).

"Relative age is the age of a rock formation or geologic event in comparison to something else; that is, the rock or event is known to be older or younger than some other feature" (Murck, 2001, p. 56). Relative age is "the geological age of a rock, fossil or event, as defined relative to other rocks, fossils or events rather than in terms of years" (MacDonald, *et al.*, 2003, p. 367). Relative dating is "the chronological arrangement of fossils, rocks or events with respect to the geological time scale, without reference to their *absolute ages*" (MacDonald, *et al.*, 2003, p. 367). However, for stronger age dating "a relative stratigraphic scale should be used that is as close as possible to the numerical geological time scale" (Gradstein, *et al.*, 2004, p. 108). In the case of the Kidd Copper Property, the numerical geological time scale can be accurately correlated with historical time records. That is the main difference between this analysis and those relying on relative age dating. There have been no estimates in the age correlations on the Kidd Copper property for the purpose of this report.

16.3 Historicity of Relative Age Dating

"The concept of relative age is based on a few simple but profound principles, which come from the study of sedimentary rock layers or *strata* (the singular form is *stratum*, from the Latin word for 'layer')" (Murck, 2001, p. 49).

"In the fifteenth century, natural philosopher Nicholas Steno took a close look at how rock layers were formed and at the curious fossils within them. He would lay down the important principles of stratigraphy that have been used to help with relative age dating of rocks. *Stratigraphy* is the study of layered sedimentary rocks. Steno's Law of superposition states that older rocks are found beneath younger rocks. This happens because when sediments settle out of a fluid, the youngest layers will be deposited on top of previous layers. According to the principle of original horizontality, these layers will initially be set down in a horizontal fashion. Any change from this orientation occurs at another time when the rocks have been tilted by tectonic forces. Early detectives of Earth's history used this technique, *relative age dating*, to place rock units into a chronological sequence" (Albin, 2004, p. 44).

"In 1869 John Wesley Powell ... realized that the evidence for an ancient Earth is concealed in its rocks" (Tarbuck & Lutgens, 1999, p. 190). "Geological events by themselves, however, have little meaning until they are put into a time perspective" (Tarbuck & Lutgens, 1999, p. 190). "It was apparent from Lyell's principle of uniformitarianism that the many thick deposits of the stratigraphic column must have taken a long time to accumulate" (Busbey, *et. al.*, 1999, p. 20).

16.4 Relative Time

"For the geologist, there are two ways of looking at time: relative time, which embraces the sequence of events that formed the rock; and absolute time, which provides the actual age of the rock in years" (Discovery Books, 1999, p.36; Busbey, *et. al.*, 1999, p. 20). "The work of early geologists William Smith, James Hutton, and Charles Lyell resulted in an understanding that rocks accumulate in layers with the oldest at the bottom, and that this process occurred in the past at the same slow rate as it does today (Lyell's principle of uniformitarianism)" (Busbey, *et. al.*, 1999, p. 20).

"In studying the Earth's history, we talk about stretches of time too long to imagine. Geologists have divided the Earth's history into periods. Each period lasted millions of years" (Williams, 1993, p. 81). "After careful mapping of all the world's known rock units, a theoretical continuous stack, called the stratigraphic column was constructed" (Busbey, *et. al.*, 1999, p. 20). To facilitate working with this cumbersome intellectual structure, the column was divided into a number of subsequences called Eras, Periods, and Epochs" (Busbey, *et. al.*, 1999, p. 20). This is known as relative time.

16.5 Units of Geological Time

Geological time is "the time extending from the end of the formative period of the Earth (i.e. after it was 'established' as a planet) to the onset of human history; it is the part of the Earth's history that is recorded and described in the succession of rocks" (MacDonald, *et al.*, 2003, p. 205). "By the beginning of the twentieth century, geologists had developed a relative geologic time scale using the fundamental principles of stratigraphy: the law of horizontality, which dictates that sedimentary rock generally forms in horizontal layers; and the Law of superposition, which holds that layers are older as they are deeper. Using these laws, they were able to correlate rock formations around the world and assemble a chronological record of the planetary past, dividing it into time units known as eons, eras, periods, and epochs" (Discovery Books, 1999, p. 36).

Eons. "The oldest radiometric dates, about 4.4 billion years, have been obtained from individual mineral grains in sedimentary rocks from Australia" (Murck, 2001, p. 62). "These ancient rocks are all from the Archean eon. Recall that the Hadean eon predates the Archean. No rocks that might provide radiometric dates are preserved from early Hadean time – none that we have yet found, at any rate" (Murck, 2001, p. 62).

Eras. "A geological (geochronological) time unit during which rocks of the corresponding *Erathem* were formed. It ranks one order of magnitude below *eon*" (MacDonald, *et al.*, 2003, p. 166).

"A *chronostratigraphical unit*, also called *chronolithological unit*, *time-rock unit* or *chronolith*, is one that was formed during a defined interval of geological time; it represents all, and only, those rocks formed during designated time span of geological history" (MacDonald, *et al.*, 2003, p. 414).

Periods. "Different plants and animals lived in different periods" (Williams, 1993, p. 81).

Epochs. "An interval of geological time (geochronological division) longer than an *age* and shorter than a *period* in which the rocks of a particular series were formed" (MacDonald, *et al.*, 2003, p. 166).

Age is a "unit of geological time, longer than a chron and shorter than an *epoch*, during which rocks of a particular stage were formed. A frequently used term for a geological time interval within which were formed the rocks of any *stratigraphical unit*" … "the time during which a particular event occurred or a time characterized by unusual physical conditions, e.g. the Ice Age … the position of any feature relative to the geological time scale: e.g. 'rocks of Eocene age'" (MacDonald, *et al.*, 2003, p. 18). This is important because "Time-stratigraphic (of rock units) having boundaries that are based on their age or time of origin" (MacDonald, *et al.*, 2003, p. 431). For example, "the *Cenozoic* began at 65.5 Ma, and spans 65.5 myr (to the present day)" (Gradstein, *et al.*, 2004, p. 6).

"The thermal history of a sedimentary layer can be used to determine whether organic material in the sediments has been converted to petroleum" (Turcotte & Schubert, 2002, p. 180). This is beyond the scope of this report except to note that the "depths to sedimentary layers deposited at times t_s as functions of time" (Turcotte & Schubert, 2002, p. 180).

16.6 The Geologic Column

A partial geologic column is shown in Table 17 - Section 16.6 (Tarbuck & Lutgens, 1999, p. 9). The units of the geologic time scale shows that the Kidd Copper mine tailings case study falls under Phanerozoic Eon, Cenozoic Era, Quarternary Period and under the Holocene Epoch according to data from Geolgocial Society of America.

"The crucial questions on the relationship of Genesis and geology, of religion and geological science nearly all hinge in some way on one's understanding of the meaning and significance of the geologic column. It is important, therefore, to understand something of its origin as a system. Most commonly the term geologic column is applied to a composite columnar section in which there is an attempt to superimpose rocks representing every period of time for the world as a whole. In this way it is thought that the column can be used somewhat like a calendar for dating rock formations and as a unifying concept, a datum in relation to which a large part of the vast fund of geological information and theory can be organized. In the various major regions of the world, locations with favorable exposures and typical deposits for individual segments of the column (epoch or series divisions) may be selected as type sections for use in correlation, as, for instance, the magnificent, little disturbed Middle Cambrian formations of the Canadian Rockies (the Albertan Series) to which Middle Cambrian rocks throughout North America are compared" (Ritland, 1981, p. Introduction).

Table 17 - Section 16.6: The Geologic Column (Tarbuck 1999; Ritland 1981-Incomplete)

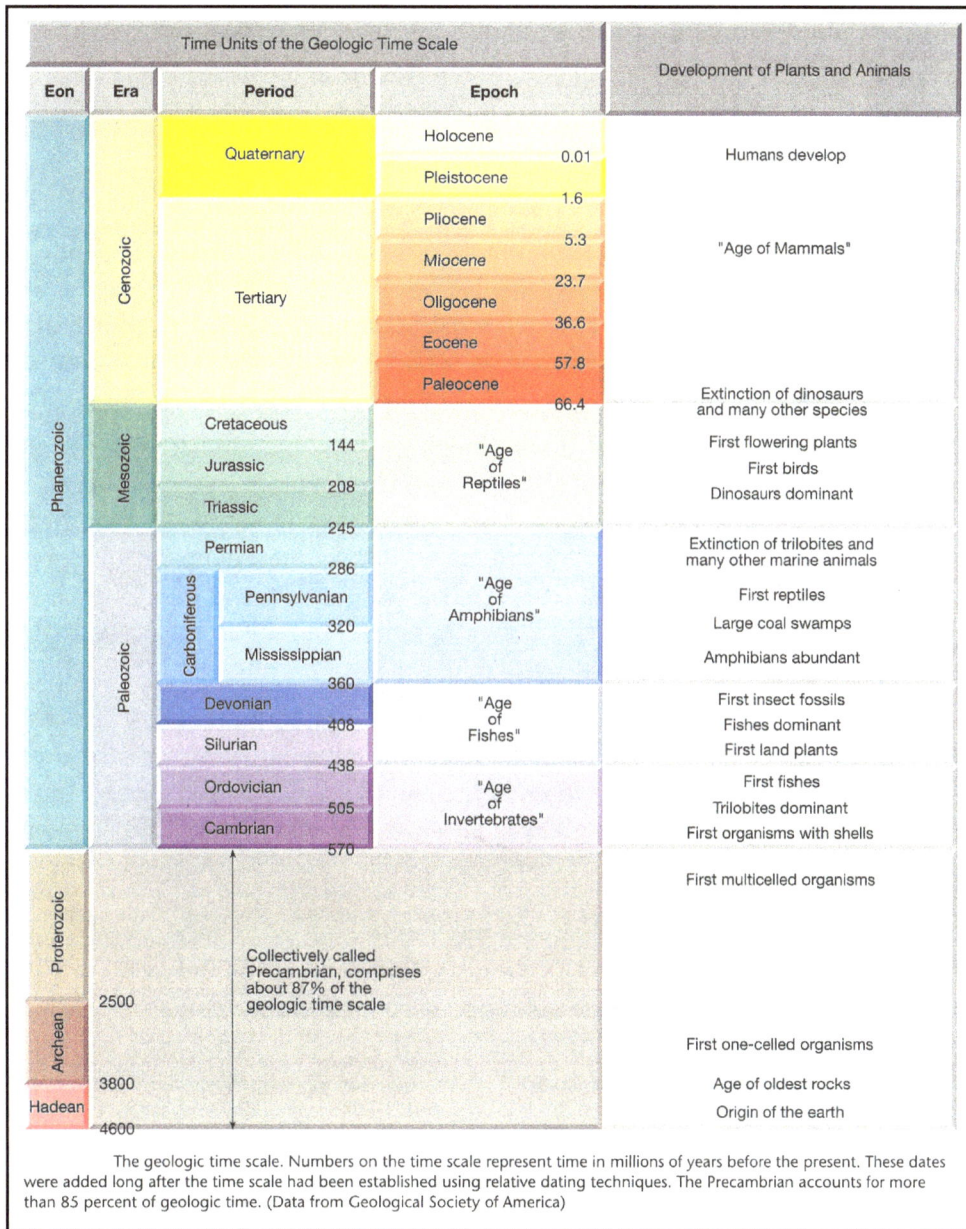

Eon	Era	Period		Epoch		Development of Plants and Animals
						Time Units of the Geologic Time Scale
Phanerozoic	Cenozoic	Quaternary		Holocene		Humans develop
					0.01	
				Pleistocene		
					1.6	
		Tertiary		Pliocene		"Age of Mammals"
					5.3	
				Miocene		
					23.7	
				Oligocene		
					36.6	
				Eocene		
					57.8	
				Paleocene		
					66.4	Extinction of dinosaurs and many other species
	Mesozoic	Cretaceous		"Age of Reptiles"		First flowering plants
			144			First birds
		Jurassic				Dinosaurs dominant
			208			
		Triassic				
			245			
	Paleozoic	Permian		"Age of Amphibians"		Extinction of trilobites and many other marine animals
			286			
		Carboniferous / Pennsylvanian				First reptiles
			320			Large coal swamps
		Mississippian				Amphibians abundant
			360			
		Devonian		"Age of Fishes"		First insect fossils
			408			Fishes dominant
		Silurian				First land plants
			438			
		Ordovician		"Age of Invertebrates"		First fishes
			505			Trilobites dominant
		Cambrian				First organisms with shells
			570			
	Proterozoic					First multicelled organisms
		Collectively called Precambrian, comprises about 87% of the geologic time scale	2500			
	Archean					First one-celled organisms
			3800			Age of oldest rocks
	Hadean		4600			Origin of the earth

The geologic time scale. Numbers on the time scale represent time in millions of years before the present. These dates were added long after the time scale had been established using relative dating techniques. The Precambrian accounts for more than 85 percent of geologic time. (Data from Geological Society of America)

16.7 Absolute Age

"Absolute age [GEOL] The geologic age of a fossil, or a geologic event or structure expressed in units of time, usually years. Also known as the actual age" (Licker, 2003, p. 1; (MacDonald, *et al.*, 2003, p. 205). "It commonly refers to ages determined by radiometric dating but other methods such as *fission track dating* and *dendrochronology* may also be used" (MacDonald, *et al.*, 2003, p. 11). "It is therefore not possible to date the formation of rocks made up from detrital grains and this excludes virtually all sandstones, mudrocks and conglomerates" (Nichols, 2003, p. 254). "The term 'absolute' is somewhat misleading, since it applies an exactness that is not always readily achieved" (MacDonald, *et al.*, 2003, p. 11).

16.8 Absolute Time

"Absolute time [GEOL] Geologic time measured in years, as determined by radioactive decay of elements" (Licker, 2003, p. 1). "Absolute time – to establish the absolute age of a rock (in millions of years), we must find and read some form of clock within the rock itself" (Busbey, *et. al.*, 1999, p. 21). "This process can take minutes or billions of years – depending on the element" (Discovery Books, 1999, p. 33).

"Today, we recognize several 'rock-clocks' that operate in different ways. One is the decay of particular types of isotopes (forms of an element) to other isotopes. In cases where the rate of decay from one to another is known, reading the ratio of the first (parent) isotope to the second (daughter) isotope tells how long that particular process of decay has been going on. There are a few restrictions on this technique. The rock-clock must be set to zero when the rock is laid down, as occurs in molten rocks were the product isotope is a gas that can bubble off" (Busbey, *et. al.*, 1999, p. 21).

"All radiochemical methods of dating have some uncertainties associated with them. Several assumptions must be made in determining an age. Perhaps the most significant assumption is the supposition that the sample was a closed system throughout its existence, that is, no parent or daughter isotope was gained or lost" (Shakhashiri, April 2008, General Chemistry).

16.9 Geologic Age

Geological age is "the age of a fossil organism or a geological event or formation, as referred to the geological time scale and expressed as *absolute age* or in terms of comparison with the immediate environment (*relative age*)" (MacDonald, *et al.*, 2003, p. 204, 205). "An age datable by geologic methods" (Bates & Jackson, 1984, p. 206).

16.10 Absolute Date

"Scientists who worked out the geologic column were tantalized by the challenge of measuring *numerical age* – that is, the actual age of rock and fossils in years (formerly called 'absolute' age)" (Murck, 2001, p. 55). "*Absolute* (or *chronometric*) *techniques* give an absolute estimate of the age and fall into two main groups. The first depends on the existence of something that develops at a seasonally varying rate, as in dendrochronology and varve dating. The other uses some measurable change that occurs at a known rate, as in chemical dating" (Isaacs, *et. al.,* 2003, p. 215). For example, chemical dating is "an absolute dating technique that depends on measuring the chemical composition of a specimen" (Isaacs, *et. al.,* 2003, p. 149). Used as an example to establish an old earth "the oldest radiometric dates, about 4.4 billion years, have been obtained from individual mineral grains in sedimentary rocks from Australia" (Murck, 2001, p. 62). See Table 22 - Section 18.2 on *Problematic Dating Methods* where dangers using dating methods are noted.

16.11 Age Equation

Age equation is "the relationship between geological time and radioactive decay, expressed mathematically as:

$$t = (1/\gamma) \times \ln (1 + d/p)$$

Where *t* is the age of the sample rock or mineral, γ is the decay constant of the particular radioactive series used for the calculation, *ln* is the logarithm to base *e*, and d/ρ is the present ratio of radiogenic daughter atoms to the parent isotope" (MacDonald, *et al.,* 2003, p. 18). Note in Section 18.3 a variance of this calculation is quoted from the same source. Radioactive decay is "the spontaneous transformation of a nucleus into one or more different nuclei by the emission of a particle or photon, by fissioning or by the capture of an orbital electron" (MacDonald, *et al.,* 2003, p. 363).

"Age of the earth – approximately 4530 Ma. This is based on analysis of the abundances of uranium and lead isotopes in meteorites and lunar rocks. The oldest meteorites give ages of 4560 Ma., and chemical similarities between meteorites and the Earth suggest that the Earth and the other planetary bodies were formed approximately 4530 Ma ago. The oldest lunar rocks give ages of 4460 Ma, and if the impact origin of the Moon from the Earth is correct then the age of lunar rocks gives the minimum age for the Earth. Zircon crystals from Australia have given ages of 4200 Ma, the oldest so far recorded from rocks on Earth" (MacDonald, *et al.,* 2003, p. 19).

17.0 Geology the Science of Telling Time

"No other field of science has found it necessary to construct its own special scale of time. There is no formalized biological time scale, or chemical time scale. All other fields of study simply use the intervals of time known to us all: seconds, minutes, hours, days, years. Geologists, on the other hand, talk about periods and epochs, eras and zones, stages and series: the subdivisions of what is known as the geological time scale" (Gribbin, 2004, p. 149).

17.1 Rock Strata Record Earth's History

"In 1788, Hutton suddenly comprehended the vast time required for the deformation of the primary strata, their erosion and subsequent deposition of secondary strata. *'The mind seemed to grow giddy by looking so far into the abyss of time'* he wrote. Here then, was proof that the Earth had not been formed instantaneously as biblical scholars argued but has a long history of strata being deposited, tilted, eroded, and buried" (Eyles, 2002, p. 15).

"In any area where rocks are laid down in layers, known as strata, common sense decrees that the lowest strata will have been laid down first, and the others laid on top. This basic principle applies regardless of whether the rock strata are formed from solidified mud or other sediments, or represent successive flows of lava, ash, or other material that has been ejected from volcanoes and spread over the surrounding landscape" (Gribbin, 2004, p. 153). "Rock strata provide a remarkable record of Earth's history, each layer representing one period. Each layer is hundreds of yards thick" (Gribbin, 2004, p. 153).

"The idea that the rock we see around us has been laid down in strata over many millions of years is the basis of stratigraphy" (Discover Books, 1999, p. 33). Geological record is "the testimony of the history of the Earth as revealed by its *bedrock*, *regolith* and *morphology*" (MacDonald, *et al.*, 2003, p. 205). "Geologists read strata like the lines of a book to gain an understanding of Earth's past. The location, thickness, texture, color, and chemistry of the strata tell them where winds blew, waters flowed, and mountains rose" (Discover Books, 1999, p. 33). In Appendix J a list of the divisions of geology and the earth sciences is provided.

"Eighteenth-century European geologists believed that a rock's type was the best clue to its age. They assumed that all rocks of a specific type, such as all granites, had been formed at the same time. By the early 1800s, however, they had come to believe, correctly, that the composition or mineral content of a rock is virtually independent of its age" (Gribbin, 2004, p. 151).

"The earth sciences, more than any other field of scientific inquiry, are inextricably linked to the passage of time. The study of

geology – the foundation of all earth sciences – has been described as
the science of telling time, and establishing the dates and sequences of
events on Earth. Sooner or later any geological investigation moves
from process to history, and to events that occurred during a particular
period in the past. So measuring time – to a geologist – is almost a
religion" (Gribbin, 2004, p. 149).

"The measurement of time is clearly indispensible to the study of geology, and it
is also a source of great error, confusion, and even heartbreak for all geologists and other
earth scientists trying to wrest the truth from our planet's rocky cover" (Gribbin, 2004, p.
149).

The best facies interpretation is based on the facts of history, science is based on
what is seen, not what is not seen. The Kidd Copper property case is very important
because textbook theory can be tested and validated with observations and known facts.
The guesswork is eliminated. Multiple layers have been used to date the age of the Earth
but if multiple layers are not indicators of ages, stages, eons or periods of time then how
is time to be correlated? If multiple layers can significantly indicate a fast process then
rapid stratification suggests a new way to correlate time and date a younger Earth.

17.2 Geological Time

"Geologic history – the history of the earth and its inhabitants throughout
geologic time. It includes all conditions, processes, and events from the beginning of the
planet to the present" (Bates & Jackson, 1984, p. 207). For example, "the total duration
of the Carboniferous just over 60 myr: the Mississippian (Early Carboniferous) lasted 41
myr and the Pennsylvanian (Late Carboniferous) 19 myr" (Gradstein, *et al.*, 2004, p.
248).

Geological time – "time scale embracing the history of the Earth from its physical
origin to the present day. Geological time is traditionally divided into eons (Archaean or
Archaeozoic, Proterozoic and Phanerozoic in ascending chronological order), which in
turn are subdivided into eras, periods, epochs, ages, and finally chrons" (Research
Machines Plc., 2003, p. 260). Refer to Figure 71 - Appendix M. "Geological Time – the
time extending from the end of the formative period of the Earth (i.e. after it was
'established' as a planet) to the onset of human history; it is the part of the Earth's history
that is recorded and described in the succession of rocks" (MacDonald, *et al.*, 2003, p.
205). "Geologic time [GEOL] the period of time covered by historical geology, from the
end of the formation of the earth as a separate planet to the beginning of written history"
(Licker, 2003, p. 135).

"Geologic-time unit – a span of continuous time in geologic history, during which
a corresponding *chronostratigraphic unit* was formed; a division of time distinguished on
the basis of the rock record" (Bates & Jackson, 1984, p. 207). Formal unit is "a

stratigraphical unit that is defined and named in accordance with the rules and guidelines of an established system" (MacDonald, *et al.,* 2003, p. 192). "Geologic-time units in order of decreasing magnitude are eon, era, period, epoch, and age" (Bates & Jackson, 1984, p. 207). Such as, "*Acme zone, epibole* or *peak zone* – A *biozone* consisting of a body of strata in which a particular species or genus of an organism occurs at maximum abundance. *Hemera* is the corresponding unit of geological time" (MacDonald, *et al.*, 2003, p. 15).

"Lithostratigraphical units, in descending order, are: *supergroup, group, formation, member, bed*" (MacDonald, *et al.*, 2003, p. 414). "Chronostratigraphical units are, in descending order: *eonotherm, erathem, system, series, stage, chronozone*" (MacDonald, *et al.*, 2003, p. 414). System is "a major unit of rank in *chronostratigraphy* that serves as a worldwide reference unit. A system is based on a type area and includes all the rocks formed during a period of geological time. In rank, system is above *series* and below *erathem*" (MacDonald, *et al*, 2003, p. 422). "Series [GEOL] (1) a number of rocks, minerals, or fossils that can be arranged in a natural sequence due to certain characteristics, such as succession, composition, or occurrence. (2) A time-stratigraphic unit, below system and above stage, composed of rocks formed during an epoch of geologic time" (Licker, 2003, p. 333). Stage is "(1) a *stratigraphical unit* ranking above *chronozone* and below *series*; it is based on *biozones* considered to approximate to deposits that are equivalent in time and includes the rocks formed during an age of geological time. (2) A major subdivision of a glacial epoch" (MacDonald, *et al.*, 2003, p. 410).

17.3 Chronostratigraphy

"The organisation of rocks into a scale of named units – *Lithostratigraphy* – is linked to time by *chronostratigraphy*, which organizes the sequence of rock units into chronostratigraphic units (*chronozones*) that correspond to intervals of time – *geochronological units*" (MacDonald, *et al.*, 2003, p. 205, 206). Geologic Time Scale defined from the Collins Dictionary Geology: "A scale that subdivides all geological time into named units and all rocks formed during geological time into a sequence from the oldest to the youngest. The organisation of rocks into a scale of named units – *Lithostratigraphy* – is linked to time by *Chronostratigraphy*, which organises the sequence of rock units into chronostratigraphic units (*Chronozones*) that corresponds to intervals of time – *Geochronological Units*" (MacDonald & Burton, 2003, p. 205, 206). "The division of rocks into units which were thought to be chronostratigraphic had been carried out long before any method of determining the geological time periods involved had been developed" (Nichols, 2003, p.255).

17.4 Geologic Time Scale

"Geologic time scale [GEOL] the relative age of various geologic periods and the absolute time intervals" (Licker, 2003, p. 135). *Geologic Time* defined from the Collins

Dictionary Geology: "The time extending from the end of the formative period of the Earth (i.e. after it was 'established' as a planet) to the onset of human history; it is the part of the Earth's history that is recorded and described in the succession of rocks. *Geological Time Scale* – A scale that subdivides all geological time into named units and all rocks formed during geological time into a sequence from the oldest to the youngest" (MacDonald & Burton, 2003, p. 205**).**

Table 18 - Section 17.4 is a 2009 representation of the GSA Geologic Time Scale assembled by "The Geological Society of America. Product code CTS004R2C. The compilers were A.R. Palmer and John Geissman" (Geological Society of America, April/May 2009, p. Time Scale).

Table 18 - Section 17.4: 2009 Geologic Time Scale, The Geologic Society of America

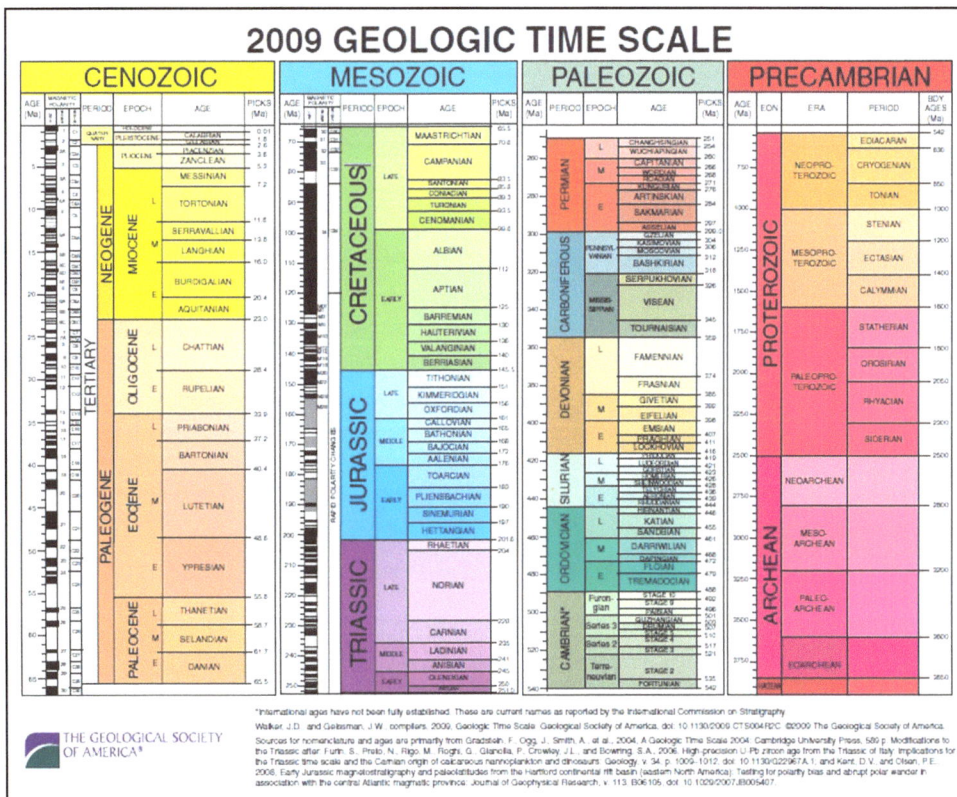

Note the changes from the earlier 1999 GSA Geologic Time Scale assembled by "The Geological Society of America. Product code CTS004. : Table 19 – Section 17.4 "International ages have not been established. These are regional only. Boundary Picks were based on dating techniques and fossil records as of 1999. Palemagnetic attributes have errors. Please ignore the paleomagnetic scale" (Geolgocial Society of America, 1999, *p. nd*).

Table 19 - Section 17.4: 1999 Geologic Time Scale, The Geologic Society of America

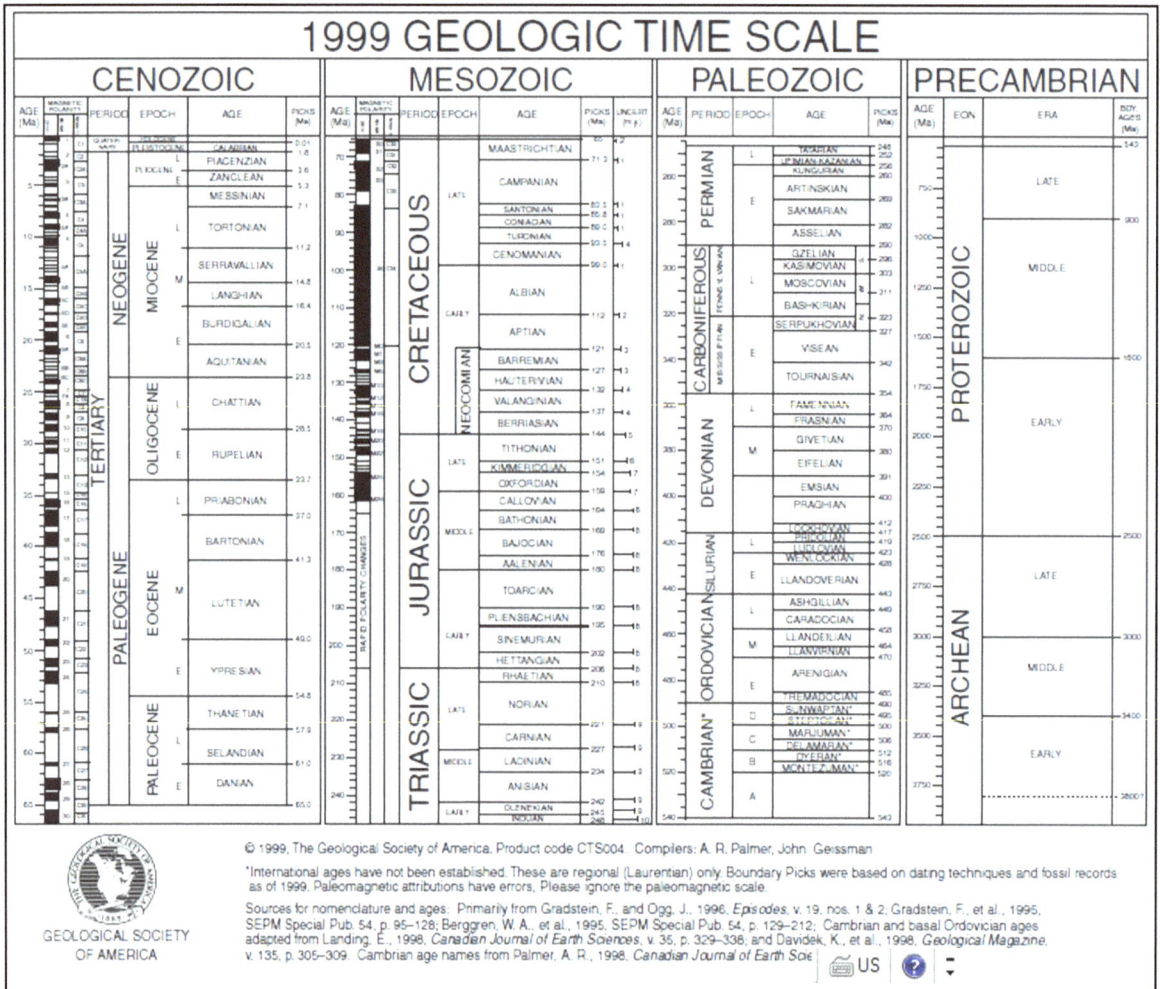

1999 GEOLOGIC TIME SCALE

CENOZOIC — MESOZOIC — PALEOZOIC — PRECAMBRIAN

© 1999, The Geological Society of America. Product code CTS004. Compilers: A. R. Palmer, John Geissman.

*International ages have not been established. These are regional (Laurentian) only. Boundary Picks were based on dating techniques and fossil records as of 1999. Paleomagnetic attributions have errors, Please ignore the paleomagnetic scale.

Sources for nomenclature and ages: Primarily from Gradstein, F., and Ogg, J., 1996, *Episodes*, v. 19, nos. 1 & 2; Gradstein, F., et al., 1995, SEPM Special Pub. 54, p. 95–128; Berggren, W. A., et al., 1995, SEPM Special Pub. 54, p. 129–212; Cambrian and basal Ordovician ages adapted from Landing, E., 1998, *Canadian Journal of Earth Sciences*, v. 35, p. 329–338; and Davidek, K., et al., 1998, *Geological Magazine*, v. 135, p. 305–309. Cambrian age names from Palmer, A. R., 1998, *Canadian Journal of Earth Sciences*...

GEOLOGICAL SOCIETY OF AMERICA

"For convenience in international communication, the rock record of Earth's history is subdivided into a 'chronostratigraphic' scale of standardized global stratigraphic units, such as 'Jurassic', 'Eocene,' '*Harpoceras falciferum* ammonite zone,' or 'polarity Chron C24r.' Unlike the continuous ticking clock of the 'chronometric' scale (measured in years before present), the chronostratigraphic scale is based on relative time units, in which global reference points at boundary stratotypes define the limits of the main formalized units, such as 'Devonian'. The chronostratigraphic scale is an agreed convention, whereas its calibration to linear time is a matter for discovery or estimation" (Gradstein, *et al.*, 2004, p. 3). Eons – Eras (Nichols, 2003, p. 229). Periods – Epochs – Ages – Chrons (Nichols, 2003, p. 230).

"Continual improvements in data coverage, methodology, and standardization of chronostratigraphic units imply that no geologic time scale can be final" (Gradstein, *et al.*, 2004, p. 3).

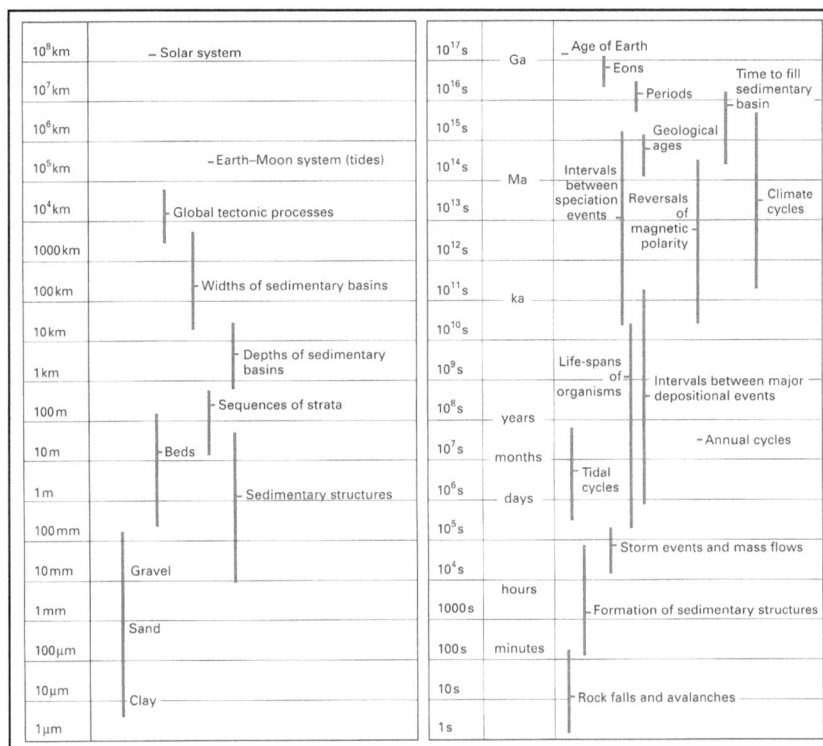

Spatial scale	Process		Time scale		Process
10^9 km	– Solar system		10^{17} s	Ga	– Age of Earth / Eons
10^7 km			10^{16} s		Periods / Time to fill sedimentary basin
10^6 km			10^{15} s		Geological ages
10^5 km	– Earth–Moon system (tides)		10^{14} s	Ma	Intervals between speciation events
10^4 km	Global tectonic processes		10^{13} s		Reversals of magnetic polarity / Climate cycles
1000 km			10^{12} s		
100 km	Widths of sedimentary basins		10^{11} s	ka	
10 km			10^{10} s		
1 km	Depths of sedimentary basins		10^9 s		Life-spans of organisms / Intervals between major depositional events
100 m	Sequences of strata		10^8 s	years	
10 m	Beds		10^7 s	months	Annual cycles
1 m	Sedimentary structures		10^6 s	days	Tidal cycles
100 mm			10^5 s		
10 mm	Gravel		10^4 s		Storm events and mass flows
1 mm	Sand		1000 s	hours	Formation of sedimentary structures
100 µm			100 s	minutes	
10 µm	Clay		10 s		Rock falls and avalanches
1 µm			1 s		

Figure 39 - Section 17.4: Magnitude of Geological Process in Space and Time (Nichols, 2003, p.3)

"Categories of environment consist of end members and points along a spectrum" (Nichols, 2003, p. 6). In Figure 39 - Section 17.4, a magnitude of geological processes in space and time are compiled displaying "the published geological time-scales (Table 20:2; Harland *et al*, 1989) have been constructed by integrating information from biostratigraphy, magnetostratigraphy and data from radiometric dating to determine the chronostratigraphy of rock units throughout the Phanerozoic" (Nichols, 2003, p. 255).

"Sediments accumulate and rocks formed in different environments pile up on top of each other to provide the stratigraphic record of these changes in environment" (Nichols, 2003, p.7). "In a simple, undeformed succession of beds, the upper beds are younger than the lower beds" (Nichols, 2003, p.7). "The geological time-scale constructed from stratigraphic information present in rocks provides us with the temporal framework for events in Earth history" (Nichols, 2003, p.7). Figure 39 - Section 17.4 depicts the magnitudes of geological processes in space and time. "Successions of sedimentary rocks indicate how regions of accumulation (sedimentary basins) have formed and filled" (Nichols, 2003, p.7). "Stratigraphy provides the record of Earth history and with it much of the evidence for the way the planet works as a physical, chemical and biological unit" (Nichols, 2003, p.7). "Global Stratigraphic Age (GSSA)" (Gradstein, *et al.*, 2004, p. 26). "The debates about the causes of the extinction of major groups such as the dinosaurs are all based on interpretation of physical, chemical and biological evidence found in the stratigraphic record" (Nichols, 2003, p.9).

The following is a direct quote from A Geologic Time Scale 2004:

"A Geologic Time Scale (GTS2004) is presented that integrates currently available stratigraphic and geochronologic information. Key features of the new scale are outlined, how it was constructed, and how it can be improved. Major impetus to the new scale was provided through: (a) advances in stratigraphic standardization and refinement of the International Chronostratigraphic Scale; (b) enhanced methods of extracting linear time from the rock record, leading to numerous high-resolution ages; (c) progress with the use of global geochemical variations, Milankovitch climate cycles, and magnetic reversals as important stratigraphic tools; (d) improved statistical techniques for extrapolating ages and associated uncertainties in the relative stratigraphic scale, using high-resolution biozonations, including composite standards, that scale stages."

GTS2004 is Problematic. "This sparse skeleton of age control, especially prior to ~30 Ma (as of 2004), leaves considerable room for interpolation in construction of a geologic time scale" (Gradstein, *et al.*, 2004, p. 455). "In some previous time scales, stages are simply assumed to be of more-or-less uniform duration or scaled according to the number of biozones that they contain (Boucot, 1975; McKerrow *et al.*, 1985; Harland *et al.*, 1990)" (Gradstein, *et al.*, 2004, p. 177).

Interpolation of geologic time or factual observeable evidence? Two choices manifest themselves to the astute scientist, choose interpolation or choose newly presented facts to determine the age of the strata. In the case of Kidd Copper extracting linear time from the multiple layers using the idea that each layer was a different time period from the rock record would give an inaccurate date. The tailings deposit did not exist prior to the 1960's, so why would it be thousands of years old? The deposits were laid in ~1970's and thus the age of the layered deposits is less than 40 years old. In the case of Kidd Copper no interpolation is required to extract the strata's age.

17.5 Assumptions Underlying the Global Chronostratigraphic Scale

"The development of new dating methods and the extension of existing methods has stimulated the need for a comprehensive review of the geologic time scale. The construction of geologic time scales evolved as a result of applying new ideas, methods, and data" (Gradstein, *et al.*, 2004, p. 3). For example, "Permian biostratigraphy has been greatly refined over the last two decades" (Gradstein, *et al.*, 2004, p. 250). "We ought to make the comparison within equal times; and this we are incapable of doing" (Darwin, 1958, p. 108).

"The geologists who followed Hutton realized that the earth is very ancient, but they lacked a way to determine exactly how old it is. All they could do was determine the *relative ages* of past events – that is, how old one rock formation or geologic feature was in comparison to another. In other words, they were concerned with establishing the chronological sequence of events in Earth's history. Nineteenth-century geologists built a geologic time scale on relative ages. In doing so, they took the essential first steps toward unraveling the history of Earth" (Murck, 2001, p. 49).

17.6 The New Geologic Timescale, Named GTS 2004

"The new geologic timescale, named GTS2004 [underwent]… "a total of 18 senior and 22 contributing authors, for a total of 40 geoscience specialists from 15 different countries, would work on the book and deal with the new time scale" (Gradstein, *et al.*, 2004, p. xv). "Information on ICS, its organization, its mandate, and its wide-ranging geoscience program can be obtained from www.stratigraphy.org." (Gradstein, *et al.*, 2004, p. xvi).

"The fascination in creating a new geologic time scale is that it evokes images of creating a beautiful carpet, using many skilled hands. All stitches must conform to a pre-determined pattern, in this case the pattern of physical, chemical, and biological events on Earth aligned along the arrow of time" (Gradstein, *et al.*, 2004, p. xv). "The development of the subdivisions of geologic time is a fascinating topic filled with philosophical concepts, dramatic personal disputes, and perceptive geological thinking" (Gradstein, *et al.*, 2004, p. 43).

"Comments received subsequent to this publication were reviewed at a meeting of the SPS in 1988, and the proposed time scale refined into a final recommendation. Subsequent postal ballots, first by SPS and then by ICS during early 1989, achieved in excess of the required majority of 60%, and the proposal was submitted to the International Union of Geological Sciences (IUGS) for ratification at the 28th International Geological Congress. Discussion was deferred to a later meeting in early 1990, at which time IUGS ratified the proposal as the recommended international time scale for the chronometric subdivision and nomenclature of Precambrian time and of the Proterozoic Eon" (Gradstein, *et al.*, 2004, p. 130).

A Geologic Time Scale 2004 – "The geologic time scale is the framework for deciphering the history of the Earth" (Gradstein, *et al.*, 2004, p. 3). "*A Geologic Time Scale 2004* (GTS2004) provides an overview of the status of the geological time scale and is the successor to GTS1989 (Harland *et al*, 1990), which in turn was preceded by GTS1982 (Harland *et al*, 1982)" (Gradstein, *et al*., 2004, p. 3).

"This study presents the science community with a new geologic time scale for circa 3850 million years of Earth history. This scale encompasses many recent advances in stratigraphy, the science of the layering of strata on Earth. The new scale closely links radiometric and astronomical age dating, and provides comprehensive error analysis on the age of boundaries for a majority of the geologic divisions of time. Much advantage in time scale construction is gained by the concept of a stage boundary definition, developed and actively pursued under the auspices of the International Commission on Stratigraphy (ICS), that co-sponsors this study" (Gradstein, *et al*., 2004, p. xv).

"Volcanic ash layers play an important role in establishing time stratigraphic correlations on a regional scale" (Gradstein, *et al*., 2004, p. 425).

"Strontium isotope stratigraphy provides a powerful tool to correlate and date marine sequences in the Cenozoic (Hess *et al*., 1989), especially when other stratigraphic tools fail, such as at high latitudes or in carbonate platforms (e.g. Ohde and Elderfield, 1992)" (Gradstein, *et al*., 2004, p. 426, 427).

Scientific Community Decisions. (1) "Decrease the age of the Rupelian-Chattian boundary by more than 1 myr" (Gradstein, *et al*., 2004, p. 408); (2) "Defining the Neogene Time Scale" … "the scientific community was left with three different options for defining the Tertiary– Quaternary boundary at the end of the nineteenth century" (Gradstein, *et al*., 2004, p. 409, 411); (3) "Notwithstanding formal ratification of the Pliocene–Pleistocene Boundary at Vrica, another decade and a half of heated discussions followed, ending with a formal attempt to lower the base of the Pleistocene in favor of the third option" (Gradstein, *et al*., 2004, p. 411); (4) "After a postal ballot within the subcommissions on Neogene Stratigraphy and Quarternary Stratigraphy (Rio *et al*, 1998), the proposal to lower the base of the Pleistocene was rejected" (Gradstein, *et al*., 2004, p. 412); (5) "*Aquitanian* GSSP, *base-Miocene Series, base-Neogene System* … Tuning of ODP Site 926 to the La2003 solution resulted in an age for the base of the Miocene of 23.03 Ma. This age is 0.77 myr younger than in the Berggren *et al*. (1995b) time scale" (Gradstein, *et al*., 2004, p. 412, 413); (6) Gelasian: Upper Pliocene – "The GSSP for the base of the Gelasian was ratified in August 1996" (Gradstein, *et al*., 2004, p. 417); (7) Macrofossil Zonations – "The uncertainty in the ages of most zonal boundaries is often considerable" (Gradstein, *et al*., 2004, p. 419). Notice how these decisions were made – postal ballot and by ratification taking into account the uncertainty in the ages. No standard could be given in which to compare the numbers of time, men had to guess.

The Timescale 2004 (GTS2004) is not standardized. "The construction of mammal-based chronological systems for other continents still lags behind those for Europe and North America… The nature of Australian mammalian record is patchy (Archer *et al*., 1995)" (Gradstein, *et al*., 2004, p. 420). Sequence stratigraphy – "rely on the assumption of a global or eustatic control on sequences, especially third-order episodes. This assumption has been strongly criticized (e.g. Miall, 1992)" (Gradstein, *et al.*, 2004, p. 424).

Timescale 2004 (GTS2004) and More Problems. "Major problems concerning the magnetostratigraphy lie in interval between 13 Ma and the Oligocene–Miocene boundary" (Gradstein, *et al.*, 2004, p. 424). "Over the last 20 years, it has become apparent that calculation of the radiocarbon age of a specific sample assuming a constant production of ^{14}C in atmospheric CO_2 is not valid, but that the activity of ^{14}C in atmospheric CO_2 has varied over time" (Gradstein, *et al.*, 2004, p. 426). "That astronomical ages for the Oligocene–Miocene boundary interval are approximately 900 kyr younger than in most recent geological time scales (e.g. Berggren *et al.*, 1995b)" (Gradstein, *et al.*, 2004, p. 433).

"The climate friction effect, due to the change of momentum of inertia of the Earth arising from the change of ice load on the polar caps during ice ages has been estimated (Levrard and Laskar, 2003), but neglected as it was found to be too small and too uncertain to be taken into account" (Gradstein, *et al.*, 2004, p. 433).

Incorporation of Zonal Schemes – 'The incorporation, for instance, the North American Land Mammal Ages (NALMAs) to the new time scale is problematical due to lack of fossil sites in long and continuous sections having a well-calibrated magnetostratigraphy" (Gradstein, *et al.*, 2004, p. 438).

"Chemical fossil – Any of various organic compounds found in ancient geological strata that appear to be biological in origin and are assumed to indicate that life existed when the rocks were formed. The presence of chemical fossils in Archaean strata indicates that life existed over 3500 million years ago" (Isaacs, *et. al.*, 2003, p. 149, 150). "The lack of a diverse and well-preserved fossil record, the generally decreasing volume of supercrustal rocks, and increasing degree of metamorphism and tectonic disturbance, as well as the uncertainties in the configuration and assembly of the continents, all contribute to making the establishment of a chronostratigraphic time scale beyond the Phanerozoic Eon problematical" (Gradstein, *et al.*, 2004, p. 129).

Classification is a Matter of Debate. "The classification and interpretation of the youngest stratigraphic sequences, variously known as Pleistocene, Holocene and Quaternary, have been, and still are, a matter of debate" (Gradstein, *et al.*, 2004, p. 441).

Origin of Short-term Events. "Dansgaard–Oeschger events occur at the high-frequency end of the sub-Milankovitch band and are therefore more difficult to explain in terms on non-linear response to Milankovitch forcing. The origin of these short-term events is still puzzling" (Gradstein, *et al.*, 2004, p. 429).

"As a consequence, we decided to build the present Neogene time scale as much as possible on astronomical calibrations, this despite the fact that the tuning has not always been independently verified and magnetostratigraphic records are still lacking for the early and middle Miocene. We think that this is the only acceptable way to proceed in view of a strong tendency to incorporate increasingly older parts of the astrochronologic framework into the standard geological time scale" (Gradstein, *et al.*, 2004, p. 432).

The question this report raises is why have the calibrations not been independently verified? Science is verifiable. Placing an assumption or a point of view into a foundation for a scale is not fact-based. Time is a measurement and requires measuring. A scale to measure time is provided by the Kidd Copper data. The deposits are around 40 years old. These can be measured against the layers found in the log profiles, which show that multiple layers can form rapidly. The result is that reading separate layers in the rock record are an inaccurate time measurement.

17.7 Limitations of GTS 2004

"The geologic time scale is the framework for deciphering the history of the Earth. Since the time scale is the tool 'par excellence' of the geological trade, insight in its construction, strengths, and limitations greatly enhances its function and its utility. All earth scientists should understand how the evolving time scales are constructed and calibrated, rather than merely using the numbers in them" (Gradstein, *et al.*, 2004, p. 3). "Increased knowledge about these relationships and refinements in the techniques of radiometric dating have led to a number of modifications over the years, and there is every reason to believe that the time-scale will be amended again in the future with the boundaries between units becoming younger or older with these changes" (Nichols, 2003, p. 255).

17.8 Calibration of Linear Time of Events in the Rock Record

"This calibration to linear time of the succession of events recorded in the rock record has three components: (1) the international stratigraphic divisions and their correlation in the global rock record; (2) the means of measuring linear time or elapsed durations from the rock record; and (3) the methods of effectively joining the two scales" (Gradstein, *et al.*, 2004, p. 3).

"The chronometric calibration of stratigraphic boundaries underpins the geologic time scale" (Gradstein, *et al.*, 2004, p. 87). "The recommendations were quickly and universally accepted and remain to this day a prime underpinning in any age determination. Uncertainties in these constants can be ignored when comparing ages derived *within a single isotopic system* because the uncertainty represents a systematic error that affects all ages equivalently. Thus analytical precision and accuracy become the sole determinants of age equivalence between samples when the same isotopic ratios are analyzed. However, the nature of this error changes when comparisons are made *between* isotopic systems ... uncertainties on the accepted values need to be considered when the result is a comparison to absolute time. Furthermore, the IUGS Subcomission on Geochronology clearly stated that the 'selected values are open and should be the subjects of continuing critical scrutinizing" (Gradstein, *et al.*, 2004, p. 87, 88).

"TIMS dates only are used here for calibrating the Cambrian period" (Gradstein, *et al.*, 2004, p. 159). Cambrian – "Cambrian fossils are all marine animals; they include trilobites, which dominated the Cambrian seas, echinoderms, brachiopods, mollusks, and primitive graptolites (from the mid Cambrian). Trace fossils also provide evidence for a variety of worms" (Isaacs, *et. al.*, 2003, p. 121).

Base of the Lower – Middle Triassic Zone: Zircon and Argon calculations differ. "The cause of the discrepancy among ages in the upper Anisian–lower Ladinian interval obtained by different U–Pb studies and the $^{40}Ar/^{39}Ar$ versus U–Pb results is not resolved" (Gradstein, *et al.*, 2004, p. 305).

"Finally, data reduction and error analysis methodologies are still undergoing review because of the relatively small number of HR–SIMS installations. All measurements of the $^{206}Pb/^{238}U$ ratio depend primarily on counting statistics, in addition to an estimate of errors related to the scatter about the Pb–U calibration line. It is uncertain whether calibration can be accurately determined using statistical error analysis on measurements" (Gradstein, *et al.*, 2004, p. 90).

17.9 Stratigraphic Column and the Stratigraphic Record

"The scientific study of rock strata is called *stratigraphy*, and it was through the principles of stratigraphy that the geologic time scale was developed" (Murck, 2001, p. 49). "Sedimentary rocks are often classified according to the way they have formed and according to the size of the particles in them" (Zim, 1961, p. 32). "The stratigraphic record provides us with only a few clues with which to build up a history of the Earth. "The record is incomplete on all scales, from the years in between the events when layers of sediment are deposited, to hiatuses within successions representing hundreds or thousands of years, to the millions of years represented by major unconformities" (Nichols, 2003, p. 326).

"During the nineteenth century, the history of Earth was encoded into a hierarchical scale of time units of varying duration. The hierarchy is known as the Stratigraphic Column" (Discovery Books, 1999, p. 36). "The chronometric calibration of stratigraphic boundaries underpins the geologic time scale" (Gradstein, *et al.*, 2004, p. 87). The new geologic time scale is known as GTS 2004 that this report references.

"After careful mapping of all the world's known rock units, a theoretical continuous stack, called the stratigraphic column, was constructed. From this, it is easy to determine the age of a particular rock unit relative to others" (Busbey, *et. al.*, 1999, p. 20). Point to ponder. If the stack in the rock record is disjointed between continents as presently supported by geologists is it due to a differentiation in time or rather is the discontinuity due to heterogeneous layering on the Earth's surface in a one-time deposition? This opens the doorway to a 'younger earth'.

17.10 Global Stratotype Section and Point (GSSP)

"This precise reference point for each boundary is known as the Global Stratotype Section and Point (GSSP)" (Gradstein, *et al*., 2004, p. 23). "Absence of precise global markers for high-precision time scale work is a key problem that was glossed over in some GSSP decisions" (Gradstein, *et al*., 2004, p. 27). "Since the global chronostratigraphic scale is ultimately defined by a complete sequence of GSSPs, the limits of chronostratigraphic units (stages) are fully defined in time" (Gradstein, *et al*., 2004, p. 41). The system "cannot define the period because the system boundaries are unknowable except at unconformities where the boundary rocks of uncertain age are missing" (Gradstein, *et al*., 2004, p. 41). "A general disadvantage of chronogram methods is that the relative stratigraphic position of the sample is generalized with respect to stage boundaries that are relatively far apart in time" (Gradstein, *et al*., 2004, p. 107). "The GSSPs cannot be located at an unconformity without introducing the problem of 'missing time'" (Gradstein, *et al*., 2004, p. 146).

17.11 Devonian-Carboniferous Discrepancies

Devonian Time Scale. "The discrepancy between U–Pb TIMS and HR–SIMS dates in the Devonian-Carboniferous is resolved by taking the youngest possible estimate of age dates with the former method and the oldest possible estimate of age dates with the latter one. Essentially, a regression line is forced through the opposite extremes of error bar values for successive age dates to interpolate stages" (Gradstein, *et al*., 2004, p. 45). "The unusually low analytical error strongly contrasts with the stratigraphic range spanning late Farsnian and possibly well into Famennian time; thus, the date does not contribute to the Devonian time scale" (Gradstein, *et al*., 2004, p. 215).

"A new ^{207}Pb / ^{206}Pb date for the Hunsrück Slate (Kirnbauer and Reischmann, 2001) gives an age of 388.7 ± 1.2 Ma for early Emsian strata. This is much younger than can reasonably be expected from the preferred regression used, and hence is rejected" (Gradstein, *et al*., 2004, p. 215). Why? Is the data not supposed to be read to formulate a theory; rather than the data made to support the theory?

17.12 Precambrian Polarity Sequence is Poorly known and Poorly Dated

Current Precambrian divisions and problems. "Correlation of such boundaries between distant sections, on the basis of even our best geochronometers (U–Pb ages on single zircons), can be no better than ±5-10 million years (relative to absolute ages), even if all other sources of uncertainty (e.g. analytical scatter, Pb loss, or cryptic inheritance) are negligible" (Gradstein, *et al*., 2004, p. 141, 142). "Precambrian polarity sequence is poorly known and poorly dated" (Gradstein, *et al*., 2004, p. 86). "The task of establishing a rigorously defined and globally acceptable time scale for the Precambrian is an exceedingly difficult, and often frustrating, exercise. The reason for this is related to the fact that studying the Earth becomes increasingly difficult and uncertain the further one goes back in geological time" (Gradstein, *et al*., 2004, p. 129).

17.13 Radiogenic Isotope Geochronology Handled Disparate Results

"High-precision intercalibration studies" … "These error levels are approximations based upon what could reasonably be expected during routine analysis and through propagation of external uncertainties. If the authors detail specialized protocols of analysis or error handling, their quoted error limits may be substituted. Disparate results were not simply averaged. From a statistical standpoint, such an approach is unwarranted, unless results pass Student's *t*-test, indicating a high probability that they represent the same population" (Gradstein, *et al*., 2004, p. 95). "Where original data were reported to other time scales, the original ages have been converted to the current time scale using the formulae of Wei (1994)" (Gradstein, *et al*., 2004, p. 100). "Furthermore, age models are ultimately based (mostly) on radiometric dates and are as accurate as those dates. Interpolation of ages between tie points, however, may be more precise, although necessarily systematically inaccurate" (Gradstein, *et al*., 2004, p. 100).

17.14 Geomathematics Assumes a Hypothetical Trial Age

Geomathematics includes statistical, chronological and graphical methods. "New challenges for construction of the GTS2004 numerical time scale include: (1) use of new age determinations that can be an order of magnitude more precise than earlier dates" (Gradstein, *et al*., 2004, p. 107). "Assuming a hypothetical trial age for an observed chronostratigraphic boundary, rock samples from above this boundary should be younger, and those below it should be older. An inconsistent date is either an older date for a rock sample known to be younger than the trial date, or a younger date for a sample known to be older. The difference between each inconsistent date and the trial age can be standardized by dividing it by the standard deviation of the inconsistent date. Thus relatively imprecise dates receive less weight than more precise dates" (Gradstein, *et al*., 2004, p. 107).

"Assuming a hypothetical trial age for an observed chronostratigraphic boundary, rock samples from above this boundary should be younger, and those below it should be older. An inconsistent date is either an older date for a rock sample known to be younger than the trial date, or a younger date for a sample known to be older. The difference between each inconsistent date and the trial age can be standardized by dividing it by the standard deviation of the inconsistent date. Thus relatively imprecise dates receive less weight than more precise dates. The underlying assumptions are that: (1) the rock samples are uniformly distributed along the time axis, and (2) the error of each date satisfies a normal (Gaussian) error distribution with standard deviation equal to that of the age determination method used. Standardized differences between inconsistent dates and trial age can be squared and the sum of squares (written as E^2) determined for all inconsistent dates corresponding to the same trial age" (Gradstein, *et al.,* 2004, p. 107).

Inconsistency and assumptions underly the construction of the GTS2004 numerical time scale and thus expose problems with the current method of dating the age of the earth. With all of these acknowledged assumptions, does this not leave a measurable amount of error in dating the age of the Earth as old?

17.15 The Chi-Square Test

"The chi-square test of residuals indicated that the 0.5-myr error bar on the 390.0 Ma monazite date (no. 6) was too narrow, and an 'average' error bar was used instead. This, together with the large error bars on dates 8 and 9 reduces the duration of the Emsian. A straight-line fit, as produced by Tucker *et al.* (1998), would substantially increase the duration of the Emsian Stage" (Gradstein, *et al.*, 2004, p. 215).

Who determines the best-line-of-fit to determine the resultant age of the Earth?

17.16 The Optimum Choice of Age Agterberg Improved

"The optimum choice of age was at the trial age where E^2 was a minimum" (Gradstein, *et al.*, 2004, p. 107). Why would the Nineteenth-century geologists choose this and what criteria would they select as their base? Is the answer to combine inconsistent dates with consistent dates? "Agterberg (1988) made the following improvement to this method. In addition to inconsistent dates, there are generally more consistent dates for any trial age selected for a chronostratigraphic boundary. The statistical maximum likelihood method can be used to combine consistent with inconsistent dates resulting in an improved estimate of the age of the chronostratigraphic boundary considered. Each standardized difference with respect to a trial age was interpreted as the fractile of the normal distribution in standard form, and transformed into its corresponding probability. Summation of the logarithmically transformed probabilities then yields the log-likelihood value of the trial date. In this type of calculation, inconsistent dates receive more weight than consistent dates. Consequently, the improvement resulting from using consistent dates, in addition to the inconsistent dates, is relatively minor" (Gradstein, *et al.*, 2004, p. 107). Is science to be supported by combining inconsistent data with consistent data or should scientists determine to be consistent, such as in the case of Kidd Copper throughout this scientific report?

Loreen Michele Sherman

235

17.17 Methods for Age Estimations

"Final estimates of the ages of the Ordovician and Silurian stage boundaries and durations are given … the method used to estimate 2-sigma values … Summarizing, it can be said that the 2-sigma value of a stage base is equal to the product of the age estimate (millions of years) and an age uncertainty factor that amounts to 0.00345 for the Orodovician-Silurian data sets … Ordovician 2-sigma values were calculated from the uncertainty factors for the Ordovican-Silurian data set. To provide a smooth transition from Orodovician to Devonian uncertainties, Silurian 2-sigma values were increased toward the base of the Devonian by linear interpolation between base-Rhuddanian (the basal Silurian stage) and base-Lochkovian (the basal Devonian stage), a process described as 'ramping'" (Gradstein, *et al.*, 2004, p. 113).

"The other Gorstian date (420.7± 2.2 Ma) is an average age taken from the K–Ar and Rb-Sr dates of Wyborn *et al.* (1982)" (Gradstein, *et al.*, 2004, p. 113). "Because dating methods have improved significantly during the past ten years, previously accepted dates may have large errors that preclude their use in the time scale" (Gradstein, *et al.*, 2004, p. 114). "The preceding correction of setting the average square of scaled residuals equal to unity results in slightly wider error bars" (Gradstein, *et al.*, 2004, p. 114). "There is the implied assumption that all *s*(*t*) values are unbiased" (Gradstein, *et al.*, 2004, p. 114).

Table 20 - Section 17.17: Carboniferous – Permian Spline-Curve Cross-Validation

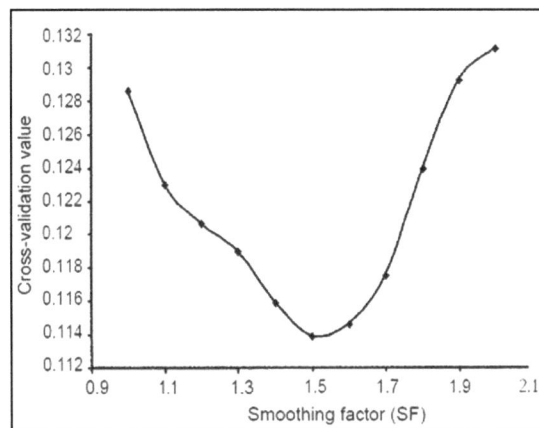

Table 20 - Section 17.17 shows the corrections to the Permian-spline-curve after statistical analysis which shows the smoothing factor of the Carboniferous-Permian-spline-curve cross-validation (Gradstein, *et al.*, 2004, p. 116).

"However, it is better to replace the base-Devonian estimate 416.7 ± or 2.9 Ma by 416.0 ± 2.8 Ma, resulting from statistical analysis of the Ordovician-Silurian data set. This replacement has already been made in calculation of the duration of the Lochkovian" (Gradstein, *et al.*, 2004, p. 115). "Permian uncertainties were based on

Copyright 2010 Star-Ting, Inc. All rights reserved. Worldwide.   1.877.896.7292   http://www.star-tingpublishing.com

Carboniferous-Permian MLFR results, but ramping was used in the Carboniferous to provide a gradual transition between uncertainties for base-Asselian and the larger Devonian uncertainties" (Gradstein, *et al.*, 2004, p. 115).

"A way out of this dilemma is to use the five dates in the immediate vicinity of the boundary … The best-fitting slpline then is the straight line … the estimate based on … The new estimate … This new estimate would imply that the rate of sedimentation remained constant between 8 cm above and 18 cm below the boundary, respectively… The slope of the line is very steep indicating that the sedimentation rate at the boundary was much less than at the first two input points and at the last input point" (Gradstein, *et al.*, 2004, p. 117). "However, this estimate was enlarged to ±0.4 myr for the following reason. On average, the China P–T dates have a 1-sigma error of 0.0537%" (Gradstein, *et al.*, 2004, p. 117). "For this reason, the TIMS dates at the P–T boundary were multiplied by approximately 1.86 to account for limiting decay-constant effect. This gives a final P–T age estimate of 251.05±0.4 Ma" (Gradstein, *et al.*, 2004, p. 117). "Cross-validation … is a good choice for the smoothing factor" (Gradstein, *et al.*, 2004, p. 118). "Because the sum of squared residuals is theoretically distributed as chi-square with (*n*-2) degrees of freedom" (Gradstein, *et al.*, 2004, p. 118).

"Final age estimates for Late Cretaceous stage boundaries as commonly assigned in the ammonite zonation of the US Western Interior and durations of stages" are given in Table 21 - Section 17.17 (Gradstein, *et al.*, 2004, p. 119).

Table 21 - Section 17.17: Carboniferous – Devonian – Silurian

Period	Stage	Base (Ma)	Duration (myr)
Carboniferous		359.2 ± 2.5	
Devonian			
	Famennian	374.5 ± 2.6	15.3 ± 0.6
	Frasnian	385.3 ± 2.6	10.8 ± 0.4
	Givetian	391.8 ± 2.7	6.5 ± 0.3
	Eifelian	397.5 ± 2.7	5.7 ± 0.2
	Emsian	407.0 ± 2.8	9.5 ± 0.4
	Pragian	411.2 ± 2.8	4.2 ± 0.2
	Lochkovian	416.7 ± 2.9	4.8 ± 0.2
Silurian			

a Estimates of uncertainty in 2-sigma.

"Only the first of these adjustments have been previously applied to Paleozoic data sets. The second adjustment is replaced by an error bar correction for relatively few individual dates" (Gradstein, *et al.*, 2004, p. 119). "It is increasingly widely recognized that the quoted errors should also include external errors associated with measurements of the K–Ar age of the fluence monitor and errors related to the determination of the decay constants. This becomes essential when ages from different systems are used together for time scale construction" (Gradstein, *et al*., 2004, p. 120).

17.18 Spline Curve Fitting Disadvantages

   Fitting Age Dates. "A possible disadvantage of spline-curve fitting is that age dates included in the data set are weighted according to their variance. Thus the size of the error bars constitutes important input for obtaining valid results" (Gradstein, *et al*., 2004, p. 124). "It is not permissible to select an older age from within the confidence interval for the top and combine it with a younger age from within the confidence interval for the base" (Gradstein, *et al*., 2004, p. 124). Fitting age dates is based on the slope of a straight line and interpolation, which means look at the dots on the graph from the findings and recognize if the data fits. The size of the error bars determines the fit. It seems strange to base a date on margins of error, such as when "the experiment seems to confirm that the estimate based on fitting $Y` = a + bx$ is reasonable although on the conservative side" (Gradstein, *et al*., 2004, p. 125).

   "A body of strata that is identified by particular fossil content is a *biostratigraphical unit*. Biozone (*biostratigraphical zone*) is the term used for any biostratigraphical unit" (MacDonald, *et al.,* 2003, p. 414). "… found basic flaws in the biostratigraphic assumptions in relation to the series boundary data points" (Gradstein, *et al*., 2004, p. 214).

   Carboniferous Time Scale… "As with the Devonian, there is a limited number of good-quality and chronostratigraphically precise dates within the period. As a result, various assumptions have been made about the duration of stages or of zones. For example, GTS89 in Harland *et al*. (1990) assigned equal duration to 50 biochrons from the foraminiferal zones in the Donetz Basin" (Gradstein, *et al*., 2004, p. 237).

   Radiometric Data. "Hence, the HR-SIMS dates are used with caution" (Gradstein, *et al*., 2004, p. 237). And even Charles Darwin stated, "From these several considerations, it cannot be doubted that the geologic record, viewed as a whole, is extremely imperfect" (1958, p. 304).

17.19 Carboniferous and Permian Composite Standard

   "There is a significant carbon excursion just below the Permian–Triassic Boundary. This excursion is intimately related to the final extinction event of the Permian, but it is unclear just how or what it completely reflects" (Gradstein, *et al*., 2004, p. 264).

   Incomplete Records are Adjusted. "Most of the sections may include potential defects, such as changes in sedimentation rate, discontinuities, or an incomplete fossil record. Hence, the resulting scale was adjusted by adding several more complete sections from the Urals, Donetz Basin, Cantabrian Mountains, and Central Asia" (Gradstein, *et al.*, 2004, p. 244). Why was a scale adjusted to accommodate the pre-set expectations by the 19th Century geologists?

17.20 Decay Constants and Their Uncertainties

Bias of Decay Constants. "A second problem that has been identified only relatively recently is that the decay constants themselves may be biased" (Gradstein, *et al.*, 2004, p. 120). "Unfortunately, it is not possible at this time to revise the 2-sigma error bars on all dates published in the early 1990s in order to incorporate the external uncertainty. Possible bias of the decay constants is unknown at present and cannot be considered at all" (Gradstein, *et al.*, 2004, p. 120). "If the published error bar of a $^{40}Ar/ ^{39}Ar$ is too large because external uncertainty was not considered, it can be adjusted to a median or average value" (Gradstein, *et al.*, 2004, p. 120).

Over-estimations are practiced. "It is noticed that duration uncertainty correction 1 is on the conservative side (slightly too large) for younger stage boundaries in the data set" (Gradstein, *et al.*, 2004, p. 120).

"Also because ^{40}K decay constants and their uncertainties will probably be revised, it is better to overestimate than to underestimate time scale uncertainties associated with $^{40}Ar / ^{39}Ar$ ages" (Gradstein, *et al.*, 2004, p. 124). If the decay constants do not work with the younger then how can we assume the older is correct?

17.21 Neoproterozoic Paleontology, Isotopes or Climate History

"Many uncertainties attend stratigraphic interpretations of Neoproterozoic paleontology, isotopes, and climate history. There is no consensus on the number of Neoproterozoic ice aes or the correlation of individual tillites" (Gradstein, *et al.*, 2004, p. 133). *Oxygen isotope stratigraphy* too has problems. "The extent to which seawater $\delta^{18}O$ can have changed over geological history is still a matter of controversy" (Gradstein, *et al.*, 2004, p. 135). With regards to the Neoproterozoic Ice Ages and chronometric constraints, "at present it is unclear how many glaciations occurred during the Neoproterozoic Era or whether all these glaciogenic deposits represent times of global glaciation (Evans 2000) " (Gradstein, *et al.*, 2004, p. 139). "At present, there is no evidence for any early Neoproterozoic, i.e. pre-750 Ma glaciation, which had been envisaged previously (Eyles and Young, 1994)" (Gradstein, *et al.*, 2004, p. 139).

17.22 Arbitrarily Defined Geochronomety

"More than seven-eighths of Earth history, between ~4500 and 500 Ma, is subdivided into segments of geological time that are more arbitrarily defined in terms of geochronometry" (Gradstein, *et al.*, 2004, p. 140). In Regional Cambrian stage suites "boundary stratotypes are therefore not designated" (Gradstein, *et al.*, 2004, p. 153). In the Pre-Ordian, "stages have not yet been designated" (Gradstein, *et al.*, 2004, p. 153).

"The magnetostratigraphic record cannot yet be precisely tied to international stage boundaries" (Gradstein, *et al.*, 2004, p. 174, 175).

"But given that the Ordovician isotopic trends are patched together from a variety of world locations, the variable rate of change in ^{87}Sr / ^{86}Sr through the interval might well be caused by correlation difficulties and artifacts of the time scale" (Gradstein, *et al.*, 2004, p. 176). "The Ordovician chronostratigraphic scale is well constrained by isotopic dating between 453 amd 457 Ma ... but elsewhere is poorly constrained. This poses a problem for interpolating stage boundaries into the linear scale of geologic time" (Gradstein, *et al.*, 2004, p. 177).

"Two alternative calibrations for the Ordovician ... until this problem is resolved, we use only the conventional TIMS dates in calibration" (Gradstein, *et al.*, 2004, p. 177).

17.23 Triassic & Jurassic Periods

Triassic & Jurassic Periods do not have a standardized ammonoid zonation. *Included but with caution.* "It has proven difficult to correlate potential definitions for the bases of the Kimmeridgian and Tithonian stages, and long-held traditions of regional equivalence have proven to be erroneous" (Gradstein, *et al.*, 2004, p. 323). "We have included the genera of the ammonite 'index' species, but with caution that these are not always the 'guide' species of the named zone" (Gradstein, *et al.*, 2004, p. 327).

Triassic has No Standard Zonation. "Despite their historical importance in subdividing the Triassic, there is not yet a standardized ammonoid zonation (or nomenclature) for Alpine regions" (Gradstein, *et al.*, 2004, p. 283). "The Alpine ammonoid scale is modified from a compilation by van Veen and Gianolla (1998)" (Gradstein, Ogg & Smith, 2004, p. 284). 'But there is no consensus on aspects of the nomenclature and inter-correlations" (Gradstein, *et al.*, 2004, p. 284).

Jurassic is Incomplete. For example, in the Lower Jurassic "widespread hiatuses or condensation horizons mark the bases of the classical Sinemurian, Pliensbachian, and the Toarcian stages" (Gradstein, *et al.*, 2004, p. 308). "There is no accepted groupings into substages of the three Hettangian ammonite zones" (Gradstein, *et al.*, 2004, p. 309). For example in the Middle Jurassic, Bajocian – "Ammonite lists of d'Orbigny indicate that he erroneously assigned species of the upper Toarcian to the lower Bajocian and vice versa" (Gradstein, *et al.*, 2004, p. 321). Bathonian – "where strata characterized by oolitic limestone are exposed in a number of quarries, but these are incomplete and lack adequate characterization by ammonites (Torrrens, 1965)" (Gradstein, *et al.*, 2004, p. 321). "The compact Bathonian–Callovian boundary interval is bounded by unconformities and may contain minor hiatuses... Therefore, this suggested GSSP has not been adopted by the International Stratigraphic Commission" (Gradstein, *et al.*, 2004, p. 322). "A suite of Bathonian and Bajocian sections in Spain displays a rapidly changing magnetic polarity pattern (Steiner *et al.*, 1987), but this has not yet been fully verified elsewhere" (Gradstein, *et al.*, 2004, p. 335).

Upper Jurassic Erroneous. "However, it has proven difficult to correlate potential definitions for the bases of the Kimmeridigian and Tithonian stages, and long-held traditions of regional equivalence have proven to be erroneous" (Gradstein, *et al.*, 2004, p. 323). Kimmeridgian and Tithonian – "this equivalency has not been incorporated into our Jurassic time scale, pending satisfactory consistency with other correlation constraints" (Gradstein, *et al.*, 2004, p. 338).

"Calibration of the radiolarian assemblages to standard geological stages and reference ammonite scales of Europe has been challenging and controversial" (Gradstein, *et al.*, 2004, p. 334). "Possible contributing factors to this divergence are diachroneity of radiolarian or ammonite datums and ranges among basins, errors in taxonomy assignments, imprecise correlation of radiolarian markers to regional ammonite stratigraphy, and miscorrelation of ammonite assemblages among paleogeographic provinces (Pessagno and Meyerhoff Hull, 1996)" (Gradstein, *et al.*, 2004, p. 335).

Albian Stage has Discrepancies. "Mutterlose *et al.* (2003) conclude that definition of 'the base of the Albian on these criteria cannot be used in a global context,' therefore an alternate biostratigraphic or non-biostratigraphic marker must be identified" (Gradstein, *et al.*, 2004, p. 360). "The base of the Upper Turonian is not yet formalized" (Gradstein, *et al.*, 2004, p. 363). "Some of these biostratigraphic markers have discrepancies in position relative to the reference carbon-13 curves from different sections (Wiese and Kaplan, 2001)" (Gradstein, *et al.*, 2004, p. 363).

17.24 Unconfirmed Data Basis of Stratigraphic Record

Manicougan Impact. "There has not yet been global recognition of the ejecta or other evidence of the Manicouagan impact in the stratigraphic record" (Gradstein, *et al.*, 2004, p. 288).

Santonian Substages "has three substages, but no markers for boundary stratotypes have yet been formalized. A possible datum … but this has not yet been confirmed in other sections" (Gradstein, *et al.*, 2004, p. 364). Why do unconfirmed sub-stages continue to be regarded as possible markers for the time scale? It seems unscientific to establish a time scale on missing information.

Why then do geologists use a time scale based on missing information and unconfirmed data to support the age of the Earth?

18.0 Problematic Dating Methods

"Scientists have discovered that k is not constant" (Fox, 2000, p. 301). … (the half-life constant .693/ k). This opens the question to what else have scientists discovered that is not constant?

18.1 Dating Methods

"Dating methods [are] various methods used to determine when a substance was formed. The particular method chosen depends on the nature of the material and the approximate magnitude of age. Some of the methods of dating rocks are listed in Figure 40 - Section 18.1.

\multicolumn{3}{c}{**MAJOR METHODS OF DATING ROCKS**}		
METHOD	**TYPES OF ROCKS**	**EFFECTIVE DATING RANGE**
RADIOMETRY		
CARBON-14	Sedimentary rock	70 thousand years or less
POTASSIUM-ARGON	Igneous rock, especially basalt	5 thousand years or more
RUBIDIUM-STRONTIUM	Igneous rock, especially granite	5 million years or more
URANIUM-LEAD	Igneous, sedimentary, and metamorphic rock	10 million years or more
STRATIGRAPHY		
INDEX FOSSILS	Sedimentary rock	700 million years or less
PALEOMAGNETISM	Igneous rock, especially basalt, and some sedimentary rock	3 billion years or less

The chart above shows some of the most common tests used to determine the age of a rock. To ensure accuracy, geologists will use different tests on a single sample. Radiometric dating methods can give the absolute age of certain rocks. Stratigraphy can tell the relative age of one rock when compared to another rock.

Figure 40 - Section 18.1: Major Methods of Dating Rocks, (Discovery Books, 1999, p. 34)

18.2 Problems with Dating Methods

"When you can measure what you are speaking about, and express it in numbers, you know something about it; but when you cannot express it in numbers, your knowledge is of a   rizon and unsatisfactory kind; it may be the beginning of knowledge, but you have scarcely, in your thoughts, advanced to the stage of Science, whatever the matter may be" (Hsü, 2004, p. 5).

Science is self-correcting and many problems are prevalent with the current scientific dating methods. Table 22 - Section 18.2 exposes some of these problems. For example problems exist in many of the radioactive dating methods, such as "Proton decay

– considerable experimental effort has been spent in looking for proton decay, so far with no success" (Isaacs, *et. al.*, 2003, p. 641). "Determine the potential for each of the sources of error, making corrections as appropriate" (Dingman, 2002, p. 125).

Many scientific searches for ways to date the Earth have significant problems as explained by Edward J. Tarbuck and Frederick K. Lutgens. Rate of sedimentation "was "riddled with difficulties, some of which are as follows: (1) different sediments accumulate at different rates under varying conditions. Thus, determining an overall rate of sediment accumulation is difficult. Further, if such a rate is determined, it does not necessarily mean that the same rate can be applied to the past. (2) Since no single locality has a complete geologic column, estimates of the total thickness of sedimentary rocks had to be compiled by adding together the maximum known thickness of rocks of each age. These estimates had to be revised each time a thicker section was discovered. (3) Sediment compacts when it is lithified; thus, a correction for compaction had to be made. Needless to say, estimates of Earth's age varied considerably each time this method was attempted" Tarbuck & Lutgens, 1999, p. 192).

How can one estimate missing data? Dingman noted "estimating missing data" (2002, p. 115). Earlier in this report it was mentioned that radiometric dating was used to establish the date for "individual mineral grains in sedimentary rocks from Australia" (Murck, 2001, p. 62). These were dated at "4.4 billion years" (Murck, 2001, p. 62). "That use of the degree of chemical and physical change from original sediments as a criterion of age is an extremely dangerous practice" (Garrels & Mackenzie, 1971, p. 67). This is because the calculations depends on "the accuracy of the assumption that the rock was formed during a relatively short time interval in comparison with its age" (MacDonald, *et al.*, 2003, p. 129). Table 22 - Section 18.2 is a compilation of many different problems in the dating methods proposed for the geologic processes of the Earth and points out that science is self-correcting. Missing data is found within the various dating methods scientists have used to re-construct the age of the Earth. A significant problem with relying on this line of reasoning and including "missing data" (Gradstein, *et al.*, 2004, p. 52) is that science facts are based on what is seen; not what is not seen.

Verification is "a procedure for checking the values of attributes for all records in a database against their correct values" (Clarke, 2003, p. 327). "Any theory that purports to be scientific must somehow, at some point, be compared with observations or experiments" (Wells, 2000, p. 2). "Many geologists expect sequence boundaries to correspond with system, series, and stage boundaries and foraminiferal zonal subdivisions even when sedimentological evidence is lacking (Izart *et al.*, 1998, 1999; Stemmerik *et al.*, 1995: Samuelsberg and Pickard, 1999)" (Gradstein, *et al.*, 2004, p. 236, 237). "If contradictory evidence turns up, the theory must be reevaluated or even abandoned. Otherwise it is not science, but myth" (Wells, 2000, p. 3).

Table 22 - Section 18.2: Problems with Dating Methods

Science is Self-correcting Problems with "early methods of dating Earth" (Tarbuck & Lutgens, 1999, p. 192).						
Uniformitarian Theory	**Fission-Track Dating**	**Magnetic Reversals**	**Radiometric Dating**	**Classification of Layers**	**Fossils**	**Geologic Time Scale**
Actualism = Principle of Uniformity = Uniformitarian sim (Licker, 2003, p. 384). The 'classical' notion as James Hutton (1726-1797) conceived uniformitariani sm "is a term that is a bit unfortunate because it suggests that changes take place at a uniform *rate*" (Plummer, *et. al.*, 2003, p. 176). "The modern earth's surface is the geologist's laboratory. Here can be studied the processes that generate sediments and their resultant deposits. By applying the principle of uniformitariansm ('the key to the past is in the present'), it is possible to deduce the origin of sedimentary rocks" (Selley, 2000, p. 181).	"Correlation of such boundaries between distant sections, on the basis of even our best geochronometer s (U–Pb ages on single zircons), can be no better than ±5–10 million years (relative to absolute ages), even if all other sources of uncertainty…co uld be reduced by defining boundaries explicitly in terms of ^{207}Pb / ^{206}Pb zircon ages or isotopic ratios, rather than absolute age numbers, but this would result in a time scale that is even less transparent" (Gradstein, *et al.*, 2004, p. 141, 142).	"Magnetic reversals are virtually useless —they are simply positive or negative binary data, like the data employed in a computer. But just as a simple plus and minus computer code can record or reveal great quantities of information when enough data is accumulated, so too can the binary code of reversal history. If it is combined with other techniques such as radiometric dating, the pattern becomes a very powerful tool" (Gribbin, 2004, p. 162).	"That use of the degree of chemical and physical change from original sediments as a criterion of age is an extremely dangerous practice" (Garrels & Mackenzie, 1971, p. 67). Reliability of the calculations depends on… "the accuracy of the assumption that the rock was formed during a relatively short time interval in comparison with its age" (MacDonald, *et al.*, 2003, p. 129).	"Mount St. Helen's initially was subdivided into nine eruptive periods. Before 4,000 years ago eruptive periods lasted for thousands of years, in contrast to younger ones that had durations of only centuries. To correct that discrepancy the eruptive periods were reclassified (Mullineaux, 1996, p 84)" (Rudd, November 2004, GE0 3E03 Research Paper, p. 3). "A theoretical continuous stack, called the stratigraphic column" (Busbey, *et. al.*, 1999, p.20).	Cannot be dated by any know technical method, their age is guessed at from the geological strata around them. "So telling time using fossils has severe limitations" (Gribbin, 2004, p. 154). "No one knew when the [Jurassic] period started or when it finished" (Gribbin, 2004, p. 154). "The fossil evidence is open to interpretation because individual specimens can be reconstructed in a variety of ways, and because the fossil record cannot establish ancestor-descendant relationships" (Wells, 2000, p. 219). "The fossil record in post-Precambrian rocks… "at any one time or place it may be quite patchy and badly biased toward one kind of organism or another" (Laporte, 2003, p. 83).	"This sparse skeleton of age control, especially prior to ~30 Ma (as of 2004), leaves considerable room for interpolation in construction of geologic time scale" (Gradstein, *et al.*, 2004, p. 455). "The concept of a geologic time-scale has been evolving for the last century and a half, commencing with a relative time-scale (mainly achieved through biostratigraphy), to which it has gradually become possible to assign dates, which are, nonetheless, subject to constant revision and refinement" (Allaby & Allaby, 2003, p. 228). "Deterministic methods assume that inconsistencies in the stratigraphic range of a taxon from section to section or well to well are due to missing data" (Gradstein, *et al.*, 2004, p. 52).

Problems with All Methods "No single locality has a complete geologic column" (Tarbuck & Lutgens, 1999, p. 192). "The record is very incomplete" (Nichols, 2003, p. 9). "The present time scale is incomplete" (Gradstein, *et al.*, 2004, p. 142).

Sources: See References

Under the heading of *Poor Data*, William D. Nesse writes, "most of the optical and physical properties reported in this and other standard sources on minerals are based on observations and reports from over a century ago" (2004, p. 127). Nesse goes on to show how these observations are cited in reference materials, which are far removed from the source data. Nesse writes, "These data are uncritically repeated in each new text or reference work, with a few additions or modifications based on new measurements gleaned from the literature" (2004, p. 127). The problem with secondary research is that "old errors are thereby perpetuated and new, often unrepresentative data added" (Nesse, 2004, p. 127). Nesse offers a sound reason as to why primary research is not used; according to him the reason is "because the time required to systematically verify and update the data on each mineral is very large" (Nesse, 2004, p. 127). Nesse then cautions. "Reader beware" (Nesse, 2004, p. 127). Unlike the secondary research that Nesse cautions his readers to be aware of, this report is founded upon first-hand, primary research with on-site observations in recent history to confirm its findings.

18.3 The Radiometric Dating Technique

Radiometric dating is "a method of calculating the age of a sample of material by knowing the half-life of a radioisotope with in it. This method has provided absolute dates for the geologic time scale and for other specific events in Earth's history" (Kump, *et. al.*, 1999, p. 342). "Radiometric methods can be used to determine the age of crystallization of an *Igneous Rock* from a *Magma*, the time of metamorphic recrystallization (with the formation of a new mineral assemblage), the time of uplift and cooling of fold mountains and the age of *Sedimentary Rocks* in which new minerals (*Authigenic* minerals) formed during deposition" (MacDonald, *et al.*, 2003, p. 129).

"The *radiometric dating technique* is based on the decay of radioactive isotopes that have very long *half-lives*. The age can be calculated from the measured proportions of parent and daughter isotopes using the *age equation*:

$$t = (1/\lambda) \ln (1 + d/p)$$

where t is the age of the sample, λ is the decay constant (which has a characteristic value for each radioactive isotope), 1_n is log to the base e and d/p is the present ratio of daughter to parent isotope. The decay constant is related to the half-life T (the time taken for half the atoms to decay) thus:

$$T = 0.693 / \lambda$$

(MacDonald, *et al.*, 2003, p. 129).
"If the age of a rock is to be calculated it is necessary to know the half-life of the radioactive decay series used, the amount of the parent and daughter isotopes" (Nichols, 2003, p. 251). Half-life is "the time required for the decomposition of one half of any given quantity of a radioactive element. It differs for various elements and isotopes but is

always the same for a particular *isotope*. It cannot be accelerated or decelerated by heat, cold, magnetic or electrical fields. A nucleus is regarded as stable if its half-life is greater than the estimated age of the Earth (about 5.5×10^9 years)" (MacDonald, *et al.*, 2003, p. 227). Refer to Figure 41 - Section 18.3.

"The reliability of the calculations… [is based on several factors]: (1) the half-life of the radioactive element is known; (2) whether any of the parent isotope and /or the daughter isotope has been removed from or added to the rock after formation; (3) whether corrections can be made for any daughter isotope present when the rock was formed; and (4) the accuracy of the assumption that the rock was formed during a relatively short time interval in comparison with its age" (MacDonald, *et al.*, 2003, p. 129). "Scientists have discovered that *k* is not constant" (Fox, 2000, p. 301). A partial list of radioactive elements used in radiometric age dating is listed in Collins Dictionary of Geology (2003, p. 130). Dr. James MacDonald and Dr. Christopher Burton list some of the parent and daughter isotopes with their respective half-life:

Parent isotope	Daughter isotope	Half-life
Carbon-14 ($^{14}_{6}C$)	Nitrogen-14 ($^{14}_{7}N$)	5730 years
Potassium-40 ($^{40}_{19}K$)	Argon-40 ($^{40}_{18}Ar$)	1250 Ma
Rubidium-87 ($^{87}_{37}Rb$)	Strontium-87 ($^{87}_{38}Sr$)	48 800 Ma
Samarium-147 ($^{147}_{63}Sm$)	Neodymium-143 ($^{143}_{60}Nd$)	108 000 Ma
Uranium-238 ($^{235}_{93}U$)	Lead–207 ($^{207}_{82}Pb$)	704 Ma
Uranium-235 ($^{238}_{92}U$)	Lead–206 ($^{206}_{82}Pb$)	4467 Ma

Figure 41 - Section 18.3: Parent and Daughter Isotopes with Respective Half-lives

Practical radiometric dating is "any alteration of the rock to be dated may result in an error in the age calculated, so it is important that the material collected for analysis is fresh, unaltered and unweathered" (Nichols, 2003, p. 254).

"For the geologic time scale… uncertainty estimates on stage duration and age of stage boundaries are derived by the maximum likelihood fitting of a function relation (MLFR) method" (Gradstein, *et al.*, 2004, p. 186). "Shaded boxes at stage boundaries indicate interval of uncertainty in correlation between stratotype points and the graptolite zonation" (Gradstein, *et al.*, 2004, p. 190). Another example, "unfortunately, the Devonian radiometric time scale, and the data records, are not yet precise enough to elucidate extinction rates in any detail" (Gradstein, *et al.*, 2004, p. 212). Again, in *cyclostratigraphy*, "the presumption has been made that dominant cycles may be precession or eccentricity cycles. The inadequacies of the radiometric scale make checking of this difficult" (Gradstein, *et al.*, 2004, p. 212).

"There are essentially no reliable radiometric ages with precise stratigraphic controls within the Late Jurassic or Early Cretaceous" (Gradstein, *et al.*, 2004, p. 371).

"… but they are always somewhat discontinuous leaving some leeway for subjective interpretations. This is particularly true for the Paleocene and the Oligocene" (Gradstein, *et al.*, 2004, p. 392).

Radiometric dating method can be used for determining the age of formation of igneous rocks and authigenic minerals in some sedimentary rocks" (Nichols, 2003, p. 251). Authigenic minerals are "minerals which grow crystals in a depositional environment" (Nichols, 2003, p. 16). The sediment on Kidd Copper was not authigenic nor was it igneous therefore the sediment could not be tested using radiometric dating.

"In the global composite … the order and spacing of events provides a proxy for a time scale. The order and spacing are determined separately, because the spacing requires simplifying assumptions that need not compromise determination of the optimal order. First, the optimal order of events is established by minimizing misfit in sequence between the composite and all of the individual sections. Misfit is determined primarily by the net number of event horizons through which the observed range ends must be extended in order to make all the observed range charts fit the same composite sequence" (Gradstein, *et al.*, 2004, p. 182).

"The scaling of the composite is therefore derived from all of the sections and it is the ratio of the thicknesses between events that is used, not the absolute thickness. The influence of aberrant sections, incomplete preservation, and non-uniform depositional rates is thus minimized" (Gradstein, *et al.*, 2004, p. 184). "We should anticipate that the composite spacings are least reliable near the base of the period" (Gradstein, *et al.*, 2004, p. 185). "The scaled composite is itself a proxy time scale … Uncertainty estimates are provided by relaxed-fit intervals, which give the range of positions within the composite at which the dated event lies, in successive runs during which 'best-fit' is relaxed by a small percentage (Sadler and Cooper, 2004)" (Gradstein, *et al.*, 2004, p. 186).

"Because radiometrically dated beds were included in the optimizing process, the composite relative time scale already contains the necessary ingredients for calibration to time … The best fit ... was calculated with a cubic-spline fitting method that combines stratigraphic uncertainty estimates…" (Gradstein, *et al.*, 2004, p. 186).

Time correlation based on the radiometric dates to support the geological time scale is problematic. Therefore, relying on conventional wisdom that many layers of sediment support an old date for the age of the Earth has a high margin of error.

18.4 The Lead-Lead Method

"The *lead-lead method*: because of the differences in half-life of the two uranium isotopes, the ratio ^{207}Pb / ^{206}Pb is time-dependent and can be used for age determinations. Corrections for the proportion of radiogenic isotopes incorporated at the time of crystallization can be made by means of an *isochron diagram*, using the relative proportion of non-radiogenic ^{204}Pb." (MacDonald, *et al.*, 2003, p. 129).

18.5 Uranium-Lead Method

"Either of the two isotopes of uranium may be used in the *uranium-lead method*, which is useful for rocks over 100 Ma old. It is less useful for younger rocks because the rate of production of ^{207}Pb is too low. Corrections may have to be made … Uranium-lead determinations are usually made on *zircon* and *sphene*, which are associated with many igneous and *metamorphic rocks*" (MacDonald, *et al.*, 2003, p. 129).

18.6 The Potassium-Argon Method

"The *potassium-argon method* can be used to date rocks ranging in age from 4600 Ma to as recently as 30,000 years. This method, too, is less accurate for younger rocks because of the relatively small amount of daughter product that has formed in them. K–Ar methods are suitable for dating igneous and metamorphic minerals such as *hornblende*, *biotite*, *muscovite* and *nepheline*, and for whole-rock specimens, particularly of fine-grained rocks – *lavas* and micaceous metamorphic rocks such as *phyllite* and *slate*. Some sedimentary rocks containing *glauconite* can be dated by this method, since glauconite forms at the time of deposition. Problems arise in the K–Ar method because of the diffusive loss of ^{40}Ar at temperatures well below those of metamorphic recrystallization" (MacDonald, *et al.*, 2003, p. 130).

18.7 The Rubidium-Strontium Method

"The *rubidium-strontium method* is rarely used for rocks younger than about 20 Ma unless they have a high content of ^{87}Rb. Minerals such as muscovite, biotite, all the potassium *feldspars* and glauconite can be dated by this method. Whole-rock specimens of igneous and metamorphic rocks that are rich in micas and potassium feldspar can also be dated. Allowance can be made for the amount of ^{87}Sr incorporated at the time of crystallisation, using an *isochron diagram* plotting the ratios of ^{87}Sr / ^{86}Sr against ^{87}Rb / ^{86}Sr (^{86}Sr is non-radiogenic). Dates can be calculated from the slope of the isochron obtained either from minerals from an individual rock (a *mineral isochron*) or from several whole-rock specimens (*a whole-rock isochron*)" (MacDonald, *et al.*, 2003, p. 130). The problem with this dating method is that "the whole-rock isochron may give a different age from a mineral isochron from the same rock because ^{87}Sr can diffuse out of the minerals in which it formed but be retained within the rock (usually in calcium-bearing minerals)" (MacDonald, *et al.*, 2003, p. 130).

18.8 The Carbon-14 (Radiocarbon) Method

"The *carbon*-14 (*radiocarbon*) *method* is often used to date materials and events that are relatively recent; it is particularly valuable for use on materials less than 50 000 years old, as this time span is too brief for the more slowly disintegrating elements to produce measurable amounts of daughter products. The radiocarbon method has been widely used in dating archaeological finds and provides the best dating for the Holocene and Pleistocene. The method is based on the assumption that the isotope ratio of carbon in the cells of living things is identical with that in air because of the balance between photosynthesis and respiration. Although several sources of error are introduced into this method, such as the inconstant supply of ^{14}C, both in time and latitude, it has proved to be enormously useful" (MacDonald, *et al.*, 2003, p. 130, 131).

Some noted problems with the Carbon-14 dating method. "Coal from Russia from the "Pennsylvanian, supposedly 300 million years old, was dated at 1,680 years" (Ham, Snelling &Wieland, 1992, p. 73). "A freshly killed seal dated by ^{14}C showed it had died 1300 years ago" (Ham, Snelling &Wieland, 1992, p. 74). "Living snails' shells showed they had died 27,000 years ago" (Ham, *et.al.*, 1992, p. 74). "The troubles of the radiocarbon dating method are undeniably deep and serious ... It should be no surprise, then, that fully half of the dates are rejected. The wonder is, surely, that the remaining half come to be accepted (Lee, 1981)" (Ham, *et.al.*, 1992, p. 75).

18.9 Fission-track Dating

"*Fission track dating* is based on the spontaneous *fission* of ^{238}U, which releases large amounts of energy. The damage this causes to the surrounding material can be revealed by etching with hydrofluoric acid. The half-life of this process is about 8×10^{15} years, and since the number of fission tracks depends only on the amount of uranium in the sample and the time elapsed since it formed, the age can be calculated. This method is useful for dating relatively recent volcanic material (up to 5 Ma), for example *volcanic glasses* and minerals that contain uranium, such as *apatite*, zircon, sphene and mica" (MacDonald, *et al.*, 2003, p. 131). "Since a constant (age) cannot be a function of random variables (counts), the fission-track 'age equation' represents only an estimator of age (Bardsley, 1982, p. 545, 546).

19.0 Data Analysis of Kidd Copper's Stratigraphy

The laws of stratigraphy would suggest that the sharp contrasts in the contacts and matrices of each Kidd Copper vertical log profile are indicative of numerous time periods; and these contrasts occurred over a long period of time for the accumulation of sediment. However, this conclusion does not coincide with the analyses in this report. The observed stratification at Kidd Copper reveals many layers of sediment deposit were a one-time event and the correlation of the deposits shows that fine-grained sediment can be distributed unevenly. The laws of stratigraphy are simply "so-called natural laws are merely the accumulation of all our observational and experimental knowledge" (Levin, 2003, p. 7). Therefore, if the accumulation of this new knowledge from the Kidd Copper case and using data to draw conclusions about the sediment depositional processes on Kidd Copper *refutes a so-called law*, then the law needs to be re-examined.

"*Data Analysis*: The process of using organized data to test scientific hypotheses" (Clarke, 2003, p. 308). Core-to-core correlation between the boreholes in the sequential lithofacies taken of both the West-North transects and the North-South transects on the Kidd Copper property reveals differences of the vertical logs in composition, color, and grain orientation. The differences according to the laws of stratigraphy should suggest different depositional conditions; however, this is not true because the depositional conditions on all of the logs were identical.

Analyzing further stratigraphy practices, "the materials are then considered in terms of depositional environments" (Nichols, 2003, p. 3). A rapid process happened as the floodwaters subsided and the water dried up. The sediment fell upon the impoundment of mine tailings and remained intact until 2004 when some Geology students uncovered several boreholes.

19.1 Rate of Sedimentation

"Based on an average pelagic sedimentation rate of 1 mm / 1000 years, it would take a maximum of 10^9 years to accumulate all of the existing deep-sea sediment mass, an interval of time equal to only about 25 percent of geologic time" (Garrels & Mackenzie, 1971, p. 199). Refer to Section 14.11 for the *sediment transport equation* which is beyond the scope of this report. Assumption: layers are added one at a time.

Sedimentation rates have been used by geologists to establish the age of the earth. Here is a brief account of some of the detailed listings from 1971-2004.

1971. The "distribution of various rock types as a function of time" (Garrels & Mackenzie, 1971, p. 272). "The constant rate required to *deposit* the 18,000 x 10^{20} g of post-Precambrian sediments, making no allowance for subsequent destruction, is 18,000 x 10^{20}/600 x 10^6 = 30 x 10^{14} g/year" (Garrels & Mackenzie, 1971, p. 260).

1972. Rates of sediment are not constant. For example subglacial outburst floods and debris flows are very rapid rate flow of sedimentation deposit. "Debris flows commonly show a general decrease in both size and amount of coarse material downflow, showing that there is a progressive, though poorly effective, elimination of the coarser materials" (Blatt, Middleton & Murray, 1972, p. 161).

1976. An interpretation of the individual layers is described by Richard C. Selley where "the bottom of each new layer buried an older part of the Earth's solid surface, and the top of each new layer became, for a short while at least, a part of the Earth's solid surface. This burial relationship means that the configuration of each layer contains information by which the details of this burying can be reconstructed" (Selley, 1976, *p. nd.*). Schuchert in Principles of Stratigraphy on page 128 states for the Paleozoic Era that 3000 years/foot and the rate of formation is 1 ft/2700 years.

1990. "A more complex way of estimating completeness of a sequence of paleosols or of sediments is by comparing rates of sediment accumulation of a sequence with rates usual for that environment and time span, as estimated from a large compilation of rates (by Sadler 1981)" (Retallack, 1990, p. 283). Some sedimentation factors are: (1) "Different sediments accumulate at different rates under varying conditions" (Tarbuck & Lutgens, 1999, p.192); (2) "Some processes of soil formation, such as podzolization, are simple and rapid enough to be studied by means of laboratory experiments (Fisher &Yam 1984). Most soil-forming processes are too slow for such an approach" (Retallack, 1990, p. 261); and (3) "Soils form over time scales ranging from ecological (days) to geological (millions of years)" (Retallack, 1990, p. 261). "The rates decline for estimates made over longer time spans because of correspondingly longer gaps in the record" (Retallack, 1990, p. 283).

1992. "Two rates of sediment accumulation have been the focus of most studies: (1) the short-term or instantaneous rate of addition of new sediment and (2) the long-term or average rate of accumulation of strata. If short-term rates are multiplied by the long time required to reach a standard for geological comparisons, say meters of sediment thickness per million years, then large numbers result. For example, from a compilation of hundreds of measured rates, using an average time base of one day, the computed number of meters per million years is 10^5 (100 km). In contrast, geologically determined rates, using a time base of 10 million years, average out to 10^{-2} m, or 1 cm, per million years. This is a difference of seven orders of magnitude" (Friedman, *et. al.*, 1992, p. 174, 175). "This implies that the stratigraphic record accumulates episodically; short-term, rapid events of sediment addition are separated by much-longer times when no sediment is added, indeed, previously deposited sediment may be removed. Because they represent times, however brief, during which deposition was interrupted, the *boundaries between strata resulting from times when sedimentation was interrupted are* called surfaces of discontinuity. The time represented by a *discontinuity* is known as a hiatus"

(Friedman, *et. al.*, 1992, p. 175). The rate of change on the exposed surface in 1992 is significantly slower from 1983-1992, than the previous rapid change from 1970-1983.

2003. "The magnitude of events and rates of processes which occur today may have been different in the past" (Nichols, 2003, p. 316). "Methods based on sedimentation rates and the rate of increase in salinity of sea water have been shown to be unreliable" (MacDonald, *et al.*, 2003, p. 203). "It now seems probable that in some senses both [Uniformitarianism and Catastrophism] are correct, with the uniformitarian view of a constancy of physical laws being logical, but with elements of catastrophism to explain some of the events of the past" (Nichols, 2003, p. 317).

2004. "When a core of an ancient sedimentary formation is obtained from a borehole or an oil well, one can now compare its sedimentary structures with a known standard, in much the same way an art historian identified a purported Rembrandt by comparing its composition, coloring, shading, and brush strokes with known Rembrandts. Sometimes the comparison is purely empirical. Other times there are good theoretical reasons why a sediment should look the way it does" (Hsü, 2004, p. 4).

The 'classical' notion as James Hutton (1726-1797) conceived uniformitarianism also implied that geological rates were also constant and 'uniform' through time. Modern geology does not hold this to be true anymore – it is now known that geological processes such as sedimentation rates are not constants as can be seen by the evolution of the scientific understanding accumulated about sedimentation rates reported from 1971-2004. Modern geology is dispelling earlier beliefs about uniformitarianism. Mainstream geologists recognize that changes do not occur at a uniform rate, "such as short-lived volcanic eruptions" (Plummer, *et. al.*, 2003, p. 176).

So, a uniform layer of sediment will not be found when the mine tailings re-settled. Therefore, can multiple layers of deposits of various sediments signify a short depositional time period? Are the processes which formed rocks and geomorphic features in geological past the same as processes operating today?

Dating of geological strata based on the prevailing theory of slow geological change. The 'classical' notion as James Hutton (1726-1797) conceived uniformitarianism also implied that geological rates were also constant and 'uniform' through time. The geologic processes at Kidd Copper provide strong evidence that this is not true – the geological processes such as sedimentation rates are not constants.

19.2 Relevance of the Rates of Sedimentation.

"Rates of sediment accumulation have been used to derive time scales. The simplest methods average fossil zone thickness in several sections and assume that thickness is directly proportional to duration (e.g. Carter *et al.*, 1980)" (Gradstein, *et al.*, 2004, p. 49). Using this derivation of time on the Kidd Copper property on four log profiles taken from Table 5 - Section 10.6 Cross Section B–B' [North–South Transect would lead to the wrong conclusion because in Point 5/5 ten different zones would be used; in Point 6/6 two different zones would be counted; Point 51/10 five different zones; and in Point 113/L2 five different zones would be designated to represent different times of deposition. Refer to Figure 42 - Section 19.2.

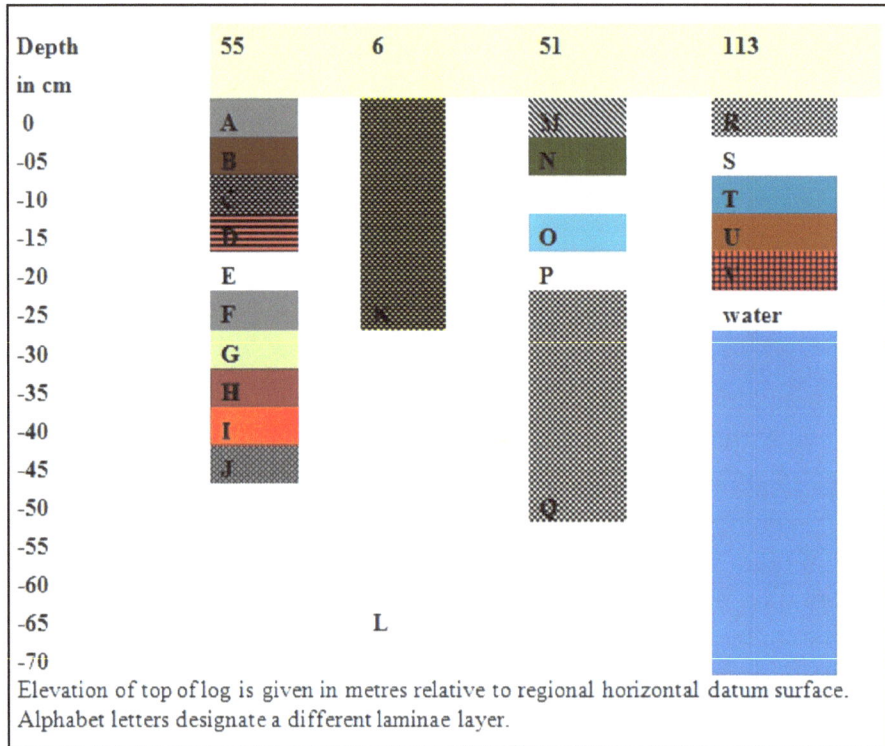

Depth in cm	55	6	51	113

Elevation of top of log is given in metres relative to regional horizontal datum surface.
Alphabet letters designate a different laminae layer.

Figure 42 - Section 19.2: Vertical Profile of Four Associations, Cross-section B–B', August 30, 2004

Sediment accumulation is not an accurate time measurement for the deposition of these sediment layers based on the Kidd Copper property site observations. The reason for the five or even ten depositional periods being inaccurate is because the sediment deposits were all placed at the same time. The rates used in this compilation include those from direct observation of sediment accumulating from floodwaters after the dam receded. Therefore, the rate of sedimentation is not constant as Gradstein, Ogg & Smith assumed (2004, p. 117). Thickness is not directly proportional to duration. Remember back to the extension of mine tailings, which happened not slowly over a long period of time but quickly. Refer to Figure 10 - Section 3.6 where 13 years later, the impoundment area was noticeably extended by "traditional panchromatic aerial photographs covering the study site were obtained from the National Air Photo Library of Natural Resources Canada" (Shang, *et al.*, 2004, p. 3).

The question to ask: Do the different alphabet letters in Figure 42 - Section 19.2 represent different time periods, sediment layers, depostional processes, stages of evolution? The answer is that none of those choices is correct. The right answer is that these alphabet layers represent one layer of sediment from one depositional process from one unit of time. They should have been marked with the same alphabet letter to mark each log profile as one mixed-layer. Each log being a heterogeneous mixture is a more accurate representation of the findings at Kidd Copper. In total, all points 55-6-51-113 in Figure 42 - Section 19.2 are representative of the same depositional process, step 1 initial deposition of mine tailings; step 2 upswelling from a flood event and step 3, redeposition.

19.3 Rates of Erosion

"Rocks on the Earth's surface are worn away into tiny grains by water, wind, ice and other natural forces. Rain washes the grains into rivers, which leave a muddy 'sediment' behind whenever they flow most slowly. Sediment piles up in layers that, over millions of years, are pressed down so hard that they become rock. Shale is sedimentary rock made from clay. Sandstone is made from sand" (Williams, 1993, p. 23). "Furthermore, rates of erosion are so fast, relative to the great lengths of geologic time through which erosion and deposition have been active, that the continents should have been reduced to featureless plains long ago unless there have been mechanisms at work in addition to buoyant restoration and sinking" (Garrels & Mackenzie, 1971, p. 29).

Some would say that the multiple layers seen on the Kidd Copper property are due to erosional processes. Consider that erosional or surface-run off processes causing these layers might be reasonable in a small area of the site but, highly unlikely to affect the entire land-site of the impoundment. This landform covered roughly 4 km^2. Erosion would mainly affect the surface layers and leave the underlying layers intact. Erosion would remove the top-layers without changing the structure of the underlying formations. Aeolian winds could cause surface changes but highly unlikely to change the entire vertical column or profiles of four associations, which were taken as a small sample of the borings. The extensive diggings throughout the site area tested a very large proportion of the mine tailings deposits found in the land at the Kidd Copper site. One hundred and twenty boreholes showed the same evidence. This makes it easy to dismiss the idea that stratification was due to erosional or surface run-off processes because the entire area of the site exposed consistent data. Every borehole noted a reference to stratification. Multiple layers of strata were documented. These remarks were candid, refutable, scientific observations from thirty-two students including the author.

19.4 Heavy Particles Fall First

"Stones and sand cannot remain long in suspension; they soon drop out and are deposited" (Hsü, 2004, p. 89). "The forces which bind soil water are related to the soil porosity and the soil water content. The forces are weakest in the case of open textured, wet soils, and greatest for compact dry soils. Thus, at a given value of S the water potential is greatest for a clay, least for a coarse sand, and intermediate for a loam" (Oke, 2001, p. 49). "Smaller stones and coarse sand grains could not remain long in suspension; they settled out of the suspension first. Finer sand, silt, and clay-sized particles settled later. Eventually all suspended particles dropped out, as the turbulence of the suspension died down" (Hsü, 2004, p. 89). Eluviation is "the downward movement of materials in suspension or solution that are being carried through the soil by descending soil water" (MacDonald, *et al.*, 2003, p. 162).

"*Graded bedding* – layers, each with a sharply distinct base, on which the coarsest grains of a *bed* lie. In each layer, the grain size decreases progressively upwards, with the finest-grained material at the top of the bed. This is followed abruptly by the coarse-grained base of the next layer. The layers are commonly interbedded, with others having parallel laminations. Graded bedding is commonly formed in deposits from *turbidites*. It is a valuable way-up *indicator* in strata that have been strongly folded or overturned" (MacDonald, *et al.*, 2003, p. 217, 218).

"You can demonstrate graded bedding using the same experimental setup you used to demonstrate the principle of original horizontality. Fill a tub (preferably clear plastic) with water and a mixture of sediment of various sizes – some pebbles, some coarse and fine sand, and some mud. Stir vigorously, and wait. The pebbles will settle out first, followed by the sand. The fine silt from the mud will settle last. You should be able to see the variation in grain sizes through the clear sides of the container" (Murck, 2001, p. 147). The rate for the sediment to form graded beds is quick. The time required for the water to settle is not eons of ages; thus supporting the evidence for rapid stratification and a repeatable process for the theory of a 'younger' Earth.

"*Sorting* [GEOL] the process by which similar in size, shape, or specific gravity sedimentary particles are selected and separated from associated but dissimilar particles by the agent of transportation" (Licker, 2003, p. 348). "Sorting coefficient [GEOL] a sorting index equal to the square root of the ratio of the larger quartile (the diameter having 25% of the cumulative size-frequency distribution larger than itself)" (Licker, 2003, p. 348).

Kidd Copper sediment was in undefined patterns because the water disassembled their components and left a fine slurry mixture. The suspension, sorting and eluviations did not leave a consistent layer of heavier particles on the bottom layer. The sediment was well-mixed and randomly, unequally distributed according to the findings from the correlated in the stratification logs of the boreholes, which records the distribution of sediment once the water evaporated. The sediment was left undisturbed until the uncovering of the 120 boreholes on August 30, 2004 to attest to the flood event.

What is seen at Kidd Copper is contrary to common thought and contrary to conventional geology. For example, a conventional answer to the question, "Why are some rocks stripy? Some rocks are made from layers of different-colored sedimentary rocks. There might be layers of shale made from mud, sandstone made from sand, and coal made from rotting trees. The different layers or stripes show whether the land was a forest, a beach, or under water when the layer was formed" (Claybourne, 2002, p. 8).

New evidence from the Kidd Copper case provides another answer to the question.

"Why are some rocks stripy?" The new answer is rocks are stripy because they were mixed by water and resettled and distributed in an unorganized, stratified manner because of a cataclysmic event that altered them.

19.5 Cross Lamination

"Beds are commonly defined as layers greater than 1 cm in thickness and layers less than 1 cm in thickness are called *laminae*" (Blatt, Middleton & Murray, 1972, p. 111).

19.6 Cross-stratification

Defined. "Cross-stratification forms either a single *set* or many sets (then termed a *coset*) within one bed. On size alone, the two principal types of cross-stratification are *cross-lamination*, where the set height is less than 6 cm and the thickness of the cross laminae is only a few millimeters, and *cross-bedding* where the set height is generally greater than 6 cm and the individual cross-beds are many millimeters to a centimeter or more in thickness" (Reineck & Singh, 1973, p. 53). What this reveals is that a set height less than 6 cm can be examined. Cross-stratification [is an] "arrangement of strata inclined at an angle to the main stratification. In modern usage, this is considered to be the general term, and to have two subdivisions: *cross-bedding*, in which the cross-strata are thicker than 1 cm, and *cross-lamination*, in which they are thinner than 1 cm" (Bates & Jackson, 1984, p. 118). "Cross-stratification deserves careful observation in the field as it is a most useful structure for sedimentological interpretations, including palaeocurrent analysis" (Reineck & Singh, 1973, p. 53). "Common types of stratification [include] Trough Cross Stratification … Interpretation. Dunes are the migrating bed forms that deposit trough cross stratification" (Harms, *et al.*, 2004, p. 46, 47). Tabular Cross Stratification "are deposited by migrating sand waves (Harms, *et al.*, 2004, p. 48).

19.7 Cross-Bedding Geological Measurements

Defined. "Cross-bedding is one of the commonest and most important of all sedimentary structures" (Selley, 1982, p. 221). "Cross-bedding forming one layer is called a set; two or more sets within a bed form a *coset*" (MacDonald, *et al.*, 2003, p.118). "Cross-bedding, as its name implies, consists of inclined dipping bedding, bounded by subhorizontal surfaces" (Selley, 1982, p. 221). "*Cross bedding* or *cross stratification* – sedimentary layering within a *bed* that is inclined at an angle to the main *bedding plane*. The steeply dipping parts of the layers are *foresets*; if the *dip* decreases towards the base of the bed, these are the *bottomset* parts" (MacDonald, *et al.*, 2003, p. 118).

"Particles are carried up and over the top of the pile by the current, accumulating on the downcurrent slope. This produces beds that are inclined. The direction in which the cross-bedding is inclined tells us the direction of flow of the water or air currents at the time of deposition" (Murck, 2001, p.147, 148). Current bedding – "any bedding or bedding structure produced by current action; specif. cross-stratification resulting from water or air currents of variable direction" (Bates & Jackson, 1984, p. 124). "Inclined bedding [GEOL] a type of bedding in which the strata dip in the direction of current flow" (Licker, 2003, p. 161).

"Turbulent flow in streams, wind, or ocean waves produce a type of bedding called *cross-bedding*, which refers to beds that are inclined with respect to a thicker stratum within which they occur" (Murck, 2001, p. 148). "It is ubiquitous in traction current deposits in diverse environments" (Selley, 1982, p. 221). "Cross-bedding is useful as a *way up indicator*: the angle between the foresets/bottomsets and the base of the bed is smaller than the angle between the foresets and the top surface, which is often an erosion surface" (MacDonald, *et al.*, 2003, p. 119).

"Cross-lamination/cross-bedding is mainly formed by the migration of *bed forms*, such as ripples, *Dunes* or *sandwaves*, as sediment is deposited on the downcurrent side but also results from the growth of *Deltas* or the migration of sand bars in a river… The nature of the cross-bedding reflects the shape of the bed form and thus the strength of the current that moved the sediment. Cross-bedding which the foresets are approximately planar is called *tabular cross-bedding* and results from the migration of straight-crested bed forms. In *trough-cross-bedding* (sometimes called *festoon bedding*), the foresets are trough-shaped, resulting from the migration of bed forms with sinuous or curved crests" (MacDonald, *et al.*, 2003, p. 118).

"*Ripple-drift cross-lamination* (or *climbing ripples*) results from the rapid deposition of sediment; ripples climb up the backs of those downstream, so that the set boundaries lie at an angle and dip in the opposite direction from the cross-layers. In some cosets the foresets within adjacent sets may dip in opposite directions: this is *herringbone cross-bedding*, produced where there are reversals of the current directions, for example in tidal areas. (MacDonald, *et al,*. 2003, p. 119). Swaley cross-bedding "a variety of trough cross-bedding with slightly curved concavities; erosional counterpart of hummocky cross-bedding; also known as flat festoon (bedding)" (Lucchi, 1995, p. 249.).

Cross-bedding is assumed to have "formed by the migration of subaqueous dunes and sand waves … or aeolian dunes" (Reineck & Singh, 1973, p. 93). Deposition: wind-blown sand along marine shorelines. "*Aeolian stratification* – The *bedding* and *lamination* formed as wind-blown sediment, usually of *sand* grade, accumulates on a dry surface" (Kearey, 2001, p. 4). However, in the case of the Kidd Copper property the cross-bedding occurred on the ground surface and not within the bed layers and so is not a primary influence on the formation of the boreholes. The aeolian factors were prevalent on the surface but the sediment was covered underneath. The mine tailings had been redeposited. Wind blown migration is a minimal factor influencing the stratification of the sub-surface layers.

19.8 Palaeocurrent Data

Paleocurrent data involves recording azimuths, which were not collected at Kidd Copper (Selley, 2000, p.169). "The azimuths are, however generally manipulated in some way to make their interpretation easier" (Selley, 2000, p.169). "In the case of Kidd Copper no distinction was made from the palaeocurrent direction or the original dip of cross bedding (Reineck & Singh, 1973, p. 94). "These techniques are only applicable to unimodal distribution of azimuthal data" (Selley, 1982, p. 245). Therefore Kidd Copper cannot be classified, "in a vector magnitude of 0% the distribution is completely random. There would be no vector mean in this case" (Reineck & Singh, 1973, p. 96). A viable model is lacking because the observations do not match the textbook theory. More on the calculating the vector mean can be read in Richard C. Selley's Introduction to Sedimentology on page 245. Table 23 - Section 19.9 lists some palaeocurrent patterns.

19.9 Palaeocurrent Patterns

"Interpretations of the palaeogeography need to be made with care" (Reineck & Singh, 1973, p. 96). Without knowing the fluvial and depositional history of the Kidd Copper property it would be easy to come to the wrong conclusion of the principal depositional environment because it could be assessed as fluvial, deltaic or a turbidite basin according to the information found in Table 23 - Section 19.9. Only because the history of the site is known that an accurate assessment is made at Kidd Copper. The correct deposition is from an upswelling and a redeposition of the mixed sediments from a flood deposit.

"Once the various facies have been differentiated, they can then be interpreted by reference to published accounts of modern sediments and ancient sedimentary facies, and by the erection of facies models" (Reineck & Singh, 1973, p. 99). "Some facies are readily interpreted in terms of depositional environment and conditions, while others are not environmentally diagnostic and have to be taken in the context of adjacent facies" (Reineck & Singh, 1973, p. 99). The best facies interpretation is based on the facts of history, science is based on what is seen, not, what is not seen. This is why the Kidd Copper property case is very important because conventional wisdom and textbook theory can be tested and validated by examining observations and known facts. The guesswork is eliminated because the parameters of beginning and end are known.

Table 23 - Section 19.9: Palaeocurrent Patterns

Palaeocurrent Patterns of Principal Depositional Environments, Together with Best and Other Directional Structures		
Environment	Directional Structures	Typical Dispersal Patterns
Aeolian	Large scale cross bedding	Unimodal common, also bimodal and polymodal: dependent on wind directions/dune type
Fluvial	Cross bedding, also parting lineation, ripples	Unimodal down palaeoslope, dispersion reflects river sinuosity
Deltaic	Cross bedding, also parting lineation, ripples	Unimodal directed offshore, but bimodal or polymodal if marine processes important
Marine Shelf	Cross bedding, also ripples, fossil orientations	Bimodal common through tidal current reversals but can be normal or parallel to shoreline; Unimodal and polymodal patterns also
Turbidite Basin	Flutes, also grooves, parting lineation, ripples	Unimodal common, either downslope or along basin axis if turbidites, parallel to slope if contourites
Source: (Reineck & Singh, 1973, p. 97).		

19.10 Key Indicators

The lithological units were shown to be inconsistent in all areas of the site, taking into account sides, slopes and interior proximities. The parallel and perpendicular cross-sections clearly show distinctive layering throughout the mine tailings deposits in a random, non-consistent spatial pattern. "An immediate problem with this definition of a sedimentation unit is the difficulty in determining the limits of such a unit. If there are no clear internal breaks, how may a bed be subdivided into sedimentation units? For example, a graded bed is an easily identifiable unit in the field but this bed was not deposited under constant physical conditions" (Blatt, Middleton & Murray, 1972, p. 7, 8). Again, the classification of which layers belong to which bed would be difficult without prior knowledge. The non-consistent spatial pattern would be likely classified as many units through time if the history of the true depositional process was not known. That misclassification would occur from not knowing the original depositional environment. Could this same mistake happen elsewhere in the rock record? High rate of probability.

19.11 Rhythmic Layering Units in Different Hierarchal Levels

"Rhythmic layering [GEOL] a type of layering in an igneous intrusion which is easily observable and in which there is repetition of zones of varying composition" (Licker, 2003, p. 303). "Rhythmic sedimentation [GEOL] a repetitious, regular sequence of rock units formed by sedimentary succession and indicating a frequent, predictable recurrence of the same sequence of conditions" (Licker, 2003, p. 303). "Rhythmic stratification [GEOL] the occurrence of sediment layers in repetitive patterns, such as a regular alternation of layers of lime and clay" (Licker, 2003, p. 303). "Rhythmic succession [GEOL] a succession of rock units showing continual and repeated changes of lithology" (Licker, 2003, p. 303).

Plates are found in geology textbooks to give descriptions of sedimentary structures. They are often used as an instructional standard for identification purposes in the field. For example, Figure 43 - Section 19.11 is of Plate 59: rhythmic stratification in fallout deposits … "the terms *band* for thin bed or lamina, and *banded deposits* in general, are often used" (Lucchi, 1995, p. 90).

Figure 43 - Section 19.11: Plate 59 Rhythmic Stratification in Fallout Deposits

"The bands in [Figure 43 - Section 19.11: Plate 59] vary in thickness from nearly 10 cm (white one at the base) to 1–2 mm. A question that can be asked is whether they represent depositional units of variable thickness but of the same rank (beds or laminae) or units organized in different hierarchical levels (laminae, beds, layers, bedsets). In other terms: do the thin bands represent small, independent events or momentary pulsations of a larger event? Answer is not easy in the present case" (Lucchi, 1995, p. 90).

Lucchi asks the question whether the bands of variable thickness represent different depositional units or are the sediment units organized in different hierarchal levels? "In other terms: do the thin bands represent small, independent events or momentary pulsations of a larger event?" (Lucchi, 1995, p. 90). "Lucchi's answer follows: "Answer is not easy in the present case. Some thick white bands have a sharp base and top, and should represent distinct beds (faint laminae inside confirm this impression). More problematic are the intermediate and thin bands. If you look at the lithology, three lithotypes are recognizable: (1) white, crystalline (saccharoidal) gypsum; (2) black shale rich in organic matter (kerogen, or bitumen), and sparse crystals are not visible to the naked eye; (3) gypsum with clayey and organic impurities (various shades of gray)" (Lucchi, 1995, p. 90-91).

"A *thick* fallout deposit with a homogeneous texture is not easily interpretable; it could derive either from a massive, instantaneous event, such as a mud flow, or from a slow settling lasting for a long time in physically stable conditions" (Lucchi, 1995, p. 91). "It can be hypothesized" as to what the depositional process was... Fallout is anyway the dominant mechanism, because reworked particles apparently traveled in suspension" (Lucchi, 1995, p. 91).

Note, that for the depostional processes without historical information the best considerations have to be "hypothesized". How can one check the accuracy, then, of a standard that is based on conjecture? Who has the knowledge to determine how truthful these assumptions are? Relying on plates is not an accurate base to determine past or unknown depositional processes.

"Geographic pattern: A spatial distribution explainable as a repetitive distribution" (Clarke, 2003, p. 313). No repetitive distribution was detected in either the parallel or the perpendicular sequential lithographic compilations on Kidd Copper, refer to Table 4 - Section 10.5 and Table 5 - Section 10.6. The key to "rhythmic sedimentation [GEOL] a repetitious, regular sequence of rock units formed by sedimentary succession and indicating a frequent, predicatable recurrence of the same sequence of conditions" (Licker, 2003, p. 303). Why, then was no repetitive patterns found in the case of the Kidd Copper sequential lithographic logs? Instead, the strata on the Kidd Copper property is more "complex in the sense of combining forms that can differ substantially in scale and orientation…"composite"…"combination of horizontal strata with cosets of cross strata" (Harms, *et al.*, 2004, p. 51).

Random stratification has been seen in other deposits. "Apparent that there are many diversely oriented subsets" (Harms, *et al.*, 2004, p. 52). The mainstream thought of interpretation is that compound cross-stratification is the results of "delta front deposits" (Harms, *et al.*, 2004, p. 53). However, the twisted orientation of the sediments deposited on Kidd Copper property is not a delta front deposit interpretation but rather a mixing and settling of deposits from a turbulent flow of a catastrophic event. Science is based on the historical evidence and not the interpretation of theoretical applications. Kidd Copper answers the question, others could not. The contrasts of color, variances in composition and differences in texture are the aftermath of a one-time depositional process.

19.12 Closely Resembles Ribbon Rock

"Ribbon rock [PETR] a rock showing a succession of thin layers of differing composition or appearance" (Licker, 2003, p. 304). "Ribbon structure [GEOL] a succession of thin layers of different mineralogy and texture often contorted and deformed" (Licker, 2003, p. 304). Some internal structures of wave-formed ripples are visible at Kidd Copper: "(a) bundled upbuilding; (b) chevron upbuilding; and (c) unidirectional crosslamination" (Reineck & Singh, 1973, p. 58).

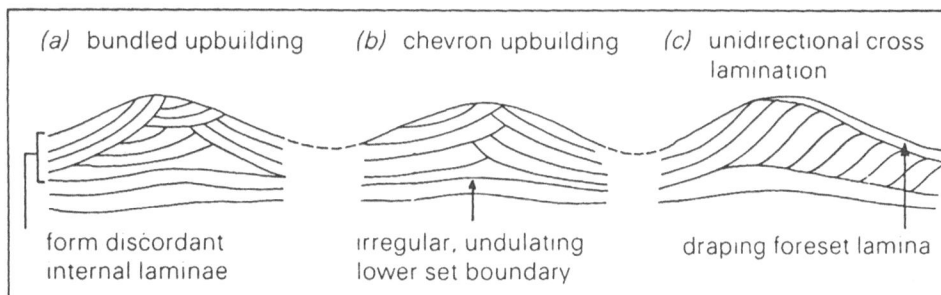

Figure 44 - Section 19.12: Bundled Upbuilding Found in Wave-Formed Ripples

Figure 44 - Section 19.12 illustrates that the bundled upbuilding that is found on the Kidd Copper property site is commonly found from the interaction of waves (Reineck & Singh, 1973, p. 58). This is evidence of water covering the impoundment area.

19.13 Event Stratigraphy

The term event stratigraphy was "first proposed by D. V. Ager (1973) for the recognition, study, and correlation of the effects of significant physical events (e.g. marine transgressions, volcanic eruptions, geomagnetic polarity reversals, climatic changes), or biological events (e.g. extinctions), on the stratigraphic record of whole continents, or even of the entire globe" (Allaby & Allaby, 2003, p. 195). Arguing "that by correlating these effects, as they are evidenced in the sedimentary record, it will be possible to define truly synchronous horizons, thus leading to greater resolution and a more accurate chronostratigraphic scale" (Allaby & Allaby, 2003, p. 195, 196). "More recently A. Seilacher (1984) has suggested the term 'event stratinomy' for the study of events at the level of individual beds" (Allaby & Allaby, 2003, p.196). The Kidd Copper data is more consistent with the "Relational model: A data model based on multiple flat files for records, with dissimilar attribute structures, connected by a common key attribute" (Clarke, 2003, p. 323).

19.14 Recurrence Time

"Recurrence time in stochastic processes, the time elapsing between the point when a system is in a certain state and the first subsequent time when it again attains that state" (DePree & Axelrod, 2003, p. 629). For example, scour marks "often form during a single erosional event" (Reineck & Singh, 1973, p. 46).

The recurrence time on the Kidd Copper Mine Tailings site was about one week for the resettling of the sediment from the dam breach to roughly 13 years where the extended area was noted for the growth of sediment in the impoundment area. Both of these start-points are in a short geologic period of time in Earth's history.

19.15 Key Time Factors

What makes the Kidd Copper property unique is that it provides a natural environment of the deposition of sediment when it has been up heaved and had to resettle. Layer is "a bed or stratum of rock" (Bates & Jackson, 1984, p. 292). If the layers seen by the 32 geology students in Section 10.3 were analyzed without recognizing that this was a one-time event occurrence it could be mistakenly rationalized that each layer was laid down separately through time; however the evidence is contrary.

It could be argued that the tailings migrated and the overland runoff and groundwater stream flow is the cause for the difference of stratified layers in the boreholes. The reason for this argument is based on conventional wisdom that the heavier particles settle leaving the finer particles to lie on top. If there was a correlated pattern throughout the waste site this argument would hold; but such is not the case. See the correlation figures and the varied mineral deposits. No such pattern exists. The random deposition of the particles is consistent with the theory that stratified layers are a quick process through time and that seeing many layers does not mean they are of different depositional processes seen elsewhere in the rock record, such as "other strata, such as layers of tephra, are deposited essentially instantaneously" (Friedman, *et. al.*, 1992, p. 175).

19.16 Data Confirming Rapid Stratification

   Datum Surface of Known Date. Stratification is clearly reported in the various documentations about the boreholes and rock profiles. On August 30, 2004 primary research was conducted by 32 geology students from the Kidd Copper property mine tailings waste site; and solely compiled by Loreen Rudd in 2004, re-married in 2008 to become Loreen Sherman, the author of this report. The use of personal names is for identification purposes only and does not constitute endorsement by the named subjects or McMaster University on the subject matter presented by the author's views on the age of the earth represented by this report. The data collected and correlated was from original field book entries from the students in *Geo3Fe3* photocopied and recorded with written permission. Refer to Appendix H for a record of permission statements.

   "The history of the Earth may be deciphered in terms of present observations on the assumption that natural laws are invariant with time" (Levin, 2003, p. 7). "Syntheses are made on the basis of direct geological observations" (Hsü, 2004, p. 19). "Field work forms the basis of all sedimentological studies. Even for experimental studies, the problems to be investigated are first defined in the field" (Blatt, Middleton & Murray, 1972, p. 4). Two areas that should support the science behind the diagenesis investigation are "either in natural environments or in a laboratory tank" (Nichols, 2003, p. 4). Natural conditions support rapid stratification and future experiments have been suggested that would further supply evidence of rapid stratification in this report.

   If the historical record was not known then the reasoning from using the conventional geologic laws and principles, refer to Section 15 in this report, would suggest a different conclusion as to the amount of time required to layer the existing strata found on Kidd Copper. However, because of the careful dating in historical records of these events, the timeline for the depositional processes is well-recorded in the archives of history so that the factors involving the sediment layering are documented and therefore difficult to refute. The multi-layers were formed rapidly; thus supporting the position that stratification occurs rapidly in the rock record.

19.17 Tilting

   In the palaeocurrent measurements, tilting is assumed to be from "tectonic processes" (Reineck & Singh, 1973, p. 91, 92). However, in Kidd Copper's case, the sediment was redeposited at the same time in the site area; and the tilting was in the original primary structures and not as a result of tectonic processes. The third step of deposition of the Kidd Copper mine tailings established non-horizontal layers and rapid stratification, even though the first initial step of mine tailings deposit began as a colluvium deposit, unconsolidated. A common pattern was not transparent either laterally, horizontally or vertically or even adjacent to the profiles. In all of the cross-sections and intersections a random distribution of the redeposited settlement was noted at Kidd Copper. The sediment pattern was irregular, without consistent form and without any repetitiveness. No tectonic processes were involved to disturb the sediment; instead the sediment was disturbed on Kidd Copper property by the dominant force of water. Thus, this shows that tilting can result from the strength of water flow.

19.18 Extensive Coverage

According to Reineck & Singh, 1973 a large number (more than 50) boreholes would be necessary to collect to determine a strong correlation and analysis (p. 91). The number of boreholes collected on the Kidd Copper property was more than double the required amount necessary for a strong analysis.

The extensive diggings throughout the site area tested a very large proportion of the mine tailings deposits found in the land at the Kidd Copper site. The natural environment is known. "An environment is considered to be any distinct geographical entity" (Walker & James, 2002, p. 8). One hundred and twenty different boreholes showed the same evidence. This makes it easy to dismiss the stratification due to erosional or surface run-off processes because the entire area of the site exposed consistent data. Every borehole noted a reference to stratification. Multiple layers of strata were documented that covered 85-90% of the surface ground area. This ratio is near the 10th percentile which means that approximately complete coverage of the area was observed at the impoundment of mine tailings on Kidd Copper.

19.19 Consistency

From the raw, primary data collected by the McMaster geology students every borehole noted a reference to stratification, refer to Section 10.2. These remarks were candid, refutable, scientific observations from thirty-two geology students. 120 boreholes confirm quick layering. There were multiple layers formed in a tiny time frame. The data is consistent in all areas of the site, taking into account sides, slopes and interior proximities.

19.20 Rhythmic Sedimentation Was Not Observed

"In many instances there is a regular repetition of the various facies to give *cycles* of sedimentation (also called cyclothems or rhythms)" (Reineck & Singh, 1973, p. 99). "The shape of the cross strata reflects the shape of the lee slope and depends on the characteristics of the flow, water depth and sediment grain-size" (Reineck & Singh, 1973, p. 53, 54).

"Rhythmic sedimentation [GEOL] a repetitious, regular sequence of rock units formed by sedimentary succession and indicating a frequent, predictable recurrence of the same sequence of conditions" (Licker, 2003, p. 303).

"Rhythmic stratification [GEOL] the occurrence of sediment layers and repetitive patterns, such as a regular alternation of layers of lime and clay" (Licker, 2003, p. 303).

Rhythmic layering [GEOL] a type of layering in an igneous intrusion which is easily observable and in which there is repetition of zones of varying composition" (Licker, 2003, p. 303).

"Rhythmic succession [GEOL] a succession of rock units showing continual and repeated changes of lithology" (Licker, 2003, p. 303).

"With many cyclic sequences there is a distinct boundary at the top (and therefore also bottom) of each cycle" (Reineck & Singh, 1973, p. 99). "Systematic changes in the character of the sediments up through a cycle, such as a gradual increase or decrease in grain-size, change in lithology and fossil content, or scale and abundance of sedimentary structures" (Reineck & Singh, 1973, p. 99).

"Simple methods involve counting the number of times each facies is overlain by the others and noting the type of boundary (sharp or gradational) between facies" (Reineck & Singh, 1973, p.100). Testimonial matrices of descriptors about the sediment on Kidd Copper property include an analysis of the fundamental and derived properties of sediments including (1) Composition; (2) Color; (3) Origin; (4) Size of grains; (5) Shapes of the grains (6) Packing of the grains; and (7) Orientation of the grains. This is in keeping with collecting information on sediment specimens. ~0% similarity between facies logs was obtained at the Kidd Copper site. Rhythmic sedimentation, rhythmic stratification, rhythmic layering or rhythmic succession was not observed on the Kidd Copper lithologic logs. The question is answered that Ricci Lucchi in Section 19.11 poses. The layers on Kidd Copper represent depositional units of multiple variables but of the same event.

19.21 Migration of Tailings

"Elemental variations in a sediment core, the pore spaces are filled with water and there is the possibility that an element will be distributed over some distance. Some elements are considered to be mobile (such as S, Mn, Fe, As, Mo, and Ba), whereas others are considered to be immobile (V, Cr, Co, Ni, Cu, Zn, Cd, Sn, Sb, Hg, PB, and Bi). In practice, element mobility should be determined for the system of interest because factors such as pH and oxidation-reduction potential play a role in element mobility" (Eby, 2004, p. 300).

It could be argued that the tailings migrated and the overland runoff and groundwater stream flow is the cause for the difference of stratified layers in the boreholes based on the reason that heavier particles settle leaving the finer particles to lie on top. "A *graded bed* is a layer with a vertical change in particle size, usually from coarse grains at the bottom of the bed to progressively finer grains toward the top. A single bed may have gravel at its base and grade upward through sand and silt to fine clay at the top. A graded bed may build up as sediment is deposited by a gradually slowing current. This seems particularly likely to happen during deposition from a *turbidity current* on the deep sea floor" (Plummer, McGery & Carlson, 2003, p. 140).

Several distinctions are required to note about the Kidd Copper property redeposited mine tailings. First, the event of the dam breaking caused upswelling and the forceful upsurge of water powerfully displaced and mixed the mine tailings; when the force of the water subsided, this rapidly redeposited the sediment as a primary structure. Second, the entire area was subject to the same depositional processes. Third, all of the lithologic logs should have shown a correlation pattern and similar observations in form, arrangement and mutual relationships. Oppositely, the Kidd Copper property *in situ* exhibits a strong random pattern. See the correlations in Table 4 - Section 10.5 and Table 5 - Section 10.6 showing varied tailings deposits. No set pattern or fixed markers exists. The random deposition of the particles is consistent with the hypothesis that stratified layers is a quick process through time and adds credibility to the 'younger earth' theory.

19.22 Peer-review Presentation

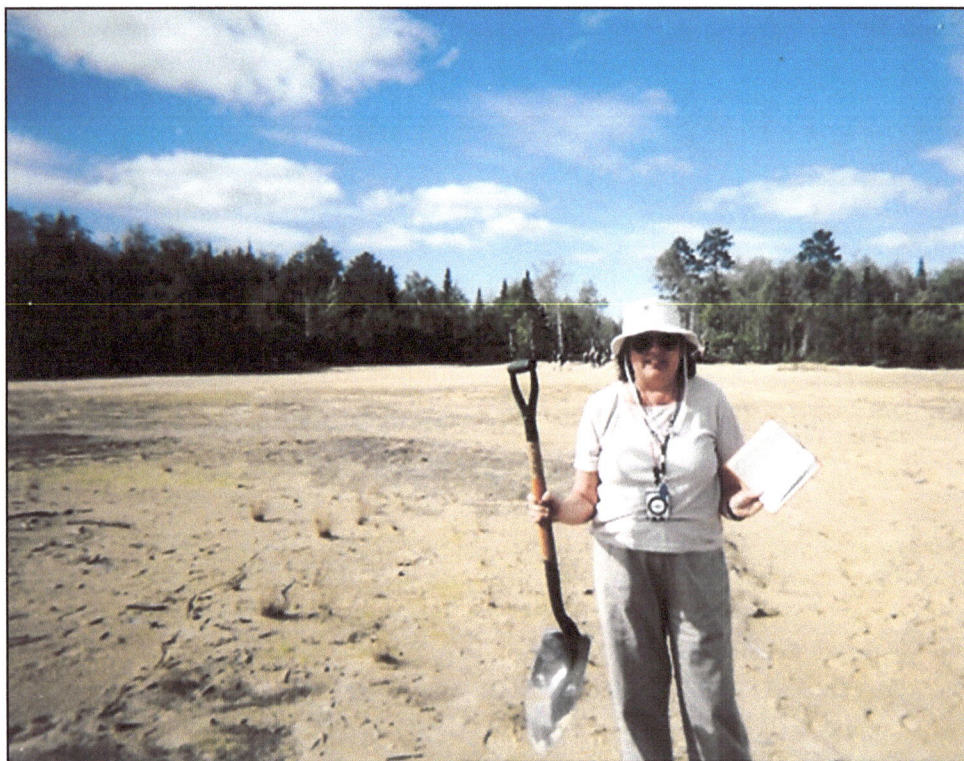

Figure 45 - Section 19.22: Loreen Rudd poses for Field Observations and Records, August 30, 2004

Figure 45 - Section 19.22 is taken on the Kidd Copper property site where Loreen Rudd is standing on the impoundment of mine tailings; and is holding one of the shovels that were used to dig up the 120 boreholes mentioned in this report. Observations and field data were entered and recorded into Forestry Supplier's Field Books, one of which Loreen is holding. This field data is the primary evidence for the main findings in this report. On Thursday, December 2, 2004, a peer-review presentation was given by Loreen

Rudd for the Sedimentary Facies and Environments' class GEO 3E03 with Professors Dr. Joe Boyce and Dr. Carolyn Eyles in attendance to introduce Loreen's observations.

See Loreen Rudd's poster used for her peer-review presentation on Thursday, December 2, 2004, Figure 46 - Section 19.22.

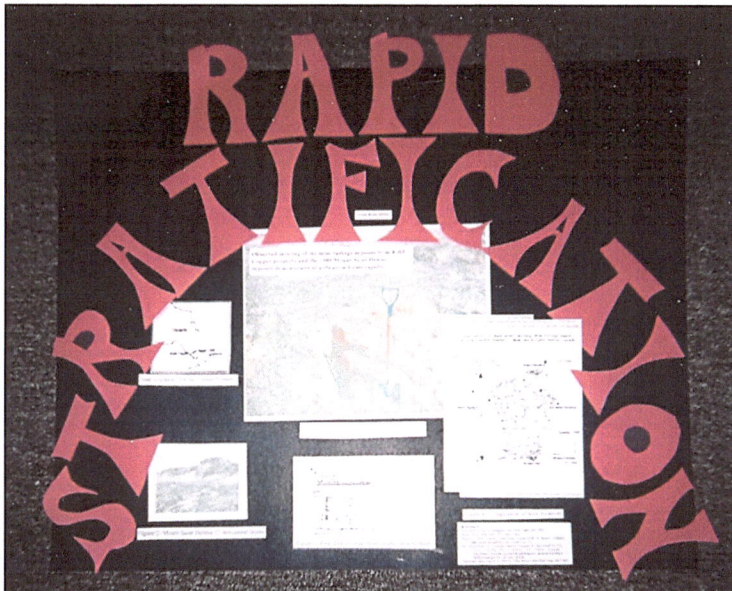

Figure 46 - Section 19.22: Peer-review Poster Presentation of Rapid Stratification

The following is the peer reviewed presentation '*Rapid Stratification*'.

"My name is Loreen. I want to thank all of you people from Field Camp that allowed me access to their field notebooks because without that data I would not have been able to put together this research paper.

Observed layering of the mine tailings deposits from Kidd Copper property and the 1980 Mount Saint Helen's deposits demonstrates stratification forms rapidly.

For my presentation, I am going to focus primarily on methodology, what I saw and how I collected the data.

The mine tailings were found on Kidd Copper property back in August 2004, when we were all at Field Camp. We were given an assignment from our professors and we had to dig holes. I saw something, then, in the holes.

When we got back to school, we were given a new assignment to do a research paper on sedimentology. I thought back to what I had seen at Kidd Copper and wondered if others had seen it too. I had taken only three holes. I needed to collect more evidence.

I got written permission to photocopy my fellow-students' field note books. I then assembled the data and noted over and over again the similar remarks: interbedded, layered, interlayers, stratified, sand layers, alternating layers, liesegang banding. I compiled 120 boreholes detailing where the samples were located.

Note, the stratification was observed to a depth of 50 cm. I dismissed the stratification due to erosional or surface run-off processes because the entire area of the mine-tailings site exposed consistent data. The stratification was noted through-out the entire area.

According to the Law of original horizontality and according to the Law of superposition the expected outcome should have been a uniform, one layer strata.

What we have at the Kidd Copper property is an outdoor observatory. Mine tailings were dumped in the 1960's onto the property and stopped when the dam broke in the early 1970's, after two weeks and the outflow lasted one hour.

When the dam broke, the water would have swept the mine tailings up; there would have been an upwelling from the increased velocity of the water. Then, when the water drained, there would have been a re-settling of sediment, the mine tailings.

According to the laws we should have expected to see uniform, one layer strata. But, what we see is multi-layering. We observe stratification.

We know that the mine tailings were redeposited within the last 20 years because they were formed in the aftermath of the dam breaking. We know this is recent history, not millions or thousands of years ago. The layers formed rapidly and were observed through a very short time in Earth's history.

Let me read from my abstract. The acceptable conventional standard of time interpretation is that the burial of each layer contains information by which the details of this burying can be reconstructed (Selley, 1976). Schuchert in Principles of Stratigraphy on page 128 states for the Paleozoic Era that 3000 years/foot and the rate of formation is 1 ft/2700 years. Expected results should be according to conventional interpretation is uniform strata when there is original deposition of sediment deposits. However, from the evidence collected from the deposits at Kidd Copper property and from eye-witness accounts from Mount Saint Helens 1980 eruption the actual, visible, observable settlement is a stratified layering that is rapidly deposited. If there had been no eye-witness accounts, it would be easy to conclude each layer was laid down slowly over thousands of years. However, this is not the case because the field data of 120 entries supports the stratification of the mine tailings deposits over a short time period, for example the mine tailings deposits were formed in the aftermath of the broken dam. So, therefore, the time of deposition of the resulting layer did not form in thousands of years but instead rapidly. This process was observed through a very short time in Earth's history" (Rudd, December 2, 2004, McMaster University Presentation Rapid Stratification).

19.23 Rapid Stratification Peer-review Presentation Questions

The room was momentarily silenced. Someone in the class asked, "How does this affect the Mesozoic or the Palaeozoic period?" Loreen's reply, "There was no one who lived during that period to actually report and tell us what the rate of settling was. We have no observable recorded information during the age of the dinosaurs. We don't know. But, what we do know is what we have from eye-witness accounts and observable information recorded from the 120 holes at Kidd Copper – Rapid Stratification" (Rudd, December 2, 2004, McMaster University Presentation Rapid Stratification).

19.24 Areal Distribution and Significance of Rapid Stratification

"The areal distribution of continents and oceans has always been a puzzle. The gross radial symmetry of the earth might lead us to expect a crust of uniform thickness spread evenly around the sphere; the oceans, in turn might form a thin shell, overlain by the atmosphere" (Garrels & Mackenzie, 1971, p. 29, 30).

The significance of rapid stratification will be considered and timelines of the geomorphological and hydrogeological events surrounding the Kidd Copper property as to the age of the planet Earth.

"One of the most remarkable aspects of the rock record is the obvious similarity of erosional and depositional processes throughout the past 3 billion years. If we do our best to eliminate postdepositional changes, it emerges that the compositions, textures, and the structures of the most ancient sedimentary rocks can be matched in detail by comparable features of modern sediments. It is perhaps this aspect of the rock record that has kept the concept of Uniformitarianism as a basic principle for almost 200 years" (Garrels & Mackenzie, 1971, p. 67).

No "post-depositional physico-chemical and chemical processes produce stylolites, solution seams and nodules" were found on the Kidd Copper property (Reineck & Singh, 1973, p. 66). The areal distribution of continents considered in context with rapid stratification reveals that the gross radial symmetry of the earth is not a crust of a uniform layered thickness spread evenly around the sphere. Huge gaps exist in the rock record to correlate the crust. Provocative thought wonders if Kidd Copper steps 1-3 are similar to the deposition of the earth's crust. The Kidd Copper event occurs within a defined biosphere and a definitive time period, which includes three specific depositional processes that parallels the flood events of the Earth's mantle. The time references projections from Kidd Copper property could account to a larger scale of world events. Kidd Copper can be validated; the present reveals answers to the past and so, opens up room to consider flood events as redepositing Earth's sediment rapidly.

The minor flood event of the mine tailings waste was redeposited on the impoundment area and the findings from this record will be used to compare the major flood event on the Earth, which uplifted the mantle and redeposited settlement to the crust of the earth. The mini flood will be likened unto the great flood. Therefore, correlating and transferring data from the natural outdoor observatory found on the Kidd Copper property is the solid basis for the system of analysis to support the hypothesis that stratification is rapid. If the rock layers are not read as many different depositional layers through time then it opens the door to the layers accumulating in a shorter time period and quicker. This, then means that the age of the Earth can be shorter. Rapid stratification supports the theory of a 'younger age' for planet Earth.

Table 24 - Section 19.24 is a *Table of Considerations* to determine whether conventional wisdom and geology principles are verified that support the rationalization of an old earth based on slow-time processes against the observations found at Kidd Copper. This table is useful to look at the possibilities of if – then. The reflection presented is the flow of thought presented throughout this report and has been substantiated within the contextual references already provided. The table is an analysis of the expectations that following the laws and principles of geology lead to. For example some of the laws that are explored include: correlation of sedimentary strata;

lateral continuity; rate of sedimentation; Law of superposition; rhythmic layering; rock layers; horizontal lamination; Law of original horizontality; Law of equal declivities; and adjacent sedimentary strata.

Are these assertions reasonable, are these assertions verifiable? What can be supported with additional information and experiments?

Table 24 - Section 19.24: Table of Considerations: If – Then Patterns

Table of Considerations Expected – Outcomes: If – Then Patterns			
Conventional Wisdom & Geology Principles Tested Against Kidd Copper Observations			
Conventional Wisdom & Geology Principles	**If Conventional Geology is to be Observed?**	**Then Conventional Geology Conclusions**	**Verified Observed Kidd Copper Results Outcomes Conclusions**
Correlation of Sedimentary Strata	Parallel structures are side by side	They were deposited simultaneously	No, not all structures were parallel when deposited simultaneously
	Parallel structures are twisted	They were deposited at different times	No, twisted structures were deposited simultaneously
Lateral Continuity	Perpendicular logs were similar	They were deposited simultaneously	No, all logs perpendicular to each other were not the same, yet deposited simultaneously
	Perpendicular logs were different	They were deposited at different times	No, perpendicular logs were different but were deposited at the same time from the same event
Rate of Sedimentation	Many layers	Slow deposition	No many layers are rapidly deposited
	One layer	Fast deposition	No, one layer was not originally deposited; instead many layers were rapidly deposited
Law of Superposition	Younger on top	Multiple depositions over long time	No, mixed layering is one time deposit with no distinction of younger and older beds
	Oldest on botton		
Rhythmic Layering	Many bands/layers	Slowly form over long time	No, the many bands indicate one time deposition, fast
Rock Layers	Flat	Not disturbed & still have their original horizontality	No, mixed layering is one time deposit and can have flat, folding or twisting at point of original deposition
	Folded or inclined	Must have been moved afterwards	
Horizontal Lamination	Initial deposition of new layer	Are the only original sedimentary structures	No, lamination is found at various levels & primary structures can be heterogeneous
Law of Original Horizontality	First sediment layer deposited horizontally originally – always horizontal at time of formation	Time of formation of sands, gravels and muds, layers form through time	No, first sediment layers were tilted and folded and the cause was not tectonic activity
Law of Equal Declivities	Rocks are maturely dissected by consequent streams	Hillside slopes of adjacent valleys will be symmetrical	No, asymmetrical profiles were produced from adjacent boreholes, unequal declivities observed in same depositional time-unit(s) of strata

Table of Considerations Expected – Outcomes: If – Then Patterns			
Conventional Wisdom & Geology Principles Tested Against Kidd Copper Observations			
Conventional Wisdom & Geology Principles	**If Conventional Geology is to be Observed?**	**Then Conventional Geology Conclusions**	**Verified Observed Kidd Copper Results Outcomes Conclusions**
Adjacent Sedimentary Strata	Significant alterations appear in the strata	A separation of time in the deposition of adjacent layers is recognized	No, alterations in strata do not represent breaks in time

Source: Kidd Copper Property August 30, 2004, Scientific Report Analysis by Loreen Sherman, MBA.

From the findings in Table 24 - Section 19.24 it can be concluded that the conventional geologic principles did not favor the expected outcomes when observed with actual data from Kidd Copper. The table shows that many of the expected outcomes from conventional geology were not substantiated by the recorded field observations. For example, the correlation of the sedimentary strata should have been if there are parallel structures side by side then they were deposited simultaneously but the case of Kidd Copper showed that not all structures were parallel when deposited simultaneously. This falsified the expectation from mainstream geologists and challenges conventional wisdom, not all original structures are deposited parallel. Twisted parallel structures are not deposited only at different times.

Lateral continuity expected to find that similar perpendicular logs would result when sediment was deposited simultaneously but again the findings were contrary to these expectations. Instead, different perpendicular logs formed when deposited simultaneously on the Kidd Copper property.

Kidd Copper reveals that many layers or one layer could be rapidly deposited. The rate of sedimentation is not determined by the layers present. Multiple layers formed on the Kidd Copper strata in a short time period. Moreover, the many bands indicate one time deposition, and fast; rather than the expectations that rhythmic layering or many bands was a slow process over time. Again, this geologic position of horizontal layers is layed at time of formation is falsified and conventional wisdom is challenged.

Mixed layering of the rock layers is a one time deposit and can have flat, folding or twisting at point of original deposition. The irregularities at Kidd Copper were not from tectonic or other later processes but occurred at point of origination.

The Law of superposition was not substantiated with its claim that younger sediment lies on top because no distinction of younger and older beds could be determined in the mixed layering at Kidd Copper. Lamination is found at various levels in the Kidd Copper lithographic logs therefore, horizontal lamination is not the only structure for original layers. Rapid stratification is an original depositional structure.

The Law of original horizontality states that at the time of formation of sands, gravels and muds the first sediment layer must be deposited horizontally originally –

always horizontal at time of formation. Furthermore, strata deposited from water "are deposited parallel to the surface… thus horizontally" (MacDonald, *et al.*, 2003, p. 273). Kidd Copper refutes that claim with observations from eyewitnesses. At Kidd Copper property the first sediment layers were not deposited originally horizontally even though conventional wisdom states "such rocks are assumed to have been horizontal at the start of their geological history" (MacDonald, *et al.*, 2003, p. 273).

The Law of equal declivities defines rocks that are maturely dissected by consequent streams and hillside slopes of adjacent valleys will be symmetrical; however in Kidd Copper asymmetrical profiles were produced from adjacent boreholes. This was contrary to the expected outcome. The density of the borehole samples should have supported some regularity; instead none was found. When significant alterations appear in the strata as in the case of Kidd Copper, the adjacent sedimentary strata was expected to be a result of a separation of time. Instead, Kidd Copper supports the claim alterations in strata do not represent breaks in time. Again, conventional wisdom is challenged.

19.25 Summation of Findings at Kidd Copper

Horizontal Layers were Not Found. The Law of original horizontality explains that sediments like sands, gravels and muds are originally deposited or settled out of water in horizontal layers and will be deposited in horizontal layers. After deposition, sedimentary layers (beds) may be further tilted or folded by tectonic activity but they are horizontal at the time they are formed. However, in the case of Kidd Copper, sediment was not disturbed during the settling process, which allowed the primary sediments to settle quietly. Contrary to the expected results from the Law of original horizontality, the Kidd Copper strata did not form horizontal layers; instead the stratum that was formed was bent, twisted layers. Conventional wisdom is challenged.

In the Kidd Copper case the date of the layering of the deposits is precise, as a result of the evaporation of the flood waters over the impoundment area. No tectonic force was involved. The strata simply came to rest. "In science, a law is a descriptive principle of nature that holds in all circumstances covered by the wording of the law. There are no loopholes in the laws of nature and any exceptional event that did not comply with the law would require the existing law to be discarded" (Isaacs, *et. al.*, 2003, p. 445). This data overturns The Law of original horizontality.

The new findings strongly conclude that original deposition and primary structures do not have to be original at their point of origin as held by The Law of original horizontality. Therefore, both horizontal and non-horizontal layers can indicate the start of geological history.

20.0 Rapid Stratification Interpretative Data

The dam broke over the mine tailings deposits in the waste disposal site and created a field observatory for the settling processes. The layers formed rapidly and were observed through a very short time in Earth's history. If there had been no eye-witness accounts, it would be easy to conclude each layer was laid down slowly over thousands of years. However, this is not the case. What is evident and observed is rapid stratification. Multiple layers are laid down rapidly.

20.1 Stratification Defined

"Stratification [GEOL] an arrangement or deposition of sedimentary material in layers, or of sedimentary rock in strata" (Licker, 2003, p. 355). "Stratification index [GEOL] a measure of the beddedness of a stratigraphic unit, expressed as a number of beds in the unit per 100 feet (30 meters) of section" (Licker, 2003, p. 355). "Stratification plane [GEOL] a demarcation between two layers of sedimentary rock, often signifying that the layers were deposited under different conditions" (Licker, 2003, p. 355).

20.2 Assertion That One-time Deposition Confirms Rapid Stratification

Rapid stratification is observed in the layering of the mine tailings deposits from Kidd Copper Property. The quick multi-layering of the deposits after the redeposition of the mine tailings is strong evidence that sediment layers are not evidence of multi depositions; rather the evidence supports multi-layered deposits are one event in time. The layers below and above were deposited the same time as the middle layer. Multi layers are a process of rapid stratification from a single event in Earth's history.

The observable facts demanded of science support rapid stratification in the multi-layering of strata. Interpreting the past can indeed be done by having a standard in which to compare the true knowledge of the present with past depositions. In other words, taking the findings from the Kidd Copper case and using them to unravel the mysteries of the past. The quick multi-layering of the flow from the dam break and the resettling of the mine tailings shows layers are not settled as one deposit on top of another through many events and much time; but rather the evidence supports multi-layered deposits are one event and through rapid stratification.

20.3 Rapid Stratification in Depositional Processes – Inferences

The new premise is that stratification is often laid down quickly in a single catastrophic event, and has no direct connection with the slow-earth time clock concept of thousands of years for these layers to form. Sediment is disassociated through original deposition processes forming multi-layer strata resulting in a time reference frame of about a day. Observed layering of the mine tailings deposits from Kidd Copper property deposits demonstrates stratification forms rapidly. The core lithofacies are not uniform in the layering of the strata; although the same mechanisms and processes of deposition affected each log. Multiple layers of strata are a one event occurrence. The slag was deposited to the site and the rock layers represented the accumulation of the slag being redeposited when the dam broke. The layers are not individual deposits but instead one deposit. Both horizontal and non-horizontal layers can indicate the start of geological history.

20.4 Not Interbedded or Interstratified

What is observed at Kidd Copper is not an interbedding of layers. Kidd Copper's layers were of the same chemical composition but settled out differently throughout the site area. They were not "Interbedded or interstratified – (of strata) being positioned between or alternated with other layers of dissimilar character, e.g. a lava flow interbedded with contemporaneous sediments" (MacDonald, *et al.*, 2003, p. 250). The Kidd Copper layers were more "Stratiform – having the configuration of a layer or bed but not necessarily being *bedded*" (MacDonald, *et al.*, 2003, p. 414). There are observations of similar stratified configurations in the rock record; Kidd Copper is not an anomaly, refer to Sections 13.10, 11 and Section 20.10.

20.5 Lamination Composition of Layers is not Anomolies

Alternating layers are seen elsewhere. Mixed-layer mineral is "a mineral whose structure consists of alternating layers of clay minerals and/or mica minerals; e.g. chlorite, made up of alternating biotite and brucite sheets" (Bates & Jackson, 1984, p. 331). "Mixed-layer mineral [Mineral] A mineral having an interstratified structure consisting of alternating layers of two different clays or of a clay and some other mineral" (Licker, 2003, p. 207). "Mixed ore [GEOL] any ore with both oxidized and unoxidized minerals" (Licker, 2003, p. 207).

"The origin of lamination has received somewhat more attention than the origin of bedding and a number of different types of laminae have been described: (a) Laminae composed of alternating light and dark layers, the light layers being generally coarser than the dark layers, which are enriched in fine grained organic matter. This type of lamination is the type most typical of glacial *varves*, i.e. lamination caused by the seasonal supply of sediment to glacial lakes … (b) Laminae composed of layers of approximately equal grain size, but of different mineral composition. Individual layers may be rich in detrital mica or heavy minerals. In some cases there is a strong segregation of different minerals in different layers. (c) Laminae composed of layers that differ in grain size characteristics. Such laminae may show sharp or gradational boundaries and in some cases they show grain size *grading*, with the coarse grains at the base of a lamina, or *reverse grading*, with the coarse grains at the top of a lamina. (d) Laminae alternately rich and poor in clay but otherwise similar in the grain size of the coarser fraction   I Laminae in mudstones or shales that appear differ in color only" (Blatt, Middleton & Murray, 1972, p. 116).

"It is known that laminae may be formed in muds by periodic (in some cases seasonal) changes in the physical or chemical conditions of deposition" (Blatt, Middleton & Murray, 1972, p. 116). "In contrast to lamination in muds, lamination in sands appears to be produced rapidly" (Blatt, Middleton & Murray, 1972, p. 117). This is further proof that the lamination compositions of the alternating layers present at Kidd Copper are not anomalies because rapidly produced layers are seen elsewhere in the rock records.

20.6 Start of Geologic History

Significance of Multiple Layers. Horizontal layers cannot be used to accurately indicate the start of geological history because twisted layers are equally indicative of the start of geological history. The horizontal layers do not indicate separate depositional process or periods. Uniform strata are not representative of the geologic breaks through time in the rock records. Field observations showed that mud stirred (a well-mixed in a blender deposit) and allowed to settle quietly yielded stratified laminae; multiple-layering is observable for a single, one-time depositional event.

"Hutton believed that 'the past history of our globe must be explained by what can be seen to be happening now" (Levin, 2003, p. 6). The relative age methods employs the simple notion that the deeper an intrusion was found in stratified (layered) sediments, the older it must be. James Hutton stated that the same geological processes that occur in the present also occurred in the past. The evidence suggests that layering of strata is not a verifiable way to get the relative age of a rock unit. The evidence collected from the deposits at Kidd Copper property proves the non-distinct, random, primary structures were formed as the aftermath of the flood; and quickly, as rapid stratification. If there had been no eye-witness accounts, it would be easy to conclude each layer was separate and was laid down slowly over thousands of years. However, this is not the case because the field data of 120 entries supports the rapid stratification of the mine tailings deposits over a short time period. Therefore, the time of deposition of the resulting layers did not form in thousands of years but instead rapidly. This process was observed through a very short time in Earth's history. Therefore, the Earth can be substantially younger than previously considered.

Science is supported on evidence that is seen.

20.7 Qualitative Data Interpretation

Geological record is "the testimony of the history of the Earth as revealed by its *bedrock*, *regolith* and *morphology*" (MacDonald, *et al.,* 2003, p. 205). "Regolith – "A layer of weathered rock fragments that lies below the soil and above the bedrock below. In the tropics it may be up to 60 meters thick. Fine regolith differs from soil in that it has no organic content (humus) and cannot therefore support plant life, although the roots of trees and other plants may penetrate it for water" (Isaacs, *et. al.,* 2003, p. 674). "Sedimentary rock is the Earth's greatest storyteller. This type of rock form in layers, and the order in which these layers are laid down – their sequence – recounts the geologic history of the region" (Discovery Books, 1999, p. 24). "Qualitative analysis involves determining the nature of a pure unknown compound or the compounds present in a mixture" (Isaacs, *et. al.*, 2003, p. 40). This aspect will be limited in this report to the nature of the adjacent sedimentary strata.

20.8 Adjacent Sedimentary Strata

If the same sediment and process occurs, what is observed? Does an equal distribution pattern settle in surrounding strata like the Law of equal declivities suggests? No, refer to Section 15.8. With the sediment and the depositional process being the same for every site borehole collected at the Kidd Copper Property, shouldn't the resulting strata be the same? The rational is to conclude that equal matter and equal processes provides equal results? The rational is that sediment from the same slope when lifted up and deposited land in the same manner? Conversely, this was not found. The adjacent sedimentary strata at Kidd Copper; instead was found unequal in the amount or type of layering distribution. Refer to Table 4 - Section 10.5 and Table 5 - Section 10.6. Each borehole showed different layers to the adjacent strata yet had equal correlation to time.

Textbook Expectations. "Conformity – (1) also called *conformability*, the stratigraphical continuity of adjacent sedimentary strata, i.e. they have been deposited in orderly series without evident time lapse, (2) a surface separating older strata from younger, along which neither erosion nor non-deposition is evident and with no significant hiatus" (MacDonald, *et al.*, 2003, p. 104). The expectations from the textbooks and mainstream geologists is that when significant alterations of appearance are revealed in the strata then a separation of time in the deposition of the adjacent layers is recognized. In other words, the story revealed in the rock layers according to conventional geology is that separations account for time lapses.

Contrary, to the expectations of the readings the adjacent sedimentary strata is not orderly laid down; and the given history is known that the age of each adjacent layer was the same time in the case of the Kidd Copper Property. The adjacent sedimentary strata did not show regularity in the sediment distribution in spatial patterns, nor in chemical compositional materials. The separations in the rock layers do not account for separate depositional times. Instead, the observable evidence supports that the disorderly layering marks the same time dispensation.

20.9 Conventional Interpretation Falsified

Expected results according to conventional textbook interpretation are that uniform strata should occur when there is original deposition of sediment deposits. However, from the evidence collected from the deposits at Kidd Copper property the actual, visible, observable settlement is a stratified layering that is deposited. If there had been no eye-witness accounts, it would be easy to conclude each layer was laid down slowly over thousands of years. However, this is not the case because the field data of 120 entries supports the stratification of the mine tailings deposits over a short time period, for example the mine tailings deposits were formed in the aftermath of the broken dam. So, therefore, the time of deposition of the resulting layer did not form in thousands of years but instead rapidly. This process was observed through a very short time in Earth's history.

Observations at Kidd Copper contradict textbook knowledge. The observation does not align itself with textbook knowledge in the following statements: (1) The Law of original horizontality explains that sediments like sands, gravels and muds are originally deposited or settled out of water in horizontal layers; (2) Beds may be tilted by tectonic activity but they are horizontal at the time they are formed; and (3) Based on these geologic principles it would suggest that uniform layering would be present and evident in the land base especially visible on the first layer or the first few inches of land surface at Kidd Copper. According to the geologic laws uniform, one-layer strata should have been observed. Instead, what is seen is multi-layering. Rapid stratification is observed.

The data provided by on-site field observations at Kidd Copper property provides evidence contrary to the geologic laws with which geologic age is based on. The outcomes based in this report challenges conventional thinking as the expected long-time rate of sediment deposits if these laws were true. Instead, evidence is such that based on the rapid stratification of the laminae layers on the Kidd Copper property a short period of time resulted in the formation of many layers.

20.10 Other Current Observations Support Young-Earth Strata Deposits

"To understand ancient rocks, we must first understand present-day processes and their results" (Tarbuck & Lutgens, 1999, p. 4). Turcotte (1986) argued that the 'marble cake structure' associated with imperfect mixing can be seen in high-temperature peridotites (also called orogenic lherzolite massifs), which represent samples of the Earth's mantle" (Turcotte & Schubert, 2002, p. 288). "In the best tradition of the uniformitarianism and the natural-history approach to geology, one should study sediment-gravity flow deposits by reading the observations of actual rockfalls by eyewitnesses" (Hsü, 2004, p. 87).

"Mount St. Helens' initially was subdivided into nine eruptive periods. Before 4,000 years ago eruptive periods lasted for thousands of years, in contrast to younger ones that had durations of only centuries. To correct that discrepancy the eruptive periods were reclassified. The first three periods were combined as stages and the last six periods were placed into a single stage. In the revised classification, each stage included repeated episodes of activity and intervening dormancy spread over more than a thousand years" (Mullineaux, 1996, p 84)" (Rudd, November 2004, GE0 3E03 Research Paper, p. 3). The re-classification of the system is problematic and leaves room for question as to the validity of the classification. How many other problems does this classification system have? If Mount St. Helens' classification was changed and corrected once already how about other reclassifications to support the evidence of new findings?

Figure 47 - Section 20.10: Seventeen Horizontal New Layers in the Enbankment, Loren Nelson

Photograph was taken Saturday, June 23, 2001 by Loren Nelson in the restricted area between Spirit Lake and Mount Saint Helens Crater along the Truman Trail (#207), Refer to Figure 47 - Section 20.10.

"Eye witness accounts verify that the current layers at Mount St. Helens were laid down in a matter of hours not a matter of millions of years. In the article, Flood Strata formed at Mt. St. Helens, Loren Nelson states that everything seen in [Figure 47 - Section 20.10] was non-existent 21 years ago. Where this picture was taken the depth of the fill is 300' because of the volcanic eruption occurring on May 18, 1980 and the subsequent landslide that followed. Fourteen miles of the north Fork of the Toutle River was buried by rocks, mud & debris to an average depth of 150'. By standard geologic measurements there are at least 17 horizontal layers in the embankment, which could represent several million years according to current scientific reasoning. Yet, most of the deposit occurred during the main eruption on May 18, 1980. If there had been no eye-witness accounts, it would be easy to conclude each layer was laid down slowly over millions of years. However, this is not the case" (Rudd, November 2004, GE0 3E03 Research Paper, p.1-3).

"If the date of the responsible volcanic expression can be determined, then a tephra layer serves as datum surface of known date" (Friedman, *et.al.*, 1992, p. 175, 176). Observed layering of the mine tailings deposits from Kidd Copper property and the 1980 Mount Saint Helens deposits demonstrates stratification forms rapidly because of a distinguishable datum surface of known date. Therefore, the Earth can be substantially younger than previously considered because the rock layers may not be interpreted correctly as one layer being laid down slowly over thousands of years.

20.11 Past Tendency to Interpret Each Laminae Separately

"Sedimentation onto the surface of a bed form produces inclined laminae. Sedimentation alone ultimately results in burial of the bed form and formation of a plane surface but in combination with some bed load movement it may result in the formation of structures such as a climbing ripple-drift cross-lamination. Accretion may form cross-lamination on the upflow (stoss) side of bed forms such as ripples, dunes, and antidunes" (Blatt, Middleton & Murray, 1972, p. 118).

"In the past there has been a tendency to interpret each lamina as produced by a separate sedimentation event, for example, a tidal cycle, the swash and backwash from a single wave, or a single bed load avalanche. It is now clear, however, that laminae may also be produced by steady flow, particularly during traction on a plane bed in the upper flow regime. The mechanism is not well understood but it appears to be related to factors that cause grains of similar size and shape to lodge together in patches on the bed. In the case of grains of mixed size, the larger grains tend to roll over smaller grains … but lodge against grains that are larger or of equal size. It is not easy to understand how a finer than average lamina is deposited but laminae of alternating grain size have been produced under conditions of steadily decelerating flow in laboratory experiments (Kuenen, 1966)" (Blatt, Middleton & Murray, 1972, p. 117).

However, today with the evidence from the Kidd Copper property alternating layers are understood as a single event, refer to Section 20.2.

20.12 Structural Geology

The study of stress and strain, the processes that cause them, and the types of rock structures that result from them are the subject of structural geology. Structural geologists study evidence of past rock deformation to help find out what kind of deformation has occurred in a particular area. This helps them understand the "stresses that prevailed at different times in the past, which in turn allows them to decipher the geologic history of the area" (Murck, 2001, p.174). The fact the Earth is very old has been established on the basis of structural geology. This report questions the foundation of geologic time based on relative time and structural geology because the findings from the data does not substantiate the underlying principles or laws as noted earlier. Instead, the multiple layers in the Kidd Copper property were laid down in a very short period of time and thus the Earth can be younger than the common answer of 4.4 billion.

20.13 Secondary Stratification

In geology the definition of "secondary clay [GEOL] a clay that has been transported from its place of formation and redeposited elsewhere" (Licker, 2003, p. 327). "Secondary stratification [GEOL] the layering that occurs when sediments that were at one time deposited are resuspended and redeposited. Also known as indirect stratification" (Licker, 2003, p. 327). The tailings deposit formed an unstratified layer as the miners deposited the slag onto the impoundment area but after the water washed up the tailings they were redeposited. The new deposit was stratified. What this does is strongly confirm the supposition of this report. The stratified layers at Kidd Copper are the result of resuspended and redeposited sediment and not the result of a long time process, which opens the doorway to consider that other stratified layers have the same depositional process as observed on the Kidd Copper property.

20.14 Complicated Geometries

"Geologists working in areas of deformed strata carefully scrutinize rocks for indications of the original tops and bottoms of beds. Features providing such information are called *geopetal structures*" (Levin, 2003, p. 75). "No very satisfactory answer to the intriguing question of why a sediment bed is molded into complicated geometries instead of simply remaining flat during sediment transport" (Harms, *et al.*, 2004, p. 8). Until now; because the findings at Kidd Copper reveal that initial deposition can be complicated geometries and that the multiple layers are not many events but rather one event can provide many layers unravels the puzzle.

21.0 Stratified Layers and Time Conclusions

Recap. The sequential layering of sediment in water and the redeposition of the sediments occur as a single-event in the case of the Kidd Copper property study. A definite random deposition in step 3 of the mine tailings is prevalent with noticeable mixing of colors and various elements. There is not a single similar pattern in the comparisons of lithographic logs correlated in Table 4 - Section 10.5 and Table 5 - Section 10.6. Rhythmic sedimentation or rhythmic stratification, where repetitive patterns or repeated changes in lithology were not observed in the stratified sediment on Kidd Copper. No repetitive pattern is distinguishable. The heterogeneous sediment found on the site occurred from rapid stratification in a short period of time. The layering of strata can occur very quickly.

21.1 Young Age of Tailings

The dam broke over the mine tailings deposits in the waste disposal site and created an outdoor, natural observatory for the settling processes. The Kidd Copper site provides a natural outdoor observatory of original sedimentation and layering of sediments. The water provided by the breaking of the dam provided the force to uplift the mine tailings and redeposit them. The settling of the deposits was without disturbance. The mine tailings were redeposited within the last forty years; they were formed in the aftermath of the dam breaking. This is a fact recorded in recent history.

A known fact is that the tailings deposit didn't exist prior to the 1960's, so why would it be thousands of years old? This is a true statement because the history of the Kidd Copper Mine's deposits confirms deposition in the 1970's and thus the age of the layered deposit is less than 40 years old. The stratification is not an event happening over thousands or millions of years. The layers formed rapidly and were observed through a very short time in Earth's history. If there had been no eye-witness accounts, it would be easy to conclude each layer was laid down slowly over thousands of years. However, this is not the case. What is evident and observed is rapid stratification.

The settling processes were uneventful: (1) mine tailings stirred; and (2) settled quietly forming stratified layers on the Kidd Copper property. The water drained off the region and redeposited the mine tailings. The mine tailings sediment has remained inactive for the last 40 years left untouched because the mining production stopped. There are no additional strata being deposited to the ground surface; yet what is seen is clearly stratified layering of sediment deposits. Interpreting the past by reference to the present is a strong fact finder. A precise date can be given for the mine tailings. The multiple layers were settled in a relatively short period of time and; thus it can be concluded that stratification of sediment can be a one event occurrence; instead of stratification marking individual time layers and denoting a long deposition time period.

21.2 Stratigraphic Correlation Defined

Stratigraphical correlation is "the procedure by which the mutual correspondence of *stratigraphical units* in two or more removed locations is shown. It is based on *fossil* content, geological age, lithographic features or some other property. The term usually implies the identification of rocks of equivalent age in different places" (MacDonald, *et al.,* 2003, p. 414).

"Stratigraphical unit or rock stratigraphical unit – A stratum or body of strata, recognizable as a unit, that may be used for mapping, description or correlation; it does not constitute a time-rock unit. A *lithostratigraphical unit* is a body of rock strata with certain unifying lithological features. Although it may be sedimentary, igneous or metamorphic, it must meet the critical requirement of an appreciable degree of overall homogeneity" (MacDonald, *et al.*, 2003, p. 414).

21.3 Time-stratigraphic Facies Defined

"The different aspect of each measurement unit is summarized by the term *facies* (from the Latin word for aspect, or 'appearance of' something)" (Walker & James, 2002, p. 1). "Time-stratigraphic facies [GEOL] a stratigraphic facies based on the amount of geologic time during which deposition and nondeposition of sediment occurred" (Licker, 2003, p. 375). "Time-stratigraphic unit [GEOL] a stratigraphic unit based on geologic age or time of origin. Also known as chronolith; chronolitholgic unit; chronostratic unit; chronostratigraphic unit; time-rock unit" (Licker, 2003, p. 375).

Sedimentary facies is "an areally confined part of an assigned *stratigraphical unit* that shows characteristics clearly distinct from those of other parts of the unit" (MacDonald, *et al.*, 2003, p. 386). "Strata-bound ore body – a *mineral deposit* of any form (*concordant* or *discordant*) that is restricted to a single stratigraphical unit" (MacDonald, *et al,* 2003, p. 413, 414).

21.4 Interpretation of Stratigraphic Unit Point 64/X1

"Infancy [GEOL] the initial (youthful) or very early stage of the cycle of erosion characterized by smooth, nearly level erosional surfaces dissected by narrow stream gorges, numerous depressions filled by marshy lakes and ponds, and shallow streams. Also known as topographic infancy" (Licker, 2003, p. 162, 163).

"Youth [GEOL] the first stage of the cycle of erosion in which the original surface or structure is the dominant topographic feature; characterized by broad, flat-topped interstream divides, numerous swamps and shallow lakes and progressive increase of local relief. Also known as topographic youth" (Licker, 2003, p. 401).

"Old age [GEOL] the last stage of the erosion cycle in the development of the topography of a region in which erosion has reduced the surface almost to base level and the landforms are marked by simplicity of form and subdued relief. Also known as topographic old age" (Licker, 2003, p. 227).

Without knowing the history of the Kidd Copper site, it would be difficult to determine by looking at the log 64/X1 (Figure 48 - Section 21.4) if the layer of mixed sediment of 30 cm was mature or an infant stage. Mark D. Licker published that topography infancy is the initial or youthful stage (2003, p. 162, 163). Determination of the stage is difficult without direct knowledge of the sedimentation processes involved. How does one know when or *if* "the initial (youthful) or very early stage of the cycle of erosion characterized by smooth, nearly level erosional surfaces ..." (Licker, 2003, p. 163). Or is the stage "the last stage of the erosion cycle in the development of the topography of a region in which erosion has reduced the surface almost to base level and the land forms are marked by simplicity of form and subdued relief ... known as topographic old age" (Licker, 2003, p. 227). These questions would be difficult to answer without knowing the history of the site. A guess may be inconclusive. Then again, one could suppose that this stratigraphic log unit 64/X1 marked "the first stage of the cycle of erosion in which the original surface or structure is the dominant topographic feature; characterized by broad, flat-topped interstream divides, numerous swamps and shallow lakes, and progressive increase of local relief ... known as topographic youth" (Licker, 2003, p. 401). Was Kidd Copper early erosion or topographic old age? The answer comes from knowing the historical depositional record at Kidd Copper.

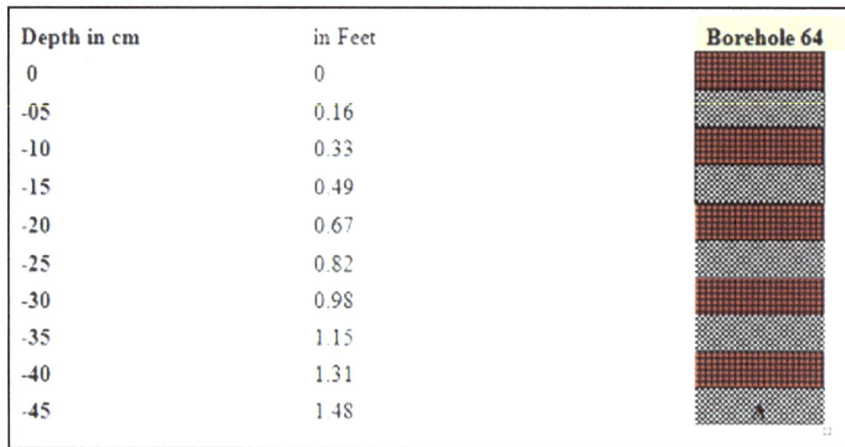

Depth in cm	in Feet	Borehole 64
0	0	
-05	0.16	
-10	0.33	
-15	0.49	
-20	0.67	
-25	0.82	
-30	0.98	
-35	1.15	
-40	1.31	
-45	1.48	

Figure 48 - Section 21.4: Stratigraphic Unit Point 64/X1

The correct interpretation of Borehole 64 stratigraphic log unit 64/X1 (Figure 48 - Section 21.4) is that the sediment was laid as slag waste on the impoundment area and a flood event stirred the sediment. No disturbance affected the quiet settling of the mixed sediment as it was redeposited on the Kidd Copper site area. Unearthing of the sediment revealed that for 45 cm a depth of about 1 ½ feet varied layers were uncovered of rust color and white. No erosional surface was noted. Since the sediments had been buried oxygenation was not a factor. The stratified layers were a result of rapid burial and do not account for different depositional times. Each layer represents the same depositional process and is one unit. The recent deposit of sediment on the Kidd Copper is the initial stage of topographic erosion. According to conventional wisdom and published by Mark D. Licker the initial (youthful) or very early stage of the cycle of erosion should be characterized by smooth, nearly level erosional surfaces.

21.5 Interpretation of Stratigraphic Unit Point 56/4

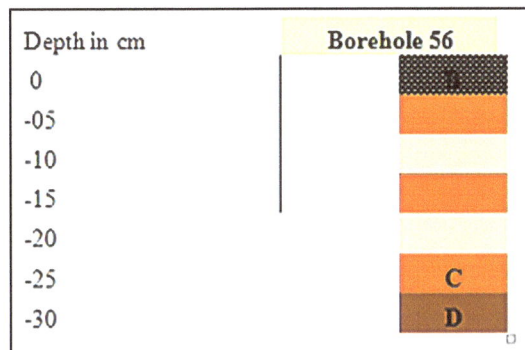

Figure 49 - Section 21.5: Stratigraphic Unit Point 56/4

The field description of Point 56/4 is: Depth 30 cm; 1 cm Dark black/beige spotted; Beige/ bright orange layers; Iron waste product oxidized, see Figure 49 - Section 21.5.

Point 56/4. Is this an *"interlayer slip – Deformation accomplished by displacement* along *bedding* or *foliation* surfaces?" (Kearey, 2001, p. 136). The deformation seen on Point 56/4 was not caused by either disturbance or by slipping; instead, it was the result of the quiet settling of sediment out of water. Point 56/4 is not an interlayer slip because the point was not displaced along bedding or foliation surfaces.

Borehole Control Point 56/4 reveals three layers of laminae (B, C, D) in Figure 49 - Section 21.5. Known fact: original time that the deposition of all three layers of laminae was simultaneously; therefore, the correct interpretation is not "time-transgressive… Diachronous [GEOL] of a rock unit, varying in age in different areas or cutting across time planes or biostratigraphic zones" (Licker, 2003, p. 96). The multi-layering in way Point 56/4 was not formed through the passage of time; therefore, time transgression is an erroneous conclusion if applied as the interpretation.

Annual layer is "any sedimentary or precipitated layer of material of variable composition and thickness, presumed to have been deposited during the course of one year. Annual layers are most frequently utilized in the study of varved deposits laid down in glacially fed lakes. Certain non-glacial deposits also exhibit annual layers, e.g. the alternating laminae of anhydrite and calcite in a salt intrusion. It is thought that the anhydrite and calcite represent deposits of dry and humid seasons respectively" (MacDonald, *et al.*, 2003, p. 33). "Diagenesis is the general term for all modifications to sediment characteristics that take place after the sediments have been deposited" (Lewis,

1984, p. 33). This sounds like a reasonable explanation to the layering of the sediments in the Kidd Copper property if the true deposition of sediment was not known; however the diagenesis is known to have occurred as the result of rapid stratification in a brief period; so this explanation is erroneous. Note, that no alternating laminae of anhydrite and calcite was found, nor was Kidd Copper property in a salt intrusion; thus an *annual layer* can be rejected as a possible interpretation.

Another possible explanation offered as an interpretation could be the aforementioned two chronostratic unit interpretations were guilty of "*overprinting* – the superimposition of a *mesoscopic* to *microscopic* younger *structure* on an older one" (Kearey, 2001, p. 192). Overprinting is not the correct interpretation because the sediments were laid down at the same time. These interpretations would be reasonable if one did not know the history of the site but science is evidenced on what is seen and observed. The facts of the site do not support these other explanations.

21.6 All Boreholes are Isochronous

"Chronohorizon or chronostratigraphic horizon – a stratigraphical surface, theoretically without thickness, that is the same age (*isochronous*) at all points. In practice it is a thin and characteristic interval that serves as a good time-correlation or time-reference zone (MacDonald, *et al.*, 2003, p. 92). The lithostratigraphic correlation at Kidd Copper revealed all the boreholes to be isochronous units to establish a time base line because they were formed at the same period. Although, each borehole was isochronous they did not share 100% similarities in stratigraphy as seen in Table 6 - Section 11.4. All the Kidd Copper property boreholes in this report can be correlated as a chronohorizon or chronostratigraphic horizon because they share the same deposition and age. The Kidd Copper data is isochronous.

21.7 Reconstruction of Palaeoenvironment of Kidd Copper Property

"The role of the sedimentologist is to interpret the history of sedimentary deposits. The scope of the subject is broad and entails both the study of processes involved in the origin, transportation, deposition, and postdepositional modification of sediments and the study of the lateral and vertical distribution of sedimentary deposits in relation to modern and ancient geographic, tectonic, and climatic settings" (Lewis, 1984, p. 1). "The first stage in a stratigraphic analysis is to determine the order in which rocks were formed" (Nichols, 2003, p. 229). "One of the objectives of sedimentological and stratigraphic analysis of rocks is the reconstruction of palaeoenvironments" (Nichols, 2003, p. 231). "In order to reconstruct the palaeoenvironment at any time in the past it is necessary to know what depositional environments existed and where they were to be found at that time" (Nichols, 2003, p. 231). The event of the redeposition of the sediment deposit occurred around the 1970's, so the reconstruction of the palaeoenvironment of Kidd Copper is definitive.

It would be easy to conclude each layer seen on the Kidd Copper property was laid down slowly over millions of years if there had been no eye-witness accounts; however, this is not the case. Observations and testimonials about the mine tailings deposits from Kidd Copper property is strong evidence that stratification forms rapidly. The three step process was observed through a very short time in Earth's history.

21.8 Extrapolations Not Needed at Kidd Copper

"It has always been a goal of geologists to determine the age of the Earth and give dates to events in Earth history. Rock relationships, lithostratigraphy and biostratigraphy provide essential information about the order in which things have happened in a relative sense, but give no idea of the absolute age of anything … In practice the whole process of dating and correlating rock relies on the integration of information from a number of different sources and techniques" (Nichols, 2003, p. 251).

The deposition of sediment and the redeposition of the sediment from the aftermath of a flood event are recorded in the annals of time at Kidd Copper. The depositions occurred in the early 19th century leaving multiple layers; rather than the layers accounting for multiple events and a long period of time to accumulate. In this brief historical period, multiple layers of sediment are observed. The evidence attained from the Kidd Copper impoundment submits that reading the rock layers is not a verifiable way to date the events in Earth's history. Often guesswork is necessary to extrapolate any time line. The extrapoliations are not reliable since no start date can be determined of the layers unless the age of the geological strata is known.

21.9 Rapid Stratification Opens the Door to a Younger Earth

Sedimentary structures and the formation of outcrops are what conventional geology looks at to consider the conditions of deposition. Sometimes the style of stratification is considered. Often the layers of the piles of strata on the earth have been used to determine the age of the earth. Conventional wisdom says that the Earth is very old based on many radiometric dates. This report strongly supports that basing dates on radiometric or other calibrations is not reliable, refer to the evidence in Section 17.

Many processes can deposit stratified sediments very rapidly. For example – volcaniclastic deposits (i.e. ash and tuff layers), glaciofluvial deposits, mass movements etc., this substantiates that rapid stratification is not an anomaly. Moreover, observed layering of the mine tailings deposits from Kidd Copper property demonstrates stratification forms rapidly.

The stratified layers found in the lithographic log associations on Kidd Copper occurred through a very short time in Earth's history. The quick multi-layering of the flow from the dam break and the undisturbed resettling of the mine tailings show that layers are not settled as one deposit on top of another through many events and much time; but rather the evidence supports multi-layered deposits are one event as rapid stratification. The data acquired through the mine tailings deposits on the Kidd Copper property is that the multiple layers were laid down in a relatively short time period and thus opening the door to the age of the Earth can be younger.

21.10 Doorway to a Younger Earth Opens

The evidence in this report is based on the following supportive facts about the Kidd Copper property research collected on August 30, 2004: (1) The site is a well-known study area; (2) The site area has remained inactive; (3) The collected data was found in a natural state in a primary depositional structure; (4) Future experiments can be synthesized in laboratory conditions to reinforce findings; (5) Objective data was collected; (6) Samples covered an extensive area of the ground cover; and (7) Log entries were marked as an assignment for McMaster geology courses; (8) The eye-witness observations provided the basis of analysis; (9) The observations were compared with conventional wisdom established on geologic laws and principles; (10) The observations from Kidd Copper provided primary research, not relying on assumptions from textbook theories. The research in this report is primary and observation-based for all findings related to the Kidd Copper property; however, extensive secondary research has been conducted to determine the accuracy of the analysis of the findings throughout this report.

This report has verified that one stratified deposit observed at Kidd Copper deposited in the last 40 years is geologically young.

The question that comes to mind is: How does one stratified deposit translate into the Earth as a whole being younger?

Many variables were analyzed in the predictors for depositional outcomes to determine the resultant structure formed in the case study, Kidd Copper property dated August 30, 2004, refer to Table 25 - Section 21.10.

Conventional wisdom predicts the pattern or distribution of the variables. For example, spherical grains of one size were assumed to be evenly distributed. This assumption was not verified and can be rejected because the Kidd Copper grains were not evenly distributed throughout the tailings. Another example, in Table 25 - Section 21.10 predicts that fine grained sediment was expected to be left behind as a horizontal deposit. Again, the Kidd Copper findings reject this prediction because the observed sediment was vertical, staggered and unstructured deposits. The findings in this report show that many predictions about several sedimentological variables using conventional wisdom can be rejected because they are not verified.

Table 25 - Section 21.10: Predictors for Depositional Outcomes

Predictors for Depositional Outcomes Case Study: Kidd Copper Property, August 30, 2004 Outcome Based on Site Observations and Testimony		
Variable	**Pattern or Distribution Predicted**	**Actual Resultant Structure Formed**
"Grains are spherical and of one size" (Coggans, *et al.*, 1991, p. 457).	"Further assumptions are that the sulfide grains are evenly distributed throughout the tailings" (Coggans, *et al.*, 1991, p. 457).	Grains were not distributed evenly throughout the tailings.
"Law of original horizontality – sedimentary rocks, especially water-laid strata, are deposited parallel to the surface on which they are deposited and thus horizontally" (MacDonald, *et al.*, 2003, p. 273).	"Such rocks are assumed to have been horizontal at the start of their geological history, although they may have been folded or tilted subsequently" (MacDonald, et al., 2003, p. 273).	Outcome was not homogeneous layers, not uniform layers. Heterogeneous structure is a primary sedimentary structure.
High solubility of chemical composition of sediment.	Predict – homogeneous mixture when stirred because "solubility is a property of a substance by virtue of which it forms mixtures with other substances that are chemically and physically homogeneous throughout" (DePree & Axelrod, 2003, p. 686).	"If a mineral reacts with an aqueous solution, and if the material dissolved from the mineral has the same composition as the mineral, it is said to dissolve *congruently*" (Garrels & Mackenzie, 1971, p. 355). Kidd Copper sediment in solution was well-mixed but redeposited as heterogeneous.
Turbulent flows capture sediment.	"Heterogeneous fluids are thoroughly mixed in turbulent flows" (Nichols, 2003, p. 38).	Heterogeneous fluids deposit as random stratified layers when mixed thoroughly.
"Deposition from strong currents" (Reineck & Singh, 1973, p. 49).	"Upper plane-bed phase lamination" (Reineck & Singh, 1973, p. 49).	Mixed layers of parallel lamination of more than one unit.
"Deposition from suspension, low-density turbidity currents or from weak traction currents" (Reineck & Singh, 1973, p. 49).	"Lower plane-bed phase lamination" (Reineck & Singh, 1973, p. 49).	Unstructured lamination.
Fine-grained sediment.	"When floodwaters recede, these fine-grained sediments are left behind as a horizontal deposit" (Plummer, McGery & Carlson, 2003, p. 237).	Vertical, staggered, unstructured deposits.
Silt, sand or granules.	"Horizontal or nearly horizontal laminae in silt, sand or granules" (Harms, *et al.*, 2004, p. 49).	Horizontal was one of many deposit structure forms.
Turbidites.	Horizontal stratification (Harms, *et al.*, 2004, p. 50).	Horizontal stratification and other bed forms similar to compound cross-stratification.
Rapid flow.	"Planar lamination produced by rapid flow" (Nichols, 2003, p. 120).	No structure or common pattern found, planar lamination was not produced; instead rapid flow produced stratified and diversified yet isochronous.
Colloid – small 0.06 mm particle.	"Evenly and stably distributed in another material" (Research Machines, 2003, p. 133).	Dispersed unevenly with a heterogeneous appearance.
Within a sequence of undisturbed sedimentary rocks.	"The oldest layers are on the bottom, the youngest on the top" (Plummer, McGery & Carlson, 2003, p. 565).	Multiple layers can reflect the same time zone for deposition.
Conclusion	**NOT VERIFIED**	**Findings Support Rapid Stratification**

Mixed layers of parallel lamination of more than one unit; unstructured lamination; and multiple layers can reflect the same time zone for deposition are the main findings from the data analysis on Kidd Copper predictors, refer Table 25 - Section 21.10. All of these support rapid stratification. These findings reject that planar lamination is produced by rapid flow and that the oldest layers are on the bottom while the youngest layers are always on the top.

Conventional wisdom specifically the Law of original horizontality expected that when turbulent flows captured sediment it would mix thoroughly and the deposits would be parallel, adversely Kidd Copper did not support this prediction. The fine grained sediment and colloids on Kidd Copper showed uneven dispersion and heterogeneous layers. Again, conventional wisdom is challenged.

21.11 Uniformitarianism Re-defined

"The acceptance of uniformitarianism meant the acceptance of a very lengthy history for Earth" (Tarbuck & Lutgens, 1999, p. 5). "It was apparent from Lyell's principle of uniformitarianism that the many thick deposits of the stratigraphic column must have taken a long time to accumulate" (Busbey, *et. al.*, 1999, p. 20). "We can examine any rock, however old, and compare its characteristics with those of similar rocks that are forming today. We can then infer that the ancient rock very likely formed in a similar environment, through similar processes, and on a similar time scale" (Murck, 2001, p. 48). Applying this to Kidd Copper, the rapid rate of layers can redefine time.

The 'classical' notion as James Hutton (1726-1797) conceived uniformitarianism also implied that geological rates were also constant and 'uniform' through time. Modern geology does not hold this to be true anymore – it is now known that geological processes such as sedimentation rates are not constants as can be seen by the evolution of the scientific understanding accumulated about sedimentation rates reported from 1971-2004. Modern geology is dispelling earlier beliefs about uniformitarianism, refer to Section 14.2. "It now seems more probable that in some senses both [Uniformitarianism and Catastrophism] are correct, with the uniformitarian view of a constancy of physical laws being logical, but with elements of catastrophism to explain some of the events of the past" (Nichols, 2003, p. 317). Just like short-lived volcanic eruptions, the case of Kidd Copper provides evidence of a fast rate of deposition. These rock units are formed rapidly, which opens the door to reconsider the age of the Earth. One uniform layer did not result on Kidd Copper and the multiple layers support one specific depositional process, which opens the door to reconsider the Earth's age. "In science, a law is a descriptive principle of nature that holds in all circumstances covered by the wording of the law. There are no loopholes in the laws of nature" (Isaacs, *et.al.*, 2003, p. 445). Therefore, if uniformitarianism is re-defined by modern geologists, then new evidence provided by the Kidd Copper case in Table 25 - Section 21.10 opens the doorway that a younger Earth is a strong alternative and worth considering.

22.0 Old Earth

"Planet Earth is approximately 4.6 billion years old. This age is established by using dating techniques – called *radiometric dating* – that measure the rate of decay in given rocks" (Firefly Publishers, 2003, p. 15)

"It wasn't until the 1800s that people began to have a sense of how old Earth really is. Before that time, everyone believed that the world was 6000 years old. This was based on a theory proposed by Archbishop James Ussher of Armagh, Ireland in 1650. He claimed that Earth was created at 9 AM on October 23, 4004 BCE. His theory was formulated by calculating the ages of all the great men in the *Bible*. Biblical scholars often dominated popular theories about Earth's origins, and early studies suggested that the planet's features resulted from events that happened over brief spans of time. For example, it was often stated that the great biblical flood of Noah's day created all sedimentary rocks. People who followed this theory were called *Neptunists* – named for the ancient Greek god of the seas" (Albin, 2004, p. 43).

"These early ideas about Earth's origins began to change, however, when Sir Charles Lyle (1797 to 1875), an English geologist, published *Principles of Geology* in 1830. In this work, Lyle proposed the theory of *uniformitarianism*, which states that physical forces such as water, heat, wind, ice, and earthquakes change Earth's surface over time. His [Lyell's] theory was based, in part, on the belief that the thickness of the strata and the number of layers – which take a great deal of time to form – had to have begun way before 4000 BCE. He [Lyell] went on to write *The Antiquity of Man*, a book that supported Charles Darwin's theory of evolution" (Albin, 2004, p. 43).

22.1 Origin of the Elements

Origin of the elements … "Although life on Earth depends upon the energy of sunlight (originating from nuclear fusion of hydrogen and helium in the solar interior), the sun did not produce the chemical elements found in the earth. Rather, it is theorized by some scientists that these first two elements and their stable isotopes emerged from the first few minutes of the early high-temperature, high-density stage of the expanding Universe (so-called big bang concept)" (DePree & Axelrod, 2003, p. 139). Refer to Appendix L. "Although it has been theorized that a small quantity of lithium (element 3 in the periodic table) was also produced during the big bang, it has been proposed that the remainder of the Li and all of the beryllium and boron (elements 4 and 5, respectively) were produced by the spallation of still heavier elements in the interstellar medium by the process of cosmic radiation. The fact that these latter elements are relatively rare is consistent with this concept" (DePree & Axelrod, 2003, p. 139).

"Huttton's use of what later was termed uniformitarianism was simple and logical. By observing geologic processes in operation around him, he was able to infer the origin of particular features he discovered in rocks" (Levin, 2003, p. 7).

22.2 Origin of Man

"… maintained that Earth and all life forms were the result of an evolutionary process" (Albin, 2004, p. 44). "The world famous astrobiologist, Paul Davies, says: It's a shame that there are precious few hard facts when it comes to the origin of life. We have a rough idea when it began on Earth, and some interesting theories about where, but the how part has everybody stumped. Nobody knows how a mixture of lifeless chemical spontaneously organized themselves into the first living cell" (Mortenson, November 2004, p. 1, 2).

The Holocene Series. "Lyell (1857) used 'Recent' for the duration of time since the appearance of 'modern' (Present and Neolithic) humans. At the third International Geological Congress (IGC) in 1885 the term Holocene (Greek *holo*, 'whole') introduced by Gervais in 1867 was adopted as a substitute for Lyell's (1857) 'Recent'. The base of the Holocene Series is informally defined at the end of the last glacial period and dated at 10 000 radiocarbon years" (Gradstein, *et al.*, 2004, p. 419). The first evidence of mammals appears in the Tertiary Period approximately 10,000 years ago (Research Machines Plc., 2003, p. 261). "Man became the dominant terrestrial species during the Quarternary… mammoth and the woolly rhinoceros" (Isaacs, *et. al.*, 2003, p. 656).

22.3 Old Age Given to Date the Earth

"The work of Hutton and Smith raised the possibility that the Earth was far older and was continually changing and recycling itself" (Busbey, *et. al.*, 1999, p. 18).

Earth estimated to be 200 million years old. Uprising of the earth and the theory of Continental drift postulates that "200 million years ago Pangaea; 135 million years ago Laurasion and Gondawana; 65 million years ago Asia, Africa, India, Antartica and Australia" (Isaacs, *et. al.*, 2003, p. 184).

5.0 Billion Years. "The age of the earth has been given as approximately 5.0 billion years" (Garrels & Mackenzie, 1971, p. 63). "The oldest rocks we know are of sedimentary origin" (Garrels & Mackenzie, 1971, p. 67). "The rocks of the Earth's continents are about 3.8 billion years old. The oldest rocks in the ocean are younger – less than 200 million years old" (Williams, 1993, p. 22). "The oldest known rocks are about 3,960 million years old" (James, 2002, p.4). "The oldest known rock yet dated is the Acasta orthogneiss from the Slave Craton of Canada, which has yielded a U–Pb zircon age of 4031 ± 3 Ma (Bowring and Williams, 1999)" (Gradstein, *et al.*, 2004, p. 131). John "Playfair also cited a 1799 record of a fossil forest in Lincolnshire, now covered by tidal flat sediments. This is the oldest record of a Quarternary fossil soil" (Retallack, 1990, p. 3).

"A tremendous amount of work during the last 20 years on the determination of rock ages by methods based primarily on the radioactive decay of uranium, thorium, potassium, and rubidium leads us to a 'firm' estimate for the age of the crust of 4.5 billion years, and a current age of about 3.5 billion years for the oldest rock yet discovered. The general agreement of the age of the crust with ages of meteorites indicates a time of earth and solar system origin between 4.5 and 5.0 billion years ago. Studies of carbon-14 provide us with a detailed chronology for the past 40,000 years" (Garrels & Mackenzie, 1971, p. 92). Refer back to Table 22 - Section 18.2 on *Problematic Dating Methods*.

22.4 Biological Evolution

"Change, with continuity in successive generations of organisms. The phenomenon is amply demonstrated by the fossil record, for the changes over geologic time are sufficient to recognize distinct eras, for the most part with very different plants and animals" (Allaby & Allaby, 2003, p. 196). Geology does not support evolution. "Nor does geology at all lead to the belief that most fishes formerly possessed electric organs" (Darwin, 1958, p. 179). Biological evolution is not the scope of this report; however comments on evolutionary evolution will be considered as they relate to the rock record.

22.5 Crustal Evolution

"The definition of GSSPs older than the base of the Cambrian has commenced in the Neoproterozoic Era and may then work progressively back in geological time as the pattern of Precambrian crustal evolution is unraveled" (Gradstein, *et al.*, 2004, p. 140).

Time units defined:
> "Geons which are defined as 100–million-year intervals counting backwards from the present (Hofmann, 1990, 1999)" (Gradstein, *et al.*, 2004, p. 130).
> 'Ka' = "thousands of years before present" (Nichols, 2003, p. 229).
> 'Ma' = "millions of years before present" (Nichols, 2003, p. 229).
> 'Ga' gigayears = "thousands of millions of years before present" (Nichols, 2003, p. 229).
> 'my', 'm.y.' or 'm.yr' = "millions of years in the past" (Nichols, 2003, p. 229).

Geology has constructed "its own special scale of time … what is known as the geological time scale" (Gribbin, 2004, p. 149). The terminology is confusing and takes dedicated study to unravel. For example, Germanic Basin – "how much time is reflected in the remaining Zechstein units is unknown; the evaporites could represent very short depositional intervals between long hiatuses or a very short interval similar to that of the Ochoan of West Texas" (Gradstein, *et al.*, 2004, p. 256).

22.6 Planetary Evolution

"It is proposed that Precambrian time should be subdivided into eons and eras that reflect natural stages in planetary evolution rather than being subdivided by a scheme based on numerical ages. The six eons can be briefly characterized as:

1. 'Accretion and differentiation:' a time span of planet formation, growth, and differentiation up to the Moon-forming giant impact event.
2. Hadean (Cloud, 1972): an extended time span of intense bombardment and its consequences, but no preserved supracrustals.
3. Archean: an episode of increasing crustal record from the oldest supracrustals of Isua to the onset of giant iron formation deposition in the Hamersley Basin, likely related to increasing oxygenation of the atmosphere.
4. 'Transition:' a time span with deposition of giant iron formations up to the first bona fide continental red beds.
5. Proterozoic: the time span of a nearly modern plate-tectonic Earth but without metazoan life.
6. Phanerozoic: Earth characterized by metazoan life forms of increasing complexity and diversity" (Gradstein, *et al.*, 2004, p. 141).

22.7 Evolutionary Rate.

"Amount of evolutionary change that occurs in a given unit of time. This is often difficult to determine, for several reasons. For example, should the unit of time be geologic or biological (the number of generations)? How should morphological change in unrelated groups be compared? In practice it is necessary to adopt a pragmatic approach, such as the number of new genera per million years" (Allaby & Allaby, 2003, p.196).

Quoting Darwin, "I further believe that these slow, intermittent results accord well with what geology tells us of the rate and manner at which the inhabitants of the world have changed" (1958, p. 111). "The mere lapse of time by itself does nothing, either for or against natural selection" (Darwin, 1958, p. 108).

22.8 Darwinism

"The theory of evolution proposed by Charles Darwin in *On the Origin of Species* (1859), which postulated that present day species have evolved from simpler ancestral types by the process of natural selection acting on the variability found within populations" (Isaacs, *et. al.*, 2003, p. 214). Darwinism is based on conjecture. "Natura non facit saltum" (Darwin, 1958, p. 184).

In Figure 50 - Section 22.9, Charles Darwin had achieved a "unique literacy status among works of the scientific imagination" (Leakey, 1979, p. 9). He wrote "*The Origin of Species*" … "for the theory of evolution" (Leakey, 1979, p. 9). Figure 50 - Section 22.9 shows Darwin in 1881, twenty-two years after *The Origin of Species* was first published.

22.9 A Geological Time Scale to Emphasize Mammalian Evolution

Darwin perpetuated mammalian evolution. "*Mammalia* – the highest class of animals, including the ordinary hairy quadrupeds, the whales and man, and characterized by the production of living young which are nourished after birth by milk from the teats (mammary glands) of the mother" (Leakey, 1979, p. 227). Darwin stated "that all animals and all plants throughout all time and space should be related to each other in groups" (1958, p. 131).

"Mr. Herbert Spencer… his law 'that homologous units of any order become differentiated in proportion as their relations to incident forces become different' would come into action. But as we have no facts to guide us, speculation on the subject is almost useless" (Darwin, 1958, p. 127). "Unexplained on the origin of species, if we make due allowance for our profound ignorance on the mutual relations of the inhabitants of the world at the present time, and still more so during past ages" (Darwin, 1958, p. 127).

Figure 50 - Section 22.9: Charles Darwin in 1881, (Leakey, 1979, p. 8)

"These breaks are imaginary, and might have been inserted anywhere, after intervals long enough to allow the accumulation of a considerable amount of divergent variation" (Darwin, 1958, p. 119).

"If my [Darwinism] theory be true, numberless intermediate varieties, linking closely together all the species of the same group, must assuredly have existed; but the very process of natural selection constantly tends, as has been so often remarked, to exterminate the parent-forms and the intermediate links. Consequently evidence of their former existence could be found only amongst fossil remains, which are preserved … in an extremely imperfect and intermittent record" (Darwin, 1958, p. 166).

Special Creation is "the belief, in accordance with the Book of Genesis, that every species was individually created by God in the form in which it exists today and is not capable of undergoing any change. It was the generally accepted explanation of the origin of life until the advent of Darwinism" (Isaacs, *et. al.*, 2003, p. 738, 739).

"The idea [Special Creation] has recently enjoyed a revival, especially among members of the fundamentalist movement in the USA, partly because there still remain problems that cannot be explained entirely by Darwinian theory" (Isaacs, *et. al.*, 2003, p. 738).

"Further back in time, historical records become less and less accurate" (Retallack, 1990, p. 261, 263). "Their order is determined from superposition of artifacts or fossils in sedimentary layers, the older ones beneath the younger ones" (Retallack, 1990, p. 263).

See Figure 34 - Section 13.13. Caution – to apply this as a standard for attaining time from the geological rock record because when the layers of sediment are misread it leads to questionable dates for the age of the Earth. "The indicators of a time interval may not have changed at the same time in all areas" (Retallack, 1990, p. 263).

Figure 51 - Section 22.9 shows the way mammalian evolution and geology are interconnected "a geological time scale, plotted on a logarithmic scale to emphasize detailed understanding of mammalian evolution, human cultural stages, and glacial chronology (based on data from Nilsson 1983, Palmer 1983, Woodburne 1987, Richmond & Fullerton 1987)" (Retallack, 1990, p. 262).

The overlapping of geology, the rock record and the evolution of time is based on reading the rock records.

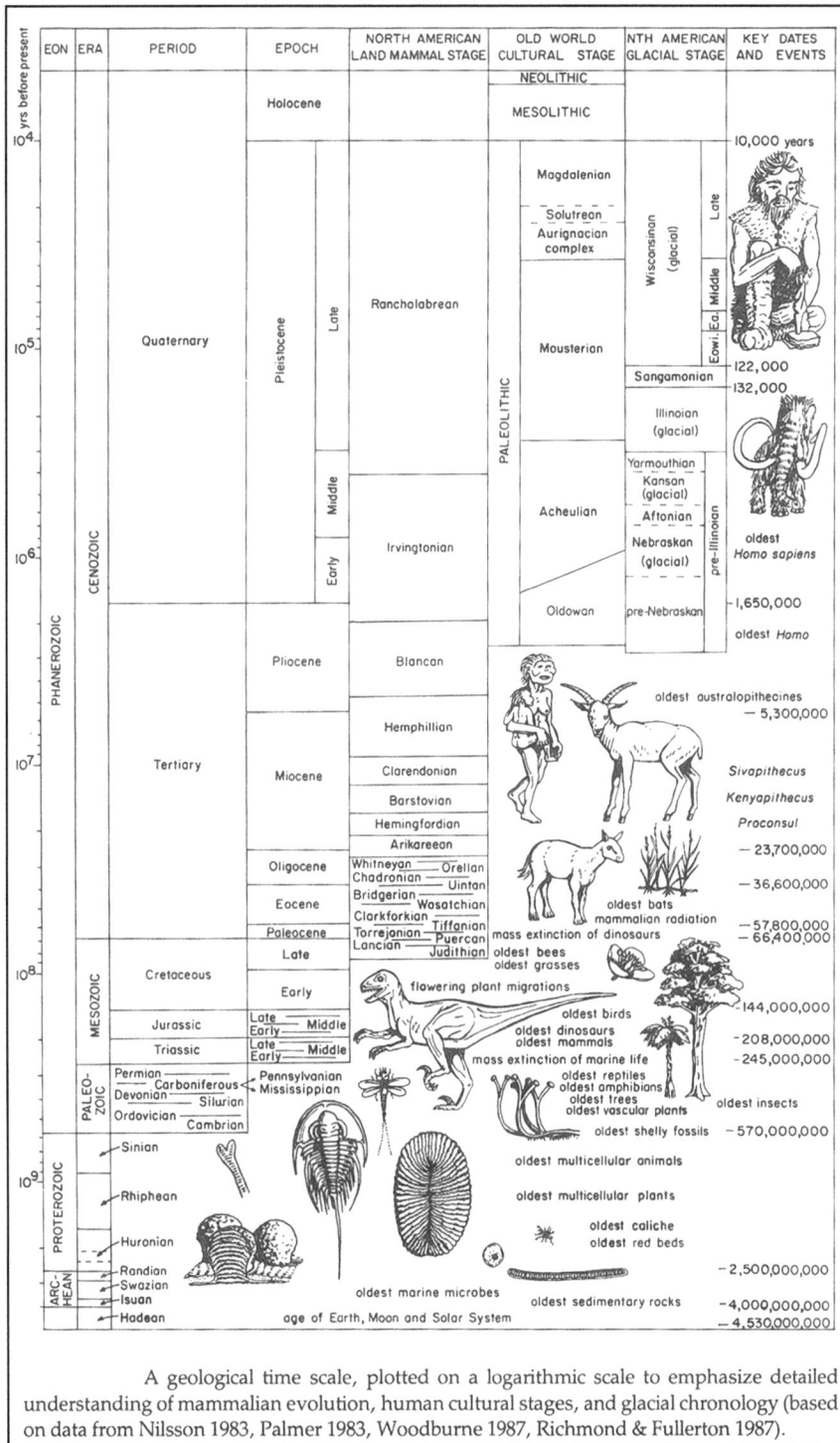

A geological time scale, plotted on a logarithmic scale to emphasize detailed understanding of mammalian evolution, human cultural stages, and glacial chronology (based on data from Nilsson 1983, Palmer 1983, Woodburne 1987, Richmond & Fullerton 1987).

Figure 51 - Section 22.9: Mammalian Evolution (Retallack, 1990, p. 262)

22.10 Young Data Is Not Readily Accepted

e.g. *Russian Platform* – "The Tatarian, basically a series of stacked soils in its type area, is difficult to correlate. It does contain the Illawarra Geomagnetic Reversal in its lower part, which ties that part to the upper Wordian–Capitanian interval. How young the upper Tatarian is, is an open question" (Gradstein, *et al.*, 2004, p. 256).

Another example is *Relative biostratigraphic position*, where young data is rejected. "A visual inspection of these two dates relative to the spline curve in Figure 52 - Section 22.10 suggests that the latter fits the time scale trend but the former is stratigraphically *too* young" (Gradstein, *et al.*, 2004, p. 221). The time scale trend is adjusted based upon the assumption of old earth even when there is evidence to support a 'younger date' for Planet Earth!

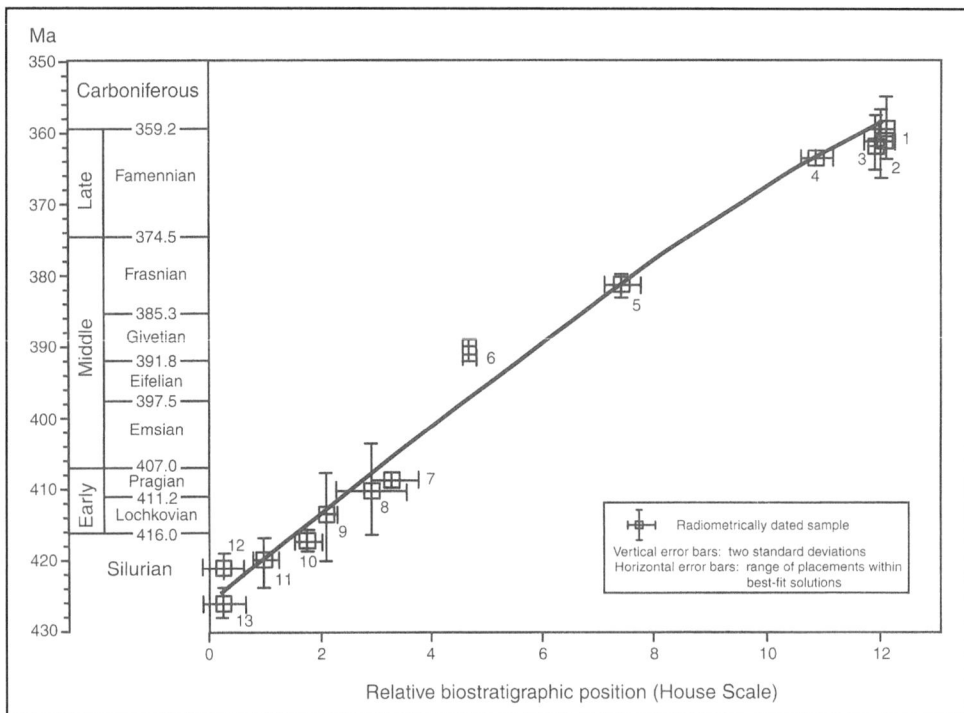

Figure 52 - Section 22.10: Relative Biostratigraphic Position Two Dates Represented

22.11 Geologic Timescale Calibration is a Sparse Reflection of Real Time

"Earth is composed of rocks and their mineral components, thus their stories are inextricably linked to the planet's history" (Discovery Books, 1999, p. 8). "The history of the Earth is recorded in a very incomplete manner in the rocks formed at different times and places" (Nichols, 2003, p. 229).

"We are still a considerable distance from the goal where geologic timescale calibration is achieved by precise direct astronomical tuning or radiometric age dating of all successive stage, zonal, or magnetic polarity chron boundaries. In fact, it is doubtful if the rock record on Earth harbors all the precise age information. This sparse skeleton of age control, especially prior to ~30 Ma (as of 2004), leaves considerable room for interpolation in construction of a geologic time scale" (Gradstein, *et al.*, 2004, p. 455).

Science facts are not based on what we don't see; but, what we do see. "Future time scales will undoubtedly re-examine and reprocess a more expanded array of Earth history data, and will undoubtedly employ even more sophisticated means of interpolation" (Gradstein, *et al.*, 2004, p. 455).

The scientific observations in this report allow for a re-examination of a 'younger age' for Planet Earth based on the rapid deposition of the mine tailings at Kidd Copper. The fine-grained sediment underwent an initial deposition, upswelling and redeposition in a very short time period. The start and end point for the depositional process is well-covered in this scientific report with conclusive evidence that the layering of the sediment as a primary structure provides a new base for the measurement of time. No longer will missing gaps be read in the rock record because the adjacent profiles on Kidd Copper provide evidence that many differences can occur in deposition. Just because they look different does not mean that they came from a different time period. For example, the profiles on Kidd Copper show a vast amount of different combinations and yet all of the borings were taken at the same moment in time, from the same source and the same mechanism of transport. Thus, rapid stratification opens the doorway to a 'younger date' for Planet Earth.

23.0 Claim for a Younger Earth

Two distinct claims for dating the Earth will be investigated in the remainder of this report: slow processes support old earth or rapid processes support a younger earth.

The evidence from Kidd Copper's lithologic logs showed many contrasts of color, variances in composition, texture differences, random distribution and non-consistent spatial pattern result from a one-time depositional process. This report has conclusively proved there are two different rates of sedimentation for primary structures. Layers can be formed in either a long period or in a short period of time as in the case of rapid stratification found on the Kidd Copper property. The Kidd Copper data in this report provides strong evidence to support that a short period of time can twist, bend and deposit sediment in layers; this is reasonable based on the sound judgment of first-hand observations. The knowledge gained from Kidd Copper covered a complete cycle of deposition, uplift and redeposition.

This new information changes the way layers are read in the rock record. Each layer of sediment is not equal to a separate time zone; thus leaving room to provide an alternative claim for the age of the Earth, from an old Earth start date to a younger Earth start date.

H_0: Possible Solution for Each Layer of Sediment = Separate Time Zone => Old Earth Date
H_1: Possible Solution for Each Layer of Sediment ≠ Separate Time Zone => Younger Date
How Are the Rock Layers Read if Stratified Layers Occur Quickly?

"The Earth was thought to have had a definite beginning, being created in the form we know at a comparatively recent time" (Busbey, *et. al.*, 1999, p. 18). "Other people too, not just scriptural literalists, remain unpersuaded about evolution. According to a Gallup poll drawn from more than a thousand telephone interviews conducted in February 2001, no less than 45 percent of responding U.S. adults agree that 'God created human beings pretty much in their present form at one time within the last 10,000 years or so'" (Quammen, November 2004, p. 6).

23.1 Age of the Earth <2 Billion Years Old

"The Hindu tradition weighed [the age of the Earth] in at slightly less than two billion years, while some Hebrew and Christian calculations resulted in values of less than ten thousand years" (Gribbin, 2004, p. 150). Earth is very old is questionable because the foundation of basing relative and geologic time has been questioned and according to this data is not substantiated. The multiple layers in the Kidd Copper property were laid down in a very short period of time and thus the Earth could be younger because the observed layering of the mine tailings deposits from Kidd Copper property demonstrates stratification forms rapidly. Each layer does not have to represent thousands of years. The reading of geologic time based on conventional wisdom is not conclusive. "Geological interpretation is an exercise in compiling all relevant data, that it produces explanations that are most probable given the set of available data, and that in many cases definitive conclusions may not be possible – several possible explanations should be provided if the data are equivocal" (Lewis, 1984, p. 5).

23.2 Time Correlation

"Time correlation [GEOL] a correlation of age or mutual time relations between stratigraphic units in separated areas" (Licker, 2003, p. 375). "To develop a geologic time scale that is applicable to the entire Earth, rocks of similar age in different regions must be matched up. Such a task is referred to as *correlation*" (Tarbuck & Lutgens, 1999, p. 195).

23.3 Time-Series with Terrestrial Matter

Parent rocks and activities on-site are relevant properties to establish a time-series with known terrestrial matter as provided in the Kidd Copper case.

Parent Rocks. Composition and initial time of deposition is known at Kidd Copper. "The bedrock is rich in goethite" (Shang, *et al.*, 2004, p. 9). Various mixtures of waste materials are observable in the field, in aerial and in subsurface imaging on the study site. The study area was sampled directly by field observations, which have been discussed in detail throughout this report.

Activities. Mining in the area led to the accumulation of mine tailings waste on the Kidd Copper property from 1936-1970. The sequence of events include the following depositional processes of the sediments/slag: (1) Initial repetitious deposition of mine tailings on the waste site, (2) Upswelling of sediment because of dam breach; (3) Resettling of particles from the turbulent flow based on real-time observations and primary field work in a well explored mine tailings waste site in northern Ontario.

23.4 Time Line

Stratotype is "the original or later designated type representative of a named *stratigraphical unit* or of a stratigraphical boundary, established as a point in a distinctive sequence of rock strata. The stratotype is the standard for the definition and recognition of a particular stratigraphic unit or boundary" (MacDonald, *et al.*, 2003, p. 414).

"Time line [GEOL] (1) is a line that indicates equal geologic age in a correlation diagram. (2) A rock unit represented by a time line" (Licker, 2003, p. 375). In the case of the geologic correlation diagram all boreholes like Point 64/X1 and Point 56/4 were able to be accurately time dated around 1970 AD. Each recorded layer in the subsequent depth measurements yielded the same time designation. 1970 was a constant date through the digging of the layers downwards, resulting in the fact that the depth did not restrain time nor did the variant layers constitute a time differential. 1970 was a fixed point that could determine what the depositional process was like in the present to be used in the past. It could be stated that the downwards layers in time have no more impact on the time variable as do the upper layers in the Kidd Copper case.

"Time-stratigraphic facies [GEOL] a stratigraphic facies based on the amount of geologic time during which deposition and nondeposition of sediment occurred" (Licker, 2003, p. 375). The time-stratigraphic facies from the Kidd Copper property bear testimony to the fact that sediment layers do not show time differentiation. Lithohorizon is "a surface showing lithostratigraphic change or a particular lithostratigraphic character; it is highly useful for correlation. It is usually the boundary of a lithostratigraphic unit but may be a thin marker layer within a lithostratigraphic unit" (MacDonald, *et al.*, 2003, p. 278). In this case control point 56/4 was randomly chosen as the correlation factor between the parallel and the perpendicular cross-sections due to the fact it intersected both the North-South transect and theWest-North transect.

A marker horizon is "a thin layer within a succession of *beds* that has distinctive characteristics such as a particular *fossil* assemblage or an unusual *lithology*, which means it can be used for stratigraphical correlation. *Marker beds* are normally assumed to be isochronous, i.e. everywhere of the same age, but this may not always be the case. As well as the more usual type of marker horizon, for example a *marine band* or a *bentonite* bed representing a layer of *volcanic* ash, beds that produce a characteristic set of seismic reflections or with distinctive electrical or magnetic characteristics can also be used as marker horizons" (MacDonald, *et al.*, 2003, p. 290). Refer to Figure 27 - Section 10.3. In this case point 56/4 was chosen as the marker bed but any bed could have been chosen because the data was taken from the same date ~1970.

In Figure 53 - Section 23.4 a time stratigraphic facies of Borehole 64 and Borehole 56 are compared. Although these points are isochronous, using conventional wisdom to date them is difficult because of the variations within these profiles. Many of the log profiles that geologists look at are similarly different. This provides much debate and differences in opinion when it comes to attaining a geologic age and leads to many changes in the building of a geologic time line, such as noted in Sections 16 and 17.

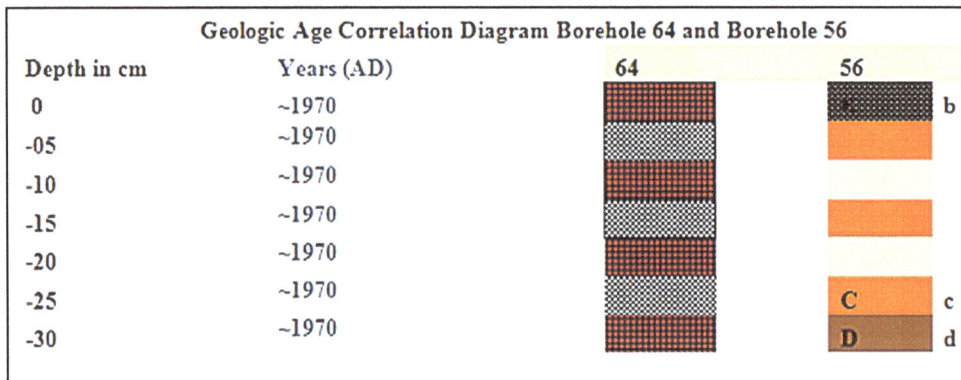

Geologic Age Correlation Diagram Borehole 64 and Borehole 56			
Depth in cm	Years (AD)	64	56
0	~1970		b
-05	~1970		
-10	~1970		
-15	~1970		
-20	~1970		
-25	~1970		C c
-30	~1970		D d

Figure 53 - Section 23.4: Time Stratigraphic Facies Borehole 64 and Borehole 56

23.5 Geologic Time Scale GTS2004 Challenged

Science looks for the best fit of evidence, just like a judge looks at the evidence to determine if the logic stands up. The Kidd Copper case seriously challenges the logic in the continuation of using a time scale that is flawed.

Geologic Time Scale. "A two-fold scale that subdivides all the time since the Earth first came into being into named units of abstract time, and subdivides all the rocks formed since the Earth came into being, into the successions of rock formed during each particular interval of time" (Allaby & Allaby, 2003, p. 228). "The geologic time scale. The absolute dates were added long after the time scale had been established using relative dating techniques (Data from Geological Society of America)" (Tarbuck & Lutgens, 1999, p. 205). Refer to Figure 72 - Appendix M. "Such a standard scale is still a long way off, but the names for geologic-time units and chronostratigraphic units down to the rank of period/system are in common use; many epoch/series and age/stage names are still regionally variable" (Allaby & Allaby, 2003, p. 228). "The branch of geology that deals with the age relations of rocks is known as chronostratigraphy" (Allaby & Allaby, 2003, p. 228). This was reported earlier in detail in Section 17.3. "The concept of a geologic time-scale has been evolving for the last century and a half, commencing with a relative time-scale (mainly achieved through biostratigraphy), to which it has gradually become possible to assign dates, which are, nonetheless, subject to constant revision and refinement" (Allaby & Allaby, 2003, p. 228). "Since the first International Geological Congress in Paris in 1878, one of the main objectives of stratigraphers has been the production of a complete and globally accepted stratigraphic scale to provide a historical framework into which all rocks, anywhere in the world, can be fitted" (Allaby & Allaby, 2003, p. 228).

The Scale of Best Fit. "Harmony: The property by which the elements of a map work together to create a balanced aesthetic whole" (Clarke, 2003, p. 314). "Unless its unit is consistently proportional to geologic time measured in millions of years, the numerical time scale resulting from spline fitting is not linearly related to the initial relative time scale. It is, however, linearly related to geologic time in millions of years. This allows for gradual changes over time in the original, hypothetical process on the basis for which the initial relative time scale is constructed. For example, deviations from a straight line on a fitted spline curve may represent corrections or changes in sedimentation rate or rate of evolutional change" (Gradstein, *et al.*, 2004, p. 108).

Straight-line Fitting for Ordovician and Silurian ages. "From the preceding analysis it follows that the spline curve is probably the best fit" (Gradstein, *et al.*, 2004, p. 112). The geologic time scale is based on corrections, estimations and a hypothetical process from which men have assigned arbitrary names and dates. Straight-line fitting is based on best-fit guesses, not accurate data often correlated on unknown parameters.

Corrections to account for time discrepencies apparent in the chronostratographic record are made with the approach to 'best-fit' the evidence and to balance all the facts as a whole. But, with no known starting and ending parameters because there is no eye-witness accounts of earth over time backwards, how can anyone know they are correct?

GTS 2004 Dating Discarded. Counting time with GTS 2004 boundaries each layer would represent a different time, which would confuse the accuracy of the date. Each layer is not a separate time zone as expected if GTS 2004 was used and the estimated date using the GTS 2004 system is in error judged with the Kidd Copper case. In its place, no estimates or extrapolations are required to determine the marker bed and its depositional time because the date of the redeposition of the sediment on the Kidd Copper property is known; therefore, the sediment layer serves as a datum surface with this known date (~1970) as an accurate date based on recent historical evidence.

23.6 Timeline of Kidd Copper Events Re-constructed

"Unravelling the nature of the surface environment through time" (Garrels & Mackenzie, 1971, p. 26). "Time plays an important role in the development of the profile" (Garrels & Mackenzie, 1971, p. 169).

The Kidd Copper site offers a current look at the present sediment deposits that can be observed and learned from real-time measurements. Real-time records of observable layering of sediment in the field; these recorded events through time can help reveal the keys to the past from referencing to present and confirmed data collection. Time lines are formulated from what is known. Speculations about past references can be overshadowed with facts from Kidd Copper. Informed information, verifiable geologic environment and depositional setting factors provide unarguable boundaries.

"In 1966 Kidd Copper commenced the construction of a 1000 tpd mill and commenced mining and milling in 1967 producing approximately 250,000 tons of ore prior to mining being concluded in 1968" (Article [2], 2008, *p. nd.*). Slag was carried by rail to the dump about 3 miles away and deposited as waste tailings to the study site area located on the topographic maps as 4116. The grain size of the mine tailings deposits is between 0.00 and 0.05 mm.

The retaining dam at the southeast corner broke in the early 1970's, creating an upsurge of deposits. Sand, liquid water and the mine tailings mix in random order due to the velocity of the water. The water provided by the breaking of the dam provided the force to uplift the mine tailings and redeposit them. The fluid lift force carried the fine-grained particles so that the mine tailings spun vigorously, and were redeposited.

Then, when the water drained, there was a re-settling of sediment, the mine tailings. The mine was closed down and remains inactive; thus the area is an ideal test site for the rate of deposition because it provides a definite initial time and the depositional processes are known. The different layers of laminae are distinctive and suggest that multi-layers are a one-time event in a short dispensation as seen by the Kidd

Copper logs. Many years were not required for this sediment to accumulate. The 120 boreholes confirm that rapid stratification is a sedimentation process to substantiate a younger earth. It can be concluded that the multiple layers are not many events but rather one event can provide many layers.

The waste site was left abandoned. In 2004, a group of *Geo3Fe3* students were requested to do a mine tailings exercise. The field books were collected and the bore holes were correlated and analyzed without the geology student's involvement in the objectives or outcomes of this study; they were blind participants. The 120 boreholes consistently revealed stratification of the mine tailing sediments through the entire waste site; instead of the expected outcome of a uniform layer for a one time occurrence. The time factor on a geologic scale is relatively young and so gives additional credence to the theory that rapid stratification supports a 'younger age' for planet Earth.

If Kidd Copper's history was not known one could erroneously conclude: "In some places, where water has moved through fractures in rocks, diffusion-controlled alteration extends for a few feet to a few tens of feet outward from the fracture and may represent tens of thousands or hundreds of thousands of years of solution flow in the fracture during which diffusion of ions into and out of the wall rock was taking place" (Garrels & Mackenzie, 1971, p. 56). The guesswork has been removed at Kidd Copper to allow for a concise interpretation of the deposition of its sediment. Kidd Copper's heterogeneous deposits do not represent sediment growth through time but are a one-time deposit. Since this is the case, a younger Earth can be reasoned by re-examining its multitude layers. Are they many deposits through time or are they like the Kidd Copper deposits of multiple layers being one event?

23.7 Rapid Stratification Supports a 'Younger Age' for Planet Earth

New Proposed Theory: rapid stratification supports a 'younger age' for Planet Earth. Stratification is "a bedded or layered arrangement of materials. The term is used especially in reference to *sedimentary rock*. Stratification can also be seen in *lava flows*, *metamorphic rocks*, masses of snow, *firn* or ice, while the waters of a lake may be arranged in layers of differing temperature or density" (MacDonald, *et al.*, 2003, p. 414). Stratification is seen elsewhere in the rock record which means that this new proposed theory is universal and not restrained to sedimentary depositional processes alone.

The hypothesis that stratification equals a fast rate of deposition in this report is based on geologic principles identified throughout that support multi layering of strata cannot be interpreted as individual time events. The evidence found by examining the primary sedimentary structure of the impoundment of mine tailings at Kidd Copper identifies stratification visibly apparent from the first few cm of land surface down to 75 cm. Field data of observed layering of the mine tailings deposits from Kidd Copper property validates stratification forms rapidly.

The claim of this report proposes the theory that the age of the Earth is relatively younger, formed under the age of 10, 000 years to correspond to the date mankind appeared on earth. This date is selected as a good starting point because of the historical records to support mankind's existence. The controversy over the evolution of mankind is beyond the pages of this report except to say that historical records for mankind's existence can be provided in other forms up to 10,000 years ago; whereas, longer dates are non-existent. The author holds to the conclusions in the interpretative data of this study showing that identifying the real record of the time events in the geologic history of the rock records is a difficult task without knowledge as to the historicity of the events. Therefore, the author suggests that the age of the earth and the age of mankind could have the same start date. This is within reason because if the age of the earth is younger then why could it not have been formed when mankind was? This report supports the theory of the formation of sedimentation deposition and sequence patterns, based on the upswelling and mixing of sediment in a one-time occurrence as it redeposited and settled in and passes upward from a mine tailings source. The observed rapid stratification at Kidd showed that sediment is disassociated through original deposition processes forming multi-layer strata resulting in a time reference frame of about a day. The premise is that stratification is often laid down quickly in a single catastrophic event, and has no direct connection with the slow – earth-time clock concept of millions of years for these layers to form. The hypothesis in this report challenges conventional wisdom like the Law of original horizontality or the slow processes of depositional layering.

Rapid Stratification. "In the past there has been a tendency to interpret each lamina as produced by a separate sedimentation event, for example, a tidal cycle, the swash and backwash from a single wave, or a single bed load avalanche. It is now clear, however, that laminae may also be produced by steady flow, particularly during traction on a plane bed in the upper flow regime" (Blatt, Middleton & Murray, 1972, p. 117). A single sedimentation event produced laminations at Kidd Copper in a steady, high velocity flow that caused the redeposition of sediment to settle quickly and in layers replacing the initial deposition of mine tailings with a well-mixed layer of mine tailings on the Kidd Copper property impoundment area of mine tailings as the floodwaters subsided and the water dried up. The sediment remained intact until 2004 when some Geology students uncovered several boreholes. This is important to recognize that scarcely any time was involved for the stratification and multiple layers to form on the Kidd Copper property. Time is not an easily determined factor in sedimentary rocks. Rocks do not always form layers through time. Rather, stratified layers are a rapid process; and a stratified layer is a rapid process. The layers do not have to wait for the process of time to accumulate these layers. It happens quickly. The multiple layers in the Kidd Copper property were laid down in a very short period of time and thus the Earth could be younger than the common answer of 4.4 billion years old because the layers seen throughout the geologic record could have similarly formed rapidly.

How Did These Layers Form? A possible explanation for the formation of the Earth's layers is that the diagenesis of the Earth was a closed system; and the explanation of the changes in thickness of the layers is due to an unbalanced distribution of resettled sediment; not tectonic causes. Nor are the thicknesses due to shifting mid-shelf carbonate depocentre. Nor are the facies changes due to eustatic sea-level fall. The new evidence is overwhelming that the end result of the dam breaking and redepositing its waste mine tailings deposits occurred in a short period of time with multiple layers; thus giving strong support to rapid stratification. Therefore, the new conclusion of the stratified layers on the Earth's crust is the result of suspended and redeposited sediment.

23.8 Factors Underlying a Good Theory

A good theory (1) identifies orderly relationships of various and diverse observations; (2) predicts future outcomes; (3) modifiable; (4) develop new direction for additional research; and (5) can be replicated. All of the factors that define a good theory have been met in this report to critique and analyze the data relating to the depositional processes of sedimentation because this report has combined all the methods to obtain its knowledge: (1) scientific method; (2) the consideration of evidence; and (3) empirical or (4) deductive reasoning.

Each factor identified to make a good theory has been used to underscore the logic in the case of the primary research and study site on the Kidd Copper property. The factors to assess the logic underneath the knowledge are as follows:

(1) *Orderly relationships are identified.* Both geographic and geologic overview of site area was included in this report. Geomorphological background and historical relationships of land and dam. Topography and mechanisms for uplift identified alongside turbidity currents and exogenetic processes. Remote sensing and a satellite site were observed to confirm relationships. The compositions of the sediment and ore mineralogy were examined. Flood deposits, field capacity and the settling velocity of the particles were carefully noted. Contrasts with other depositional processes exhibiting rapid stratification were carefully looked at, such as: sturzstroms and the Grand Banks sediment.

(2) *Textbook laws, theories and hypothesis are defined.* Orderly relationships were carefully considered in relation with conventional textbook anticipated outcomes in many areas like: recurrence time, parallel lamination, and the expectation of horizontal lamination. Much effort went into this report to analyze the findings of Kidd Copper with relationship to uniformitarianism, the Law of original horizontality; and the Law of superposition and others.

(3) *Predictions of future outcomes.* Determining depositional environments from the stratigraphic record provided from the data in the Kidd Copper case allows room to assert that the Earth is younger. The sediments settled and

redeposited as a one time occurrence as stratification and this assertion is seen in other strata formations. Kidd Copper's analysis of the data and outcomes reveal that rapid stratification is heterogeneous and not uniform in primary sediment structure contrary to the conventional belif that one uniform layer will be deposited as a result of each new mixture.

(4) *Modifiable*. Many methods for age estimations and dating methods were briefly exposed as to the problematic areas they contained. The future may yet reveal a more stable and established method to date the Earth then those discussed in this report. But, conventional wisdom found in the geological laws at present with their assumptions, missing data and interpolative data that is presently offered has been seriously challenged according to the findings in this scientific report. Identified problems with current age estimations opens the doorway to an alternative 'younger date' for Planet Earth over the date of an old age based on rejecting these conventional problematic methods.

(5) *Develop new direction for additional research*. Refining the existing Geologic Time Scale to a 'younger date' could be extensive. The data collected from other 'catastropic' events could provide valuable clues. Suggestions would be to synthesize the findings in this report more carefully with the newly formed exposed volcanic ash deposited in the Mount Saint Helen's eruption in the 1980s. Refining school texts and modifying the current assumption that an old age for Planet Earth is scientifically sound will be a great endeavour. Providing an alternative to the theory of rock dating that layers are multiple units of time and proclaiming that a younger Earth is possible based on the fact that multiple layers represent one unit of time requires time and additional research.

(6) Opening the doorway to a 'younger' Earth opens the reasonable pathway with respect to special creation is an alternative answer for the start of time. This will require additional information to support the topic of debate. Special creation has been a point of debate and controversy but provides a reasonable explanation as to how the layers on the Earth formed so quickly. Over time biological life forms are said to have evolved. The geologic time scale is used to place this evolution into a time-scale but as pointed out in the research of this report the Geologic Time Scale needs to be re-evaluated based on its many inconsistencies and unfounded assumptions, recall Section 17.

(7) *Can be replicated*. Many models for future reproducible experiments were given for example, highlighting a further flume study and research on the variations of a spigot dam flooding another site area. The stratification seen in the case of Kidd Copper has been replicated in other sedimentary and igneous deposits; thus the information in this report is universal and not only applicable to this one specific case.

The science in this report has "summarized what is known or can be deduced about the mass, origin, composition, and age distribution of sedimentary rocks and then have proposed some simple models to account for theses distributions in terms of rates of deposition and destruction of sediments throughout earth history" (Garrels & Mackenzie, 1971, p. 5). Thus, the research in this report provides valuable new knowledge.

23.9 Opening the Doorway Back to Special Creation

No one living today was alive to re-count the actual creation of the Earth firsthand. The layering of the crustal sediment that is called the earth's crust and underlying structure, the mantle, was laid without eyewitnesses to account for the actual formation of it in time or mechanism. Commonly the story goes "because the mantle is the bulk of the Earth, its story is fundamental to geology. The mantle began, during Earth's birth, as an ocean of magma atop the iron core. As it solidified, elements that didn't fit into the major minerals collected as a scum on top—the crust. After that the mantle began the slow circulation it has had for the last 4 billion years, with at least the upper part being cooled, stirred and hydrated by the tectonic motions of the surface plates" (Alden, 2009, Aspect 6).

"The complex interactions of chemical, physical and biological processes which control depositional environments have produced situations in the past which are quite different from anything we see today" (Nichols, 2003, p. 316). The first question to ask concerning this statement is 'why?" What makes this statement true if no one could testify to situations in the past? What makes the Kidd Copper property relevant is that many testimonies were provided as to the depositional processes. The sight observations included: chemical, physical and biological processes. By studying the present laws of nature, a scientist can determine facts. If today one proves that gravity exists and items fall; then is it not safe to say that this happened yesterday? If one proves today that multiple layers are not the result of a slow process but rather as the result of rapid stratification; then is it not safe to say that the same happened yesterday or millions of years ago? So, the question is reversed, does the past depositional environment need to be so radically different then that which is seen today? Is it not probable that the geologic sediment structures of today reflect the past as a mirror? For example, if stratified layers are not a measurement of multiple time units in the Kidd Copper layers, then what about elsewhere on Earth or in the Earth's past? Therefore, the new knowledge attained from the case of Kidd Copper can reasonablybe integrated into an interpretation of the genesis and history of the dating of the Earth.

The evidence in this report supports the claim for the defence of a 'younger' Earth and the doorway is re-opened to examine special creation.

24.0 Rapid Stratification Findings & Implications

Kidd Copper property Geo3Fe3 Geology Assignment and the resultant aggregate map of the Kidd Copper major mine tailings deposit, shown as a detailed sketch map in plan view in Figure 27 - Section 10.3, compilation of the n =120 boreholes where the samples and cross-sections were located with exposed surface depths up to 75 cm is the main subject matter of this report.

"Owned by Crowflight Minerals Inc. and is known as the Kidd Copper Property" boundary conditions were set by McMaster staff and these steps were followed: (1) A visual inspection of the site by having each student "walk over the entire site" to ensure an encompassing breadth of information on the site would be observed; (2) Each student reported his/her geological observations to create "a map that details the different processes that [they] observe in the different areas"; (3) Subsurface was uncovered with "a spade at several locations to observe what is occurring in the subsurface"; (4) Aerial photos were provided "to describe the changes that have occurred at the site over time" and; (5) Submit field notes with "the rough map that you drew in the field, identifying processes that are occurring in the different areas" (McMaster University, August 2004, Assignment #9 Mine Tailings).

Primarily based on the recordings from each of the field books an analysis and interpretation of the findings has been assembled and reported in this research paper, solely by Loreen (Rudd) Sherman, who later earned a double degree from the School of Geography and Geology and is an Earth and Environmental Scientist. Loreen received a Bachelor of Arts by the authority of the Senate the Chancellor conferred on her the 8th day of June, 2005 at Hamilton, Ontario from McMaster University.

So in conclusion, the author, Loreen Sherman, who attained her Masters of Business Administration from the University of Phoenix in September, 2009 completed the research started on August 30, 2004 and asserts the following twenty-four geologic findings from the evidence and observations on Kidd Copper property and its depositional processes as related to time in this scientific report and analysis.

24.1 Finding One: Historicity and Source Rock Well-established

Finding 1: In the case of the Kidd Copper property "sedimentological and stratigraphic analysis of the rock record using the material exposed at the surface … in the subsurface, provides … the means to reconstruct the history of the surface of the Earth as far as the data allow" (Nichols, 2003, p. 328). In the case of the Kidd Copper property, a reconstruction of the depositional processes has been used to correlate the sediment structures with the numerical geological time scale provided from the geological and historical time records using data provided from air photos and ground reports.

The depositional processes were based on first-hand observation and testimonials collected, entered and recorded into Forestry Supplier's Field Books in 2004. For example, the mine sits on a mineral intrusion; as a result its source rock underlying the waste material is goethite. This is the main differentiation between the stratigraphic correlations in this report versus relying on contemporary wisdom using relative age dating. No guesswork was required to come up with dates or numbers because both the age and the matter content of the original parent source rock was known; this resulted in a reliable analysis of the diagenesis of the first layer strata from step three of redeposition.

Three steps of sediment deposits are recorded in the Kidd Copper site from 1960-1970. First step is an accumulation of slag deposited on the ground level. The grain size analyses of tailings at piezometer nests reveal that a typical grain size for the Kidd Copper Area was between 0.01 and 0.06 mm. (Coggans, *et al.*, 1991, p.450). On the Udden-Wentworth grain size scale, refer to Figure 59 - Appendix B, this would translate between very fine sand to coarse silt (Nichols, 2003, p. 12).

Knowing the fluvial history of the Kidd Copper property the principal depositional environment could not be wrongly assessed as found in Table 23 - Section 19.9, the result of a fluvial, deltaic or a turbidite basin; because according to historical information the depositional environment was an upswelling and a redeposition of the mixed sediments from a flood deposit. The best facies interpretation is based on the facts of history which are known at Kidd Copper and the recorded account of what happened.

The Kidd Copper property continued its operation until the early 1970's when the mine site became abandoned because the "man made dam broke" (Belan, 2004, p. 41). A strong geological overview was presented, for example the coordinates of the site area recorded by Justin Grenier in his *Geo3Fe3* field book on August 30, 2004: "UTM 0467209 5137455" (p. 81) and the elevation is 850 feet, refer to Figure 4 - Section 1.3. The historic authenticity of the Kidd Copper property is well established and documented from 1888-2010, which makes it an ideal study site for dating sedimentation processes.

24.2 Finding Two: A Cataclysmic Event Changes Landscape in Hours

Finding 2: "Water can change our landscape in the span of just a few hours" (Albin, 2004, p. 34). The velocity or the force of the flood was strong enough to create an upsurge of sediments when "some time between 1977 and 1983, the retaining dam at the southeast corner of the site gave way" extending the tailings deposits (Shang *et. al.*, 2004, p. 4, 5). The retaining dam was not re-built leaving the displaced sediment intact. This dramatic change was prominent on the Kidd Copper surface in the 1983 photograph; whereas, little change in the tailings distribution was noted in the 1992 photograph. The landform was changed dramatically over a short period of time.

24.3 Finding Three: Exogentic Processes are Minimal

Finding 3: In the case of Kidd Copper property little change occurred on the ground surface and the bed layers were left untouched. The aeolian factors were detected on the surface but the mine tailings sediment was covered underneath as it had been redeposited. Wind blown migration is a minimal factor influencing the stratification of the sub-surface layers. Without prior knowledge to the deposition of the mine tailings classifying these layers would be difficult and would have a large margin of error. For example, in error the non-consistent spatial pattern could be classified as many units through time if the history of the true depositional process was not known. However, the margin of error is <0 because of the eye witness accounts and known historical records to validate the claim. A significant amount of sedimentation was deposited in a very short period of time, roughly ~13 years (1970-1983). The Kidd Copper site area has virtually remained untouched (1983-1992-2004); therefore the exogenetic processes discussed in this report played a relatively small factor in the reshaping of its landform. Kidd Copper property displays almost the same distribution as the present day; therefore erosional factors are minimal influencers.

24.4 Finding Four: A Base-line of Geological History at Kidd Copper

Finding 4: Geological history was observed and recorded so that the deductions from the lithographic logs and stratigraphic sequences of rocks are interpreted in the context of processes now operating on earth. These present processes provide a base-line to measure the sequence of the rock layers against time. Having the background and hands-on observations of the Kidd Copper property is relevant to modern day geology because a track record of the phenomena found at the site can be traced and analyzed against other records collected. "To understand the earth's history, it is necessary to have some sequential framework in which to place our observations; that is, the relative or absolute ages of the rock units must be known" (Garrels & Mackenzie, 1971, p. 63). No guessing is involved as to when the process started at Kidd Copper or its end date because the historical records of the Kidd Copper property are known and because the cataclysmic event of the dam breaking was chronicled.

All the Kidd Copper data is isochronous so that all the boreholes in this report can be correlated as a chronohorizon or chronostratigraphic horizon because they share the same deposition and age. Kidd Copper establishes a base-line of geologic history.

24.5 Finding Five: Heterogeneous Sediment Deposited

Finding 5: The average grain sized detected on the Kidd Copper property is *Very Fine Sand* 0.06 – 0.135 mm, particles small enough to be dispersed evenly after mixing in solution and to maintain a homogenous appearance. The Kidd Copper silty tailings are small and the minerals dissolved but contrary to the homogeneous predictions the sediment did not distribute itself in uniform deposits. The assumptions that C.J. Coggans, D.W. Blowes, W.D. Robertson and J.L. Jambor made that the sulphide grains would be evenly distributed in the concentration of dissolved material because the grains were spherical and one size did not occur.

The surface deposits on the Kidd Copper property looked uniform to the naked eye but the underlying layer revealed multiple layers of sediment deposits. The exposed surface layers of 120 Boreholes collected with site reports from the Kidd Copper property on August 30, 2004 revealed successive rock units when unearthed. Alternating bands of sediment were described in the soil profiles from 32 Geology students. The deposited sediment is recorded throughout the field entries of geology students as layers. Other words used to describe the 11 vertical logs in Cross Section A-A' include: band, interbedded, interlayers, interbedding all the way down – 44 cm (Borehole 46/6), layer(s), and layering. The total depth of Cross Section A-A' [West – North Transect] was ~75 cm or ~2.46 ft.

The shovel in Figure 26 - Section 9.12 was one of the 120 boreholes unearthed that exposed stratified layers when the earth on the Kidd Copper property mine tailings site was exposed. The uncovered sediment deposit was not uniform in appearance in either the upper or lower zone of transition. Andy Weller's shovel exposed a stratified layer of sediment, showing a depth of four feet down slope bank. Andy's description of the borehole is: "Top two feet yellow clay; Two feet below is gray and looks thick potter's clay with green running through which is copper" (Borehole 103, 2004, 29-43-A). Colored layers are visibly detected. Chemical weathering over a short period of time had little influence in the layers below the initial surface as they had not been exposed to the direct atmosphere. Remember "weathering [occurs] at the earth's surface" (Garrels & Mackenzie, 1971, p. 149).

The sediment on the Kidd Copper property was thoroughly and well-mixed in the water as the water dissolved the sediment, metals and constituents. The turbulent flow uplifted the tailings waste deposits into a clear and homogeneous mixture and according to Gary Nichols "Heterogeneous fluids are thoroughly mixed in turbulent flows" (2003, p. 38). This mixing of sediment became a blended mixture, smoothing out and dissolving the sediment particles like chocolate in milk. Heterogeneous sediment deposited ensued on Kidd Copper after homogeneous blending.

24.6 Finding Six: Primary Sedimentary Structure can be Heterogeneous

Finding 6: The dam breach resulted in an upsurge of sediment of sand, liquid and mine tailings. This process was the second step at Kidd Copper of the sediment depositional process. "Redeposition [GEOL] formation into a new accumulation, such as the deposition of sedimentary material that has been picked up and moved (reworked) from the place of its original deposition, or the solution and reprecipitation of mineral matter" (Licker, 2003, p. 293). The third step at Kidd Copper was the deposition and resettling of the particles; and their distribution in beds predominantly as laminae.

Examination of redeposited sediment in its diagenesis context reveals heterogeneous sediment layers formed as the primary structure, on the site area. The Kidd Copper deposits after the flood event clearly show stratification in all of the vertical log profiles. No similar match throughout the cross-sections for color was found; even though the sediment was laid at the same point in time, as the aftermath of the flooding of the dam. Each borehole on Kidd Copper represents a sedimentary structure and each borehole is considered a primary sedimentary structure because "collectively, all features formed during deposition or shortly thereafter" thus are "designated as primary sedimentary structures" (Friedman, *et.al.*, 1992, p. 159). Stratification was found in all the vertical log profiles which were deposited together from one event; therefore, rapid stratification can be a primary structure.

Two interpretations of the Kidd Copper stratigraphic columns are available: (1) Diversified strata were laid over time as various sediment deposits; or (2) The diversified strata is a result of one event and rapid depositional processes resulting in stratification. Undeniably, in the case of Kidd Copper property the second interpretation is the only interpretation that is verifiable because historical information confirms that all of the sediment layers were deposited at the same time. The evidence is overwhelming that the end result of the dam breaking and redepositing its waste mine tailings deposits had stratified results: (1) Deposition of the sediment occurred in a short period of time; (2) Multiple layers; (3) Randomness of sediment distribution seen in adjacent logs; and (4) Rapid stratification was the end result of a one-time occurrence. Stratification is found in primary structures. Heterogeneous deposits are a primary structure.

24.7 Finding Seven: Twisted Rocks are found at Initial Deposition

Finding 7: The Kidd Copper case is an optimal place to correlate each lithographic log as each was found in short proximity of one another. From the walk-about on August 30, 2004, 120 soil samples were taken in an area encompassing 4 km^2 about 85-90% of the surface ground area covered. This ratio is near the 10th percentile which means that approximately complete coverage of the area was observed. The deposition of each borehole point within the spatial distribution of data points is referenced at the same start point in time. Kidd Copper mine tailings aggregate map of 120 borehole deposits with samples and cross-sections of exposed surface depths up to 50 cm, refer to Figure 27 - Section 10.3. Each point had the same initial deposition and was subject to the same catastrophic event which led to its upswelling of sediment, surprisingly and contrary to expectations the sediment did not fall in a uniform layer.

"Finding *physical continuity* – that is, being able to trace physically the course of a rock unit – is one way to correlate rocks between two different places (Plummer, McGery & Carlson, 2003, p. 183). In other words, if the rock unit has similar characteristics to another then they are said to be correlated. The idea is that similar rock units can be correlated. "Correlation between two regions can be made by assuming that similar rock types in two regions formed at the same time" (Plummer, McGery &

Carlson, 2003, p. 184). The idea that only similar characteristics can be correlated is challenged with the random, twisted, irregular sediment deposits found on Kidd Copper. These bent, irregular, heterogeneous, stratified deposits were laid on initial deposition and not after. Furthermore, the tilting was not due to tectonic forces. Correlation of bent or twisted rocks in the Kidd Copper report was done based on similiarities of age and the type of depositional process that they underwent such as the catastrophic flood event.

24.8 Finding Eight: No Uniformity in the Adjacent Structures

Finding 8: "Law of original continuity [is] a general law of geology: a water-laid stratum, at the time it was formed, must continue laterally in all directions until it thins out as a result of nondeposition or until it abuts against the edge of the original basin of deposition" (Bates & Jackson, 1984, p. 291; MacDonald, *et al.*, 2003, p. 273).

Accordingly, the Law of original continuity states that the expected outcome of the adjacent boreholes would be laterally conformed to the marker and what is seen in the one unit should be seen next to it. Contrarily, there was no uniformity in the adjacent structures of the 120 borehole samples in the Kidd Copper property. Refer to Table 15 - Section 15.4, the architectural elements were not identical as expected from mixing the same elements together. The most common correlation was the fact that mixtures of color existed within each layer. Even the surface texture of each profile was different. The color and mineral compositions clearly showed random and unconformable sediment deposits in sampled holes from the Kidd Copper boreholes lithographic logs observed.

The water-laid deposits did not continue laterally as predicted. Instead when the parallel and perpendicular cross-sections were taken the boreholes were not similar; refer to Table 8- Section 11.5. The units on the Kidd Copper stratigraphic logs hold few, if any, commonalities and would be interpreted and possibly indexed as separate units in time without knowing that they came from the same sediment deposit.

Herein lays a perplexing fact about the basic unit of sampling. If the sample is taken from two different beds it cannot be properly correlated because it would lack meaning and so different findings are to be discarded; perhaps because they have no relevancy or relationship. This interpretation based on conventional wisdom is incomplete because different architectural units were found on the same bed in the case of Kidd Copper. The units marked as Boreholes 13, 17, 19, 49, 68, 114 and 117 in Table 15- Section 15.4 are all samples of the architectural elements found in vertical logs on August 30, 2004. Each of these log descriptions describes the same sediment layer deposited with the same sequence of events from the impoundment on the Kidd Copper Property. One descriptor was 'dark brown layers', another 'ferrihydrite', another 'high precipitated salt', another 'interbedded grey and dark red', another 'moss covered', another 'sand', another 'silt', and another 'no vegetation'. It sounds like they were from different regions although all these vertical logs were taken from the Kidd Copper impoundment surface and sub-surface area. The adjacent environments are the same although the architectural elements are varied and not uniform.

The average sediment size found on the Kidd Copper property is *Very Fine Sand* according to the findings in Table 10 - Section 11.9. Since no variations in grain size appear in the sediment on the Kidd Copper property the expectation based on conventional wisdom is that the layering would be homogenous and to predict the strata layers at Kidd Copper to be uniformly deposited. "Foresets may be termed heterogenous if the layering is due to variations in grain size or homogenous if it is not" (Selley, 1982, p. 221). Furthermore, since no variations in grain size appeared in the sediment on the Kidd Copper property it was expected that the layering would be homogenous. The prediction based on conventional wisdom was that the layering of the strata at Kidd Copper would be uniformly deposited but this report does not validate these expectations.

24.9 Finding Nine: Units Cycled are Not Proportional

Finding 9: Conventional reason supports that differences in proportions are seen in the rock record due to time factors, through time. "The cycing of sediments through time should lead to differences in the proportions of rock types of a given age as observed today and also those proportions at the time of deposition" (Garrels & Mackenzie, 1971, p. 275, 276). Conventional reasoning leads one to speculate that the proportional differences seen in the sediment layers is caused by a time factor. The Kidd Copper data reveals that the differences in proportions are not due to different times but rather because the settling process caused a variant distribution. "The sediments deposited at any given time in the earth's history are the same as those deposited at any other time" (Garrels & Mackenzie, 1971, p. 276). Therefore, since science is based on what is seen, not, what is not seen it can be deduced that the variant proportions in the units cycled are not from time differentiations and cannot be used to determine cyclic changes.

24.10 Finding Ten: The Law of Equal Declivities Falsified

Finding 10: The stratification on the Kidd Copper property revealed asymmetrical, heterogeneous deposits were settled out of fluid as stratified layers. No, symmetrical profiles were produced from adjacent boreholes on Kidd Copper, refer to Table 4 - Section 10.5 and Table 5 - Section 10.6 for supporting evidence. Instead, these layers had unequal declivities and different slope angles with no symmetrical profile contrary to the Law of equal declivites which states that "slopes of the valleys cut by the streams tend to develop at the same slope angle, thereby producing symmetrical profiles" (Bates & Jackson, 1984, p. 290). The same depositional process affected all the sediment observed *in situ* at Kidd Copper but had dissimilar profiles in the resultant layers.

Conventional wisdom holds that if the outcrop is similar it can be correlated. "Another thing geologists know to be true – no matter where they are – is that if the same sequence is visible in the rock outcrops in different places, it is likely that these beds are or once were continuous" (Discovery Books, 1999, p. 40). In fact, Kidd Copper had many different units or outcrops even when these visible distinctions were uncovered from a continuous bed of sediment. Contrary to conventional wisdom a different sequence of sediment was found to be continuous from Kidd Copper's bed. Kidd Copper's unequal and discontinuous sediment rejects the Law of equal declivities.

24.11 Finding Eleven: Lateral Continuity Not Found Kidd Copper Case

Finding 11: The sediment layers at Kidd Copper did not continue laterally in all directions until thinning out as a result of all redeposition processes.

Field geology "that part of *geology* that is practiced by direct observation in the field" (MacDonald, *et al.*, 2003, p. 182). The Kidd Copper property provided much information, which based upon eye-witness testimony and observations which eliminated guesswork. For example, in the case of the architectural elements found on the Kidd Copper property lateral continuity is not found. Walther's Facies Law called *lateral continuity*, "states simply that different kinds of sediment deposited side by side in nature will be preserved on top of one another in a sedimentary sequence" (Retallack, 1990, p. 119). The observations of eye-witnesses did not support this expectation.

The *principle of original lateral continuity* was the third of Steno's stratigraphic axioms. It pertains to the fact that, as originally deposited, strata extend in all directions until they terminate by thinning at the margin of the basin, end abruptly against some former barrier to deposition, or grade laterally into a different kind of sediment" (Levin, 2003, p. 4). This is not the case at Kidd Copper, where the abruptions found in the lithographic logs and sediment profiles are disjointed as the settling process was abrupt and rapid leaving a heterogeneous distribution as sediment cover; instead of the thinning suggested by Steno. The third of Steno's stratigraphic axioms is not supported.

24.12 Finding Twelve: Some Major Geologic Laws Do Not Hold

"Law. In science, a formal statement of the invariable and regular manner in which natural phenomena occur under given conditions; e.g. the 'Law of superposition'" (Bates & Jackson, 1984, p. 290). "In science, a law is a descriptive principle of nature that holds in all circumstances" (Isaacs, *et. al.*, 2003, p. 445).

Finding 12: After the dam broke, The Kidd Copper sediment was redeposited once simultaneously over the impoundment area. Applying the Law of original horizontality, Law of superposition, Law of equal declivities and Walther Facies Law would propose that a single, uniform and horizontal layer to be found when the earth was uncovered. These laws suggest that a similar pattern would exist in the borehole lithographic logs and that no tilting of the sediment would be found in the original deposition. These descriptive laws did not hold in the case of Kidd Copper. Moreover, the evidence at Kidd Copper is contrary to the expectations of these geologic laws. Science is based on what is seen, "otherwise it is not science, but myth" (Wells, 2000, p. 3). The findings at Kidd Copper challenge conventional wisdom and their predictions refer to Table 25 - Section 21.10. If contradictory evidence turns up, the theory must be re-evaluated or even abandoned. "Any exceptional event that did not comply with the law would require the existing law to be discarded" (Isaacs, *et. al.*, 2003, p. 445).

24.13 Finding Thirteen: Concept of Uniformitarianism Further Refined

   Finding 13: Concept of iniformitarianism and the conclusion of old Earth is based on slow processes through time is based on Charles Lyell's assertions. "Principle of uniformity" (Licker, 2003, p. 384) = uniformitarianism = uniform rate. However, mainstream geologists like Charles C. Plummer, David McGery and Diane H. Carlson are re-defining the concept of uniformitarianism "because it suggests that changes take place at a uniform rate" (2003, p. 176). Fast processes are being recognized. This report considers these fast rates of deposition, like the Kidd Copper case in its relevance to geologic time. No longer must the concept of geologic time be based solely on slow processes because many rock units have a fast sediment deposition rate.   To accept uniformitarianism means to accept a lengthy history for Earth based on the very slow processes that allegedly formed the rock units, which now is known can be formed rapidly and do not require long periods of time. Therefore, this report opens the doorway to re-define uniformitarianism and to re-define the history of the Earth.

24.14 Finding Fourteen: The Principle and Law of Original Horizontality Falsified

   Finding 14: "The first fundamental principle of stratigraphy is the *principle of original horizontality*, which says that waterborne sediments settle out and are deposited (laid down) as horizontal layers. The principle of original horizontality is important because it means that whenever we observe water-laid strata that are bent, twisted, or no longer horizontal, we know that some tectonic force must have disturbed the strata after they were deposited" (Murck, 2001, p. 49). It was expected that "most sediment settle in water in horizontal layers with flat upper and lower surfaces" (Busbey, *et. al.*, p. 50). Applying this principle and using conventional wisdom, it was expected that sediments that are laid down in water were deposited in horizontal or near horizontal layers.

   > "Law of original horizontality" as defined by Dr. James MacDonald, Dr. Christopher Burton, Dorothy Farris Lapidus, Donald R. Coates and Isobel Winstanley state "*sedimentary rocks*, especially water-laid strata, are deposited parallel to the surface on which they are deposited and thus horizontally. Such rocks are assumed to have been horizontal at the start of their geological history, although they may have been folded or tilted subsequently" (MacDonald, *et.al.*, 2003, p. 273).

   "According to the principle of original horizontality, these layers will initially be set down in a horizontal fashion. Any change from this orientation occurs at another time when the rocks have been tilted by tectonic forces" (Albin, 2004, p. 44). The principle of original horizontality "means that layers of sediment are generally deposited in a horizontal position. Thus, if we observe rock layers that are flat, they have not been disturbed and still have their *original* horizontality. But if they are folded or inclined at a steep angle they must have been moved into that position by crustal disturbances sometime *after* their deposition" (Tarbuck & Lutgens, 1999, p. 191). Applying the principle and the law of original horizontality would lead to the conclusion that the first deposition of the Kidd Copper sediment on ground zero must be deposited horizontally, which was not observed and not corroborated in this report, refer to Table 12 - Section 13.5. In contrast to the prediction of a horizontal layer, waterborne sediments settled on Kidd Copper randomly without order. The lithographic logs yielded various composites and distribution patterns. The actual observation confirms sediment laid down in water is deposited in variant layers and not in horizontal layers.

24.15 Finding Fifteen: Falsification of the Law of Superposition

"The power of modern science lies in the concept of falsification; that is, any scientific theory someone might propose is as good as any other theory unless or until there is experimental evidence to the contrary. Thus, in the modern scientific tradition, one does not "prove" a theory; rather, one can only disprove it. **As Albert Einstein observed, "No amount of experimentation can prove me right; a single experiment can prove me wrong"** (Simple, 2002, p.44, 45).

Finding 15: "Steno's law of superposition states that older rocks are found beneath younger rocks. This happens because when sediments settle out of a fluid, the youngest layers will be deposited on top of previous layers" (Albin, 2004, p. 44).

The Kidd Copper case provides strong evidence contrary to the stated law of superposition which is defined as "a principle or law stating that within a sequence of undisturbed sedimentary rocks, the oldest layers are on the bottom, the youngest on the top" (Plummer, McGery & Carlson, 2003, p. 565). What that means is when multiple layers are seen in the rock layers a time sequence is measured with the oldest layers on the bottom of the pile. For example, mainstream geologists have used this rationalization to date the Grand Canyon. "Applying the Law of superposition to these layers exposed in the upper portion of the Grand Canyon, the Supai Group is oldest and the Kaibab Limestone is youngest" (Tarbuck & Lutgens,1999, p. 191). To arrive at the conclusion that a multiplicity of layers represents multiple dispensations of time one has to assume that continuous, harmonious layering signifies a one-time event. Any disruption or association that is different in appearance means that a different depositional occurrence happened. The reason to support this conclusion is that "there is no way of knowing" (de Raaf, *et al.*, 1965) (Walker & James, 2002, p. 6). In contrast, the Kidd Copper evidence provides a way of knowing; the mine tailings sediment on the top layer was the same age as the mine tailing sediment on the bottom layer in the case of Kidd Copper; thus the Law of superposition can be disproved.

Indeed, the Law of superposition is falsified because the layering of sediments recorded throughout this report clearly showed that layering can occur within the same deposition of sediment. The younger sediments can be found on the bottom as easily as on the top. This is contrary to the Law of superposition but is evidenced by observation and direct witness accounts from the Kidd Copper case study. Table 5 - Section 10.6 display inclined and cross-laminae but no two lithographic logs are identical and it would be hard to determine which was the younger or older sediment based on the layers. However, in the case of Kidd Copper one does not have to worry about figuring out which of the layers is the oldest because all were deposited in the same moment of time thus all are the lithographic logs on Kidd Copper are the same age. The Law of superposition is falsified based on the fact the layers on top at Kidd Copper are not the youngest, nor are the layers on the bottom the oldest. The lithologic logs at Kidd Copper cannot be correctly correlated applying the Law of superposition.

The redeposited sediment was examined in its diagenesis context and did not meet the suggested results or expected outcome from textbook theory or conventional geologic expectations. Conventional geologic wisdom is not only challenged but falsified.

24.16 Finding Sixteen: Some Calibration Methods are Problematic

Finding 16: "calibration to linear time is a matter for discovery or estimation "(Gradstein, *et al.,* 2004, p. 3). The old age calibration methods examined in this report are problematic such as astronomical tuning or radiometric age dating (Gradstein, *et al.*, 2004, p. 455). Futhermore, the Kidd Copper Case covered three steps showing different processes of deposition in a short period, which contradicts the estimations for a long period of time required to make the many sediment deposits seen in larger scale rock profiles. Calibration for age dating cannot be dependent on extrapolation of data or trying to fit data into best-fit.

Science that discovers time parameters and historical information is required for accurate dating. The rock record is problematic because no method has been found that provides the same information when tested independently continuously. The margins of error are high in the chi-square test; spline curve fitting and looking at decay constants to name a few of the methods looked at in Section 17. To date, these calibration methods are problematic and based on the findings in this report should be rejected to accurately date the age of the Earth.

24.17 Finding Seventeen: Sedimentary Rock Order Falsified

Finding 17: "Geologists read sedimentary sequences – the order in which the rock layers were laid down. Their interpretation depends on two key facts: sedimentary rock is always laid down horizontally and if the rock layers have been undisturbed, the oldest rock will lie on the bottom. So it follows, for example, that if the rock layers are not horizontal, then the rock must have changed position over time. It also follows that the more bent and fractured and tilted the rock layers, the more change that has taken place –

and, therefore, the older they are. Sometimes the layers are buckled over so far that, all these millions of years later, geologists have to figure out which way is up, an exercise in deduction called younging" (Discovery Books, 1999, p. 40). Contrary to the rock order mainstream geologists have used to read the rock layers, Kidd Copper findings did not support the interpretation that sedimentary rock is always laid down horizontally. Conventional wisdom and logic was broken when the observations at Kidd Copper did not establish that bent or tilted rocks were changed over time after the initial deposition. Geologists reading sedimentary sequences did not corroborate that the stratified layers were resultant of many changes over much time in the Kidd Copper case. The exercise of 'younging' was not required on the Kidd Copper chronistratigraphy because the age of the layers was known to have occurred at the same point in time, thus this evidence falsifies the conventional and established rock order and reading sedimentary sequences.

24.18 Finding Eighteen: Supports All Scales Incomplete Geologic Record

Finding 18: "Early detectives of Earth's history used this technique, *relative age dating*, to place rock units into a chronological sequence" (Albin, 2004, p. 44). "In 1869, John Wesley Powell … realized that the evidence for ancient Earth is concealed in its rocks" (Tarbuck & Lutgens, 1999, p.190). Relative dating is "the chronological arrangement of fossils, rocks or events with respect to the geological time scale, without reference to their *absolute ages*" (MacDonald, *et al.*, 2003, p. 367). However, for stronger age dating "a relative stratigraphic scale should be used that is as close as possible to the numerical geologic time scale" (Gradstein, *et.al.*2004, p. 108). The stratigraphic record provides us with only a few clues with which to build up a history of the Earth. "No other field of science has found it necessary to construct its own special scale of time. There is no formalized biological time scale, or chemical time scale. All other fields of study simply use the intervals of time known to us all: seconds, minutes, hours, days, years. Geologists, on the other hand, talk about periods and epochs, eras and zones, stages and series: the subdivisions of what is known as the geological time scale" (Gribbin, 2004, p. 149). "The record is incomplete on all scales, from the years in between the events when layers of sediment are deposited, to hiatuses within successions representing hundreds or millions of years represented by major unconformities" (Nichols, 2003, p. 326). "All radiochemical methods of dating have some uncertainties associated with them" (Shakhashiri, April 2008, General Chemistry).

This report on Kidd Copper covered in detail problems with various rock dating methods found in radiometric dates and stratigraphy. Refer back to Sections 16 and Section 17. For example, problems exist in many of the radioactive dating methods, such as "Proton decay – considerable experimental effort has been spent in looking for proton decay, so far with no success" (Isaacs, *et. al.*, 2003, p. 641). "There are essentially no reliable radiometric ages with precise stratigraphic controls within the Late Jurassic or Early Cretaceous" (Gradstein, *et al.*, 2004, p. 371). "Problems arise in the K-Ar method because of the diffusive loss of ^{40}Ar at temperatures well below those of metamorphic recrystallization" (MacDonald, *et al.*, 2003, p. 130). The problem with Rubidium-Strontium dating method is that "the whole-rock isochron may give a different age from a mineral isochron from the same rock because ^{87}Sr can diffuse out of the minerals in which it formed but be retained within the rock (usually in calcium-bearing minerals)" (MacDonald, *et al.*, 2003, p. 130).

Relative dating cannot tell us how long ago something took place, only that it follows one event and preceded another" (Tarbuck & Lutgens, 1999, p. 191). "The troubles of the radiocarbon dating method are undeniably deep and serious … It should be no surprise, then, that fully half of the dates are rejected; the wonder is, surely, that the remaining half comes to be accepted" (Lee, 1981)" (Ham, *et.al.*, 1992, p. 75). Moreover, the observations at Kidd Copper supports the geologic record is incomplete on all scales.

24.19 Finding Nineteen: Assigning Ages to the Rock Layers is Unreliable

Finding 19: "The second principle of stratigraphy is the *principle of stratigraphic superposition*, which states that any sedimentary rock layer is younger than the stratum below and older than the stratum above. In other words, sedimentary rock strata are like newspapers laid in a pile day by day, providing a record from the time of deposition of the bottom (oldest) stratum to the time of deposition of the top (youngest) stratum. Newspapers, of course, have dates on them, so it is possible to determine the exact age of a given newspaper within the pile. Rock strata lack dates, so on the bases of stratigraphy we can only assign relative ages to them" (Murck, 2001, p. 49). The findings in this report clearly support that assigning ages and guessing at the ages of the rock depositions in the rock record are problematic when "using these laws, … to correlate rock formations around the world and assemble a chronological record of the planetary past, dividing it into time units known as eons, eras, periods, and epochs" (Discovery Books, 1999, p. 36). Assigning ages is not a repeatable or a conclusive method to verify the age of the Earth. Great caution should be extended when relying on this method.

24.20 Finding Twenty: The Geologic Time Scale is Insufficient

Finding 20: Time correlation based on the geological time scale is problematic; evidenced by the many errors and missing data compiled in Table 22 - Section 18.2. The geologic processes of the Earth defined by contemporary theories of time are problematic because they ignore or extrapolate data and are insufficient to account for the random mixing of the sediment on Kidd Copper. Science is self-correcting.

Section 17.7-9 dealt with the limitations of GTS2004 – its uncertainties, error changes, uncertainties on the accepted values, causes of the discrepancy… is not resolved, uncertain whether calibrations can be accurately determined, record is incomplete on all scales, a theoretical continuous stack called the stratigraphic column was constructed, rock record is disjointed between continents, absence of precise global markers, cannot define the period because the system boundaries are unknowable, and boundary rocks of uncertain age are missing.

Unlike the missing information and uncertainties this report correlated units in the rock record from the known dates on the Kidd Copper property. The observations made at Kidd Copper had boundaries and therefore are accurately determined.

Scientific judgment: Is the data not supposed to be read to formulate a theory; rather than the data made to support a pre-existing theory?

24.21 Finding Twenty-one: A 'Younger Date' for Planet Earth is Reasonable

Finding 21: Originally, "the acceptance of uniformitarianism meant the acceptance of a very lengthy history for Earth" (Tarbuck & Lutgens, 1999, p. 5). "The accumulations of deep units of sedimentary rock were formed by the same processes of sediment deposition, and at the same rate, as those that can be observed in the modern world. Since most geological processes were shown to be very slow, and the rock units that were formed by these processes were often very large, Lyell concluded that the Earth was extremely old. He was the first person to talk of its age in millions of years" (Busbey, *et. al.*, 1999, p. 19). Later on geologists revised Lyell and Hutton's position to mean that the principle of uniformitarianism "does not imply the invariance through time of conditions now prevailing; what is meant is that changes or modifications of the Earth's crust are produced only in accordance with the laws of physics and chemistry" (MacDonald, *et al.*, 2003, p. 442).

James Hutton "realized that the great thicknesses of sediment rock we find on the continents are products of sediment removed from land and deposited as mud and sand in seas. The time required for these processes to take place had to be incredibly long. Hutton upset conventional thinking (the world was believed to be less than 6,000 years old) by writing in 1788, 'We find no sign of a beginning – no prospect for an end" (Plummer, McGery & Carlson, 2003, p. 176). Hutton's problem was that he had no way to test his assumption that the Earth layers related directly to a magnitude of time; whereas, the present-day transformation and redeposition of sediment on the Kidd Copper property occurred in a relative short period of time, with knowing the beginning and the end of the geologic process. This new information allows the length of Earth's history to be re-considered within these boundaries revealed from Kidd Copper's data.

24.22 Finding Twenty-two: Old Age for Planet Earth Seriously Questioned

Finding 22: The accepted age of the planet Earth being 4.4 billion years old can be reasonably questioned. The author established the fact that stratification is a rapid process based on the interpretation of reconstruction of the paleoenvironment of Kidd Copper, which allows a 'younger' Earth date to be deliberated. The author established the fact that many time-correlations are erroneous. Modern geology is dispelling earlier beliefs about uniformitarianism and re-defining its original meaning. The author establishes the fact that the Law of superposition has been falsified in this report by not withstanding the scrutiny of site observation. With careful analysis of the findings in this report the author has established that each layer of sediment does not equal a separate time zone. Horizontal layers cannot be used to accurately indicate the start of geological history because twisted layers are equally indicative of the start of geological history. The horizontal layers do not indicate separate depositional process or periods. Uniform strata are not representative of the geologic breaks through time in the rock records. Field observations showed that when mine tailings were stirred (a well-mixed in a blender deposit) and allowed to settle quietly yielded stratified laminae, multi-layering is observable for a single, one-time depositional event. The author has provided ample reason to look carefully at the alternative solution for the age of the Earth; and support the assertion that the Earth is younger.

The Kidd Copper property provides a natural real-time small-scale event that can be factually correlated and observed without guessing to time references. This is because this event happened in modern history and not in ancient times. No guesswork is required to account for dates.

The complexity is that scientists have been trying to correlate sediment along indistinguishable paths where soil boundaries are missing and no commonalities exist. They come to the wrong conclusion that each of these multiple, sharp, ferrrunginized or truncated boundaries are the result of separate depositional events. Contrary, to this rationalization the multiple, sharp, ferrrunginized or truncated boundaries deposited rapidly in the case of Kidd Copper. No particular pattern in the layering of the sediment is discernible on the Kidd Copper Property. Instead, the mixing of colors reflects a variety of patterns and a variety of layers. The conclusive findings support that multiple layers are found deposited after a flood event. The laws of stratigraphy would suggest that the sharp contrasts in the contacts and matrices of each vertical log profile should suggest that numerous time periods occurred for the accumulation of sediment; however, these conclusions are not accurate analyses. The fact is that the many layers of sediment deposit were a one-time event and the correlation of the deposits shows that fine-grained sediment can be distributed unevenly.

The diversity of the re-distribution of the sediments as uncovered in the borehole logs was witnessed by all 32 students in their field records. Stratification is seen in every entry. Noticeable colored layers of mineral deposits at different depths to a total depth of <75 cm: clay 10 cm, orange 1 cm, etc varying layers; separates into different colored layers; 26 cm interbedded of white clay and yellow/orange sand; orange, 10 cm light clay, orange with clay interbedding (darker orange); alternating layers of dark brown sand and light grey sand; alternating grey and orange layers; interbedding orange and red with grey; and successive bands of yellowy-orange and grayish-white interbands are some of the remarks noted from the recorded field book entries describing the uncovered boreholes on the Kidd Copper property on August 30, 2004. Even more, this feature provides support that rapid stratification bears evidence of a younger earth. The layers do not require a long period of time to form. Based on layers forming in a shorter time period, 4.4 billion years can be shortened.

24.23 Finding Twenty-three: New Interpretation Given for Geologic Events

> "The record of these processes in sedimentary rocks allows us to interpret the rocks in terms of environments. To establish lateral and temporal changes in these past environments a chronological framework is required. This time framework is provided by different aspects of stratigraphy and allows us to interpret sedimentary rocks in terms of dynamic evolving environments. The record of tectonic and climatic processes through geological time is held in these rocks along with evidence for the evolution of life on Earth" (Nichols, 2003, p. 1).

Finding 23: "Geological events by themselves, however, have little meaning until they are put into a time perspective" (Tarbuck & Lutgens, 1999, p. 190).

If there had been no eye-witness accounts, no established boundaries and no supportive data about Kidd Copper it could be mistakenly concluded that the mine tailings sediment was laid down slowly over millions of years; however, this is not the case. *In situ* field data of 120 entries collected on 85-90% of the impoundment area of mine tailings covers a substantial portion for on-site observations. The lateral, parallel and perpendicular aspects of the lithographic logs consistently recorded stratification.

The quick multi-layering of the flow from the dam break and the resettling of the mine tailings shows layers are not settled as one deposit on top of another through many events and much time but rather the evidence supports multi-layered deposits are one event and through rapid stratification. The conclusive data acquired through the mine tailings deposits on the Kidd Copper property is that the multiple layers were laid down in a relatively short time period. The record of the processes on Kidd Copper establishes a chronological sequence of events to provide a framework for multiple layer deposits. The time frame was rapid and the rock record supports a shorter period. The new interpretation from the time framework established at Kidd Copper allows the possibility of a 'younger' earth to be examined.

Old Earth was contrasted to young Earth in origination. Men like Sir Charles Lyle introduced "the belief that the thickness of the strata and the number of layers – which take a great deal of time to form – had to have begun way before 4000 BCE." (Albin, 2004, p. 43). Quoting Darwin, "I further believe that these slow, intermittent results accord well with what geology tells us of the rate and manner at which the inhabitants of the world have changed" (1958, p. 111).

In contrast to Darwin's belief, the strong evidence obtained from the collective data at Kidd Copper results in fast, continuous deposits of many layers that now translate into the geologic record. This new knowledge tells geologists that the rate and manner at which the inhabitants of the world have changed in a quicker, faster, more rapid rate than Darwin exerted. Was Darwin wrong? The possibility exists.

24.24 Finding Twenty-four: New Time Parameters from Rock Readings

Finding 24: Although, biological evolution was not the scope of this report "the world famous astrobiologist, Paul Davies, says: It's a shame that there are precious few hard facts when it comes to the origin of life. We have a rough idea when it began on Earth, and some interesting theories about where, but the how part has everybody stumped. Nobody knows how a mixture of lifeless chemical spontaneously organized themselves into the first living cell" (Mortenson, November 2004, p. 1, 2).

Charles Darwin had achieved a "unique literacy status among works of the scientific imagination" (Leakey, 1979, p. 9). He wrote *"The Origin of Species"* … "for the theory of evolution" (Leakey, 1979, p. 9). Darwin perpetuated mammalian evolution. "These breaks are imaginary, and might have been inserted anywhere, after intervals long enough to allow the accumulation of a considerable amount of divergent variation" (Darwin, 1958, p. 119). "But as we have no facts to guide us, speculation on the subject is almost useless" (Darwin, 1958, p. 127).

The overlapping of geology, the rock record and the evolution of time is based on reading the rock records. "Earth is composed of rocks and their mineral components, thus their stories are inextricably linked to the planet's history" (Discovery Books, 1999, p. 8). "The history of the Earth is recorded in a very incomplete manner in the rocks formed at different times and places" (Nichols, 2003, p. 229). Layers of sediment when misread definitely lead to questionable dates for the age of the Earth.

Gradstin, Ogg and Smith recognize that calibration of the geologic time scale with precise dates is not easily achieved because of the inaccuracies in the testing methods and the insufficient data accumulated to date (2004, p. 455). "In fact, it is doubtful if the rock record on Earth harbors all the precise age information" (Gradstein, *et al.*, 2004, p. 455). These professors recognize that the present way time is estimated has limited control and open to much speculation, "especially prior to ~30 Ma (as of 2,004), leaves considerable room for interpolation of a geologic time scale" (Gradstein, *et al.*, 2004, p. 455).

The stratified layers set new time parameters. Radically, the Kidd Copper property case research established known time parameters that support a 'younger' earth date from the physical boundaries that determined precisely the geological depositional process considering three different steps for the depositional process on the impoundment area. For example, the true interpretation of Point 64/X1 (Figure 48 - Section 21.4) is that the sediment was laid as slag waste on the impoundment area and a flood event stirred the sediment. No disturbance affected the quiet settling of the mixed sediment as it was redeposited on the Kidd Copper site area. Unearthing of the sediment revealed that for 45 cm to depths of about 1 ½ feet varied layers were uncovered of rust color and white. No erosional surface was noted. Since the sediments had been buried oxygenation was not a factor. The stratified layers were a result of rapid burial and do not account for different times or different depositional processes. Each layer represents the same depositional process and is one unit. The depositions occurred in the early 20th century leaving multiple layers; rather than the layers accounting for multiple events and a long period of time to accumulate. In this brief historical period, multiple layers of sediment are observed. This report has verified that several stratified deposits deposited in the last 40 years observed at Kidd Copper are geologically young.

The 24 findings at Kidd Copper is a significant contribution to the academic body of knowledge. This opens new time parameters to be considered when reading the rock records. Now, when layers are looked at they will not be seen as multiple depositions through time but rather as a single unit. Stratified layers will be read as primary structures. Heterogenous layers will be looked at as original units. The rock record will be dramatically shortened. Discontinuous units will be seen as continuous because the adjacent structures do not have to be similar to result from the same deposition. Much of the GTS 2004 will be re-written as extrapolations and guesses are eliminated. New parameters from the rock record will open new possibilities for the date of the Earth, and credibly consider the 'younger date' for Planet Earth.

25.0 Hypothesis Supported for Rapid Stratification

The 'classical' notion as James Hutton (1726-1797) conceived uniformitarianism implied that geological rates were constant and 'uniform' through time. Modern geology does not hold this to be true anymore, as can be seen by the evolution of the scientific understanding accumulated about sedimentation rates reported from 1971-2004. If uniformitarianism has changed in meaning, then what about the conclusions that Lyell "proposed that past events occurred at the same rate as they do today… Since most geological processes were shown to be very slow, and the rock units that were formed by these processes were often very large, Lyell concluded that the Earth was extremely old" (Busbey, *et. al.*, 1999, p. 19).

Today, Lyell's conclusions can be seriously challenged with new evidence provided by the Kidd Copper case. Primary research was collected in Section 10 of this scientific report and correlations were made using both parallel and perpendicular cross-sections of the West-North transect and a North-South transect looking at the boreholes dispersed on the impoundment of tailings at the Kidd Copper site. The aggregate base map of borings provides referential data with established parameters. Secondary research was collected to support the data findings in this report, for example Copper Cliff satellite site data confirmed information about Kidd Copper's grain size D_{50} (mm) of the sediment between "0.01 and 0.06 mm" (Coggans, *et al.*, 1991, p. 450).

Rapid stratification was looked at from many different angles to accumulate a strong defense that rapid stratification is a primary structure. The findings in Section 24 provide strong evidence from numerous sources to substantiate this author's claim that conventional geologic wisdom can be challenged in defense of a 'younger' Earth.

25.1 Hypothesis Accepted that Stratification Equals a Fast Rate of Deposition

The claim in this report for a 'younger earth' is not based on what we don't see but on what we do see. The claim to re-open the doorway to a 'younger earth' has been established. "Time is determined by a sequence of events" (Garrels & Mackenzie, 1971, p.63). The claim that layers in sediment are a rapid process has been tested.

The hypothesis that stratification equals a fast rate of deposition was supported by the strong evidence reported in the findings of this scientific report. The null hypothesis that stratification occurred slowly was rejected based on the evidence provided by first-hand observations. The lithologic logs at Kidd Copper were correlated with no symmetrical or pattern distribution; instead an uneven and random distribution pattern was noted; although all the boreholes had undergone the same depositional process. The adjacent boreholes did not exhibit similar facies characteristics and if one did not know that they had settled simultaneously it would be hard to find a commonality.

The differences in the adjacent or associated units at Kidd Copper did not reflect differences in origin but rather different units can reflect the same origin. The hypothesis of rapid stratification supports a time-stratigraphic correlation of heterogeneous material can be referenced by the same point in time even when the distribution patterns are dissimilar. Time is not measured on multiple rock layers being equal to multiple units of time. Rather multiple layers of rock can be one unit of time as this report based on rapid stratification verifies. The Kidd Copper redeposited sediment corresponds directly to rapid stratification.

The null hypothesis is $H_0 \neq$ rapid and the alternative hypothesis $H_1 =$ rapid; so that, the null hypothesis can be rejected; therefore proving the hypothesis. The alternative hypothesis is accepted that rapid stratification equals a fast rate of deposition.

Rapid stratification opens the door for a 'younger' earth. This is because the units in the rock record do not need to be correlated along similar patterns thus nullifying the GTS 2004 correlations for time references as the evidence in this report indicates. Future interpretations to the age of the earth will be assisted by the data contained within this report, where science is supported by evidence that is seen. Science is self-correcting. This report established future replicable and experimental models for the accounting of the age of the earth as younger. Substantial proof exists for a 'younger' earth. This report confirms that a reasonable alternative well worth considering is that a younger Earth is viable.

25.2 Weighting of Evidence Supports Claim for a Younger Earth

The significance of this report has ethical, existential, intellectual and practical reasons for close examination. The historical geography and knowledge of the depositonal processes that stem from this research provide a basis for humans to develop a measurable time line. The exploration of the world is more fully understood by the recognition that the depositional processes in the rock record could be the result of the Great Deluge. This opens the doorway to new scientific findings as the study of science continues to understand and answer the questions concerning Planet Earth and its intrinsic nature. As in all studies, knowledge stimulates curiosity and challenges the mind to re-evaluate the conceptions of the world. As further science opportunities arise, the policies to protect the Kidd Copper property as a scientific natural wonder may offer others the examination of the depositional processes as discovered.

Summation of the Case of Kidd Copper's 24 Findings: In summation, the author noted 24 findings at the Kidd Copper property that seriously question conventional geologic laws. Many were falsified.

The historic authenticity of the Kidd Copper property is well established and documented from 1888-2009 with the source rock known. A base-line of geological history was found at Kidd Copper. The dam breach resulted in the mixing of the mine tailings and the redeposition almost simultaneously. Examination of redeposited sediment in its diagenesis context reveals heterogeneous sediment layers deposited as the primary structure, which was unexpected based on many current Geologic laws and principles that taught primary structures were homogeneous. Instead, Kidd Copper establishes that primary sedimentary structure can be heterogeneous. Rapid stratification is one outcome for a primary structure as an original deposition.

Exogentic processes had a minimal effect on the Kidd Copper Property. Stratification was found in the entire vertical log profiles found in Kidd Copper's isochronous and primary structures. The correlation of bent or twisted rocks appeared at initial deposition and not afterwards in the case of Kidd Copper, verified by the examination of 120 boreholes from original excavations *in situ*.

The sedimentary rock order was falsified. No uniformity in the adjacent structures was noted and the Law of equal declivities was falsified. Lateral continuity was not found in the Case of Kidd Copper. Some major geologic theories need re-evaluation and some geologic laws are problematic. The Law of original horizontality, the Law of superposition and the Walther Facies Law are falsified.

The Geologic Time Scale is problematic. Assigning ages to the rock layers is inconsistent. The data and the information in this report support the point that the Geologic Record is incomplete on all scales. Time correlation based on the geological time scale is incomplete. A 'younger date' for Planet Earth is reasonable because the Old Age for Planet Earth based on conventional wisdom has many problematic areas and raises serious questions. For example, to accept uniformitarianism means to accept a lengthy history for Earth based on the very slow processes that allegedly formed the rock units, which today are known to form rapidly and do not require long periods of time. The accepted age of the Planet Earth being 4.4 billion years old now is questioned. A new interpretation can be given for geologic events and new possible rock readings are within reach. The findings that supported the hypothesis that stratification equals a fast rate of deposition are based on facts, observations and testable, reproducible models. Original excavations *in situ* provided real and historical information in contrast to missing data or imagination to make this report. The weighting of the evidence supplied by the new information from the Kidd Copper case confirms there is substantial evidence to support a 'younger date' for Planet Earth.

25.3 Conclusion: Strong Evidence to Support a 'Younger Date' for Planet Earth

This report was fully documented and covered the historicity of the Kidd Copper area. All of the lithological logs were of known diagenesis context and their results were compared to the expected outcome(s) from the Geologic laws and principles. The documentation provided historical data to determine the actual events, which established a base-line in which geologic theory and laws were compared. The overtopping and the dam breach resulted in rapid stratification of sediment as a primary structure. Contrary to mainstream thought, rapid stratification was established as one-layer.

The scientific evidence for dating the Earth based on the sediment layers of 120 boreholes consistently revealed that multiple layers were not the result of multiple deposits. Instead the *in situ* excavations and first-hand observations repeatedly supported that stratification was the result of a one time event or occurrence. The accepted hypothesis in this report is that stratification in the rock record supports a fast rate of deposition. The evidence in the land base at Kidd Copper identifies stratification visibly apparent on the first layer or the first few inches of land surface. Eye-witness testimony confirms that these boreholes were deposited in a short period of time; the stratification reported throughout this report confirms the fast rate of deposition for multiple layers.

The theory that rapid stratification supports a 'younger' Earth is supported as the findings verified that multiple layering of strata cannot be interpreted as individual time events. The time factor on a geologic scale is relatively young and so gives credence to the theory that rapid stratification can support a 'younger age' for planet Earth. This has been validated by the evidence that rapid stratification supports each layer of sediment is not equal to a separate time zone; thus leaving room to provide an alterative to be reasonably considered, from an old Earth start date to a young Earth start date.

The claim of this report supports the reasonable alternative to an old age earth. "The father of modern geology is the English scientist Sir Charles Lyell (1797-1875)" (Busbey, *et. al.*, 1999, p.18). "Lyel concluded that the Earth was extremely old. He was the first person to talk of its age in millions of years" (Busbey, *et. al.*, 1999, p.19). "Another key geological discovery has been that of *geologic time*. In the three hundred years between 1650 and 1950, the known age of planet Earth increased more than 1 million times from 6,004 years to 4.6 billion years!" (Eyles, 2002, p. 4). The scientific findings and observations in this report open the doorway to support the earlier suppositions that the Earth is 'younger', based on the evidence that layers are formed rapidly. The geologic time scale must be re-calibrated.

Editor's Note: the use of personal names or brand names in a peer-reviewed paper is for the identification purposes only and does not constitute endorsement by the named subjects or McMaster University on the subject matter presented by the author's views on the age of the earth represented by this report.

References

Abulnaga, Baha E. (2002). Slurry systems handbook. *Technology & Engineering*. Retrieved February 28, 2009, from http://books.google.ca/books? Isbn=0071375082.

Albin, E.F. Ph.D. (2004). *Earth Science Made Simple*. New York: Random House, Inc.

Alden, A. (2009). Six things to know about the earth'ss mantle. *About*. Retrieved October 11, 2009, from http://geology.about.com/od/mantle/tp/mantleintro.htm.

Allaby, A. & Allaby, M. (2003). *Oxford Dictionary of Earth Sciences*. Toronto: Oxford University Press.

Article [1]. (2008). Cracking Dams Advanced Level. *Simscience*. Retrieved March 27, 2008, from http://simscience.org.

Article [2]. (2008). Crowflight Minerals Inc.. Sudbury Basin: AER Kidd Property – Friday [Online]. Retrieved March 27, 2008, from http://www.crowflight.com/s/SudburyBasin.asp.

Bardsley, W.E. (1982). Random errors in fission-track dating. *Mathematical Geology*. 14(5), 545-546. University of Waikato, Hamilton, New Zealand. *Plenum Publishing Corporation*. Retrieved June 4, 2009, from http://www.springerlink.com/index/W4003777H7712357.pdf.

Bates, R.L. & Jackson, J.A. (1984). *Dictionary of Geologic Terms* (3rd ed.). New York: Anchor Books.

Behe, M.J. (2003) *Darwin's Black Box: The Biochemical Challenge to Evolution*. New York: The Free Press.

Belan, K. (August 2004). *Geo3Fe3 Field Camp. Forestry Suppliers Field Book*. August 30, 2004.

Benson, A. (August 2004). *Geo3Fe3 Field Camp. Forestry Suppliers Field Book*. August 30, 2004.

Bharti, S. (November 2004). Crowflight Receives mega TEM results on Thompson Nickel Belt Properties from Falconridge. *Crowflight Minerals Inc*. Retrieved August 20, 2009, from http://www.crowflight.com.

Bishop, A.C., Woolley, A.R. & Hamilton, W.R. (2001). *Cambridge Guide to Minerals, Rocks and Fossils*. United Kingdom: Cambridge University Press.

Blatt, H., Middleton, G. & Murray, R. (1972). *Origin of Sedimentary Rocks*. New Jersey: Prentice-Hall, Inc.

Boggs, S. (2006). *Principles of Sedimentology and Stratigraphy. (4th ed.)*. Pearson Education, Inc. Canada.

Brady, N.C. and Weil, R.R. (2002). *The Nature and Properties of Soils*. (13th ed.). Canada: Pearson Education, Ltd.

Brown, R. (August 2004). *Geo3Fe3 Field Camp. Forestry Suppliers Field Book*. August 30, 2004.

Buckman, H.O. & Brady, N.C. (1968). *The Nature and Property of Soils* (6th ed.). New York: The MacMillan Company.

Busbey III, A.B., Coenraads, R.R., Willis, P. and Roots, D. (1999). *The Nature's Company Guides Rocks & Fossils*. Sydney: Time Life Books.

Carroll, D. (August 2004). *Geo3Fe3 Field Camp. Forestry Suppliers Field Book*. August 30, 2004.

Charbonneau, Y.P. (2003). *Worthington Ontario Ghost Town*. [Online]. Retrieved October 15, 2004, from http://www.ontarioghosttowns.com/yworth.html.

Clarke, K.C. (2003). *Getting Started with Geographic Information Systems* (4th ed.). Toronto: Pearson Education, Inc.

Claybourne, A. (2002). *Rocks and Minerals*. UK: Parragon Publishing Book.

Coggans, C.J., Blowes, D.W., Robertson, W.D., and Jambor, J.L. (1991). The Hydrochemistry of a Nickel-Mine Tailings Impoundment — Copper Cliff, Ontario. *Reviews in Economic Geology*. The Environmental Geochemistry of Mineral Deposits. Vol. 6B.

Coupe, S. (2004) *Visual Dictionary*. San Fransisco: Fog City Press.

Daily Planet. (October 2009). Discovery Channel TV: Comments about impact physics. Viewed on Channel 35, October 9, 2009 at 21:00 hours.

Dams. (2008) Topographic and Geologic Conditions for Different Types of Dams. [Online]. Retrieved March 27, 2008, from http://www.dur.ac.uk/~des0www4/cal/dams/geol/topo.htm.

Darwin. C. (1958). *The Origin of Species*. A Signet Classic. New York: Penguin Books Ltd. 2003.

Darwin, C., and Jones, S. (2001). *The Voyage of the Beagle*. NewYork: Random House, Inc.

De Medeiros, K. (August 2004). *Geo3Fe3 Field Camp. Forestry Suppliers Field Book*. August 30, 2004.

DePree, C.G. & Axelrod, A. (2003). *Van Nostrand's Concise Encyclopedia of Science*. New Jersey: John Wiley & Sons, Inc.

Dingman, S. Lawrence. (2002). *Physical Hydrology* (2nd ed.). New Jersey: Prentice Hall.

Discovery Books, Consultants: Brobst, D., Kampf, A.R., Kaylor, J., Jones, R., McCourt, G.H., and Wilson, W.E. (1999). *Discovery Channel Rocks & Minerals*. New York: Random House.

Eby, G.N. (2004). *Principles of Environmental Geochemistry*. Canada: Thomson Learning Inc.

Enders, E. (August 2004). *Geo3Fe3 Field Camp. Forestry Suppliers Field Book*. August 30, 2004.

Environmental Science 1A03/1B03/1G03 (September 2002). *Handbook for the Earth and Environmental Sciences Student*. McMaster University Custom Courseware Production Services.

Eyles, C.H. & Boyce, J. (September 2004). GEO 3E03: Sedimentary Facies and Environments: School of Geography & Geology. *McMaster University* Course Outline.

Eyles, C.H. & Boyce, J. (September 2004). GEO 3E03 Sedimentary Facies and Environments: School of Geography & Geology. *McMaster University* Custom Courseware.

Eyles, N. (2002). *Ontario Rocks*. Toronto: Fitzhenry & Whiteside Limited.

Eyles, N. and Young, G.M. (1994). Geodynmaic controls on glaciations in Earth history. In M.Deynoux, J.M.G. Miller, E.W. Domack, *et al.*, *Earth's Glacial Record*. Cambridge: Cambridge University Press, pp. 1-28.

Farndon, J. (2002). *Science*. UK: Parragon Publishing Book.

Farruggia, C. (August 2004). *Geo3Fe3 Field Camp. Forestry Suppliers Field Book*. August 30, 2004.

Firefly Publishers. (2003) Firefly Guide to Fossils. Toronto: Canada.

Fox, M.R. (2000). *A Study of the Biblical Flood*. Five F Publishing Company. Oklahoma City: Oklahoma.

Fresques, L. (2004). [Online]. Student mine remediation project now full-fledged research. Retrieved October 27, 2004, from http://www.nmsu.edu.

Friedman, G.M., Sander, J., & Kopaska-Merkel, D. (1992). *Principles of Sedimentary Deposits*. Toronto: Macmillan Publishing Company.

Gabriel, J. (August 2004). *Geo3Fe3 Field Camp. Forestry Suppliers Field Book*. August 30, 2004.

Garrels, R.M. & Mackenzie, F.T. (1971). *Evolution of Sedimentary Rocks*. New York: W.W. Norton & Company, Inc.

Gradstein, F., Ogg, J., and Smith, A. (2004). *A Geologic Time Scale 2004*. New York: Cambridge University Press.

Grenier, J. (August 2004). *Geo3Fe3 Field Camp. Forestry Suppliers Field Book*. August 30, 2004.

Gribbin, J. (2004). *A Brief History of Science*. China: The Ivy Press Ltd.

Gyatt, S. (August 2004). *Geo3Fe3 Field Camp. Forestry Suppliers Field Book*. August 30, 2004.

Ham, K., Snelling, A., & Wieland, C. (1992). *The Answers Book*. Australia: Creation Science Foundation Ltd.

Hamilton, E. (August 2004). *Geo3Fe3 Field Camp. Forestry Suppliers Field Book*. August 30, 2004.

Harms, J.C., Southard, J.B., Spearing, D.R., and Walker, R.G. (2004). Depositional Environments as Interpreted from Primary Sedimentary Structures and Stratification Sequences. *ProQuest Company*. Michigan.

Harper, N. (August 2004). *Geo3Fe3 Field Camp. Forestry Suppliers Field Book*. August 30, 2004.

Hawley, J.E., and Stanton, R.L., (1962). The facts, the ores, their minerals, metals and distribution in the Sudbury Ores – Their mineralogy and origin; *Canadian Mineralogist*. V.7, pp. 30-145.

Hodgson, D.M. *et al*. (March 2006) Stratigraphic evolution of fine-grained submarine fan systems. *Journal of Sedimentary Research*. An International Journal of SEPM (Society for Sedimentary Geology). 76(1&2), 25.

Hsü, K.J. (2004). *Physics of Sedimentology: Textbook and Reference*. (2nd ed.). New York: Springer.

Hughes, J.D. (1984). Geology and Depositional Setting of the Late Cretaceous, Upper Bearpaw and Lower Horseshoe Canyon Formations in the Dodds-Round Hill Coalfield of Central Alberta – A Computer-

Based Study of Closely-Spaced Exploration Data. *Geological Survey of Canada.* Minister of Supply and Services Canada 1984. Bulletin 361. Ottawa: Canadian Government Publishing Centre.

Hynes, E. (August 2004). *Geo3Fe3 Field Camp. Forestry Suppliers Field Book.* August 30, 2004.

Isaacs, A., Daintith, J. and Martin, E. (2003). *Oxford Dictionary of Science.* New York: Oxford University Press.

Jack, T. (August 2004). *Geo3Fe3 Field Camp. Forestry Suppliers Field Book.* August 30, 2004.

Kearey, P. (2001). *The New Penguin Dictionary of Geology* (2nd ed.). New York: Penguin Books Ltd.

Kilmenko, D. (August 2004). *Geo3Fe3 Field Camp. Forestry Suppliers Field Book.* August 30, 2004.

Knopf, A.A. (1979). *National Audubon Society Field Guide to Rocks and Minerals.* NewYork: Random House, Inc.

Kump, L.R., Kasting, J.F., & Crane, R.G. (1999). *The Earth System.* New Jersey: Prentice-Hall, Inc.

Laporte, L.F. (2003). *The Fossil Record and Evolution. Readings from Scientific American.* University of California: Santa Cruz. W.H. Freeman and Company.

Leakey, R.E., (1979). *The Illustrated Origin of Species by Charles Darwin.* Essex, England: The Rainbird Publishing Group Ltd.

Lee. R.E., (1981). Radiocarbon, Ages in Error. *Anthropological Journal of Canada.* 19(3), 9. Retrieved from, Ham, K., Snelling, A., and Wieland, C. (1992) *The Answers Book.* Australia: Creation Science Foundation Ltd.

Lengert, J. (2004). *Geo3Fe3 Field Camp. Forestry Suppliers Field Book.* August 30, 2004.

Levin, H.L. (2003). *The Earth Through Time* (7th ed.). U.S.A.: John Wiley & Sons, Inc.

Lewis, D.W. 1984. *Practical Sedimentology.* New York: Hutchinson Ross Publishing Company.

Licker. M.D. (2003). *Dictionary of Geology & Mineralogy* (2nd ed.). New York: McGraw-Hill.

Li, D. (August 2004). *Geo3Fe3 Field Camp. Forestry Suppliers Field Book.* August 30, 2004.

Lucchi, F.R. (1995). *Sedimentographica: A Photographic Atlas of Sedimentary Structures.* (2nd ed.). New York: Columbia University Press.

MacDonald, J., Burton, C., Winstanley, I., and Lapidus, D.F., and Coates, D.R. (2003). *Collins Dictionary of Geology.* Glasgow: HarperCollins Publishers.

McGinley, S. (2002). Mine tailings restoration: method uses high grade biosolids to revegetate land. *The University of Arizona College of Agriculture and Life Sciences.* Retrieved from www.arizona.edu.

McGonagle, D. (September 1999). *Collins Gem Science. Basic Facts. Blast Furnace.* UK: HarperCollins.

McMaster University, staff [TA] Donato, Simon. GEO 3E03 Sedimentary Facies and Environment McMaster University Class, Personal communication indicated Nov 8, 2004.

McMaster University. (August 2004). *Field Notebooks. Geo3Fe3 Field Camp McMaster University Class. Assignment # 3 Due Thursday, October 7, 2004 by 4:00 pm.*

McMaster University. (August 2004). *Mine Tailings. Geo3Fe3 Field Camp McMaster University Class. Assignment # 9 Due Wednesday, September 1, 2004 by 10:00 pm. McMaster University.* Professors: Dr. Bill Morris GSB302, Susan Vajoczki BSB313, and Dr. Darren Gröcke GSB305.

Melymuk, L. (August 2004). *Geo3Fe3 Field Camp. Forestry Suppliers Field Book.* August 30, 2004.

Miall, A.D., (1996). *The geology of fluvial deposits: sedimentary facies, basin analysis, and petroleum geology.* New York: Springer.

Moorhouse, WW. (1959). *The Study of Rocks in Thin Section.* New York and Evanston: Harper & Row.

Mortenson, T., (6 November 2004). National Geographic is wrong and so was Darwin. *AIG.* USA. Retrieved March 27, 2010, from http://www.answersingenesis.org/docs2004/1106ng.asp.

Mottana, A., Crespi, R. & Liborio, G. (1978). *Simon & Schuster's Guide to Rocks and Minerals.* New York: Simon & Schuster's Publishing Inc.

Mullineaux, D.R. (1996). Pre-1980 Tephra-fall Deposits Erupted From Mount St. Helens, Washington, *U.S. Geological Survey.* U.S. Geological Survey Professional Paper.

Munoz, A. (2009). Recovery of nickel and copper from mineral wastes using bacterial leaching. *Institute of Chemistry.* Retrieved August 24, 2009, from http://www.metsoc.org/COM2009.

/PDFs/Program-posters.pdf.

Murck, B.W. (2001). *Geology – A Self-Teaching Guide.* New York: John Wiley & Sons, Inc.

Nelson, L. (2004). [Online]. *Flood Strata Formed at Mt. St. Helens.* Retrieved October 25, 2004, from http://www.nwcreation.net/caps/articles/floodstrata.html.

Nesse, W.D. (2004). *Introduction to Optical Mineralogy* (3rd *ed.*). New York: Oxford University Press, Inc.

Nichols, G. (2003). *Sedimentology & Stratigraphy.* Malden, MA: Blackwell Science Ltd.

Oke, T.R. (2001). *Boundary Layer Climates.* (2nd *ed.*). New York: University Press.

Oldershaw, C. (2003). *Firefly Guide to Gems.* Toronto: Canada. Firefly Books Ltd.

Palmer, A.R., and Geissman, J. (April/May 2009). GSA Geological Time Scale. 2009. *The Geological Society of America.* http://www.geosociety.org./science/timescale.

Palmer, A.R., and Geissman, J. (1999). GSA Geological Time Scale. 1999. *The Geological Society of America.* http://www.geosociety.org./science/timescale.

Parker, S.P. (1997). *Dictionary of Earth Science.* New York: McGraw-Hill.

Pearl, R.M. (1955). *How to Know the Minerals and Rocks.* New York: McGraw-Hill Book Company, Inc.

Pham, V. (August 2004). *Geo3Fe3 Field Camp. Forestry Suppliers Field Book.* August 30, 2004.

Plummer, C.C., McGery, D. & Carlson, D.H. (2003). *Physical Geology* (9th *ed.*). New York: McGraw-Hill Companies, Inc.

Quantec Geoscience. (2002). OMET HIS Data Collection Report: Airborne Hyperspectral Surface Mapping Over Exposed and Covered Mineral Belts in Ontario, 30pp.

Quammen, D., (November 2004). Was Darwin Wrong? No. The evidence for evolution is overwhelming. National Geographic. *National Geographic Society.* 206(5), 6.

Reineck, H-E., and Singh, I.B. (1973). *Depositional sedimentary environments with reference to terrigeneous clastics.* New York: Springer-Verlag.

Research Machines Plc. (2003). *Collins Dictionary of Science.* Glasgow: Harper Collins Publishers.

Retallack, G.J. (1990). *Soils of the Past – An Introduction to Paleopedology.* London: Unwin Hyman, Ltd.

Ritland, R. (1981). Introduction to the column. *Andrews University*: Berrien Springs, Michigan. Retrieved June 4, 2009, from http://www.grisda.org/origins/08059.htm.

Roberts, D.C. (1996). *Peterson Field Guides Geology Eastern North America.* New York: Houghton Mifflin Company.

Robertson, H. (August 2004). *Geo3Fe3 Field Camp. Forestry Suppliers Field Book.* August 30, 2004.

Rosamond, M. (August 2004). *Geo3Fe3 Field Camp. Forestry Suppliers Field Book.* August 30, 2004.

Rothwell, R.G., (2005) "Deep Ocean Pelagic Oozes. Vol 5 of Selley, Richard C., L. Robin McCocks and Ian R. Pilmer, *Encyclopedia of Geology, Oxford*: Elsevier Limited.

Rudd, L. (August 2004). *Geo3Fe3 Field Camp. Forestry Suppliers Field Book.* August 30, 2004.

Rudd, Loreen. (2004). *Geo3Fe3 Mine Tailings Exercise Marking Sheet & Paper, Evaluation.* McMaster University School of Geography and Geology.

Rudd, Loreen. (December 2004). *Geo3e03 Sedimentary Facies and Environments, Presentation: Rapid Stratification.* McMaster University School of Geography and Geology, Instructors Dr. Carolyn H. Eyles and Dr. Joe I. Boyce.

Rudd, Loreen. (November 2004). Geo3e03 Research Paper: Layering of mine tailings deposits from Kidd Copper property and the 1980 eruptions of Mount Saint Helens deposits shows comparative time rates for sediment settling processes. *School of Geography & Geology*, McMaster University.

Saha, T. (August 2004). *Geo3Fe3 Field Camp. Forestry Suppliers Field Book.* August 30, 2004.

Schlumberger. (2008). *Oilfield Glossary*. Retrieved 13/08/2008, from, http://www.glossary.oilfield.slb.com/Display.cfm?Term=uniformitarianism.

Schuchert. *Principles of Stratigraphy.* Dunbar and Rogers.

Selley, R.C. (1976). *An Introduction to Sedimentology.* New York: Academic Press.

Selley, R.C. (1982). *An Introduction to Sedimentology* (2nd ed.). New York: Academic Press Inc.

Selley, R.C. (2000). *Applied Sedimentology.* San Diego: Claifornia. AP Professional.

Shang, J., Morris, B. and Howarth, P. (2004). [Online]. *Surface Mapping of Mine Waste using Hyperspectral Imagery of the Kidd Copper Mine Site near Sudbury, Ontario: Preliminary Results.* Retrieved November 7, 2004, from http://www.casi.ca/papers/2-11.pdf.

Shakhashiri, Prof. (April 2008). *Chemical of the week Uranium: a radioactive clock.* Retrieved June 4, 2009, from http://scifun.chem.wisc.edu/CHEMWEEK/PDF/Uranium_Clock.pdf.

Simple, S. (2002). *The Contrarian's Guide to Leadership.* San Fransisco: Jossey-Bass A Wiley Company.

Singh, S. (August 2004). *Geo3Fe3 Field Camp. Forestry Suppliers Field Book.* August 30, 2004.

Stockley, C., Oxlade, C. Wertheim, J. (2001). *The Usborne Illustrated Dictionary of Science.* Great Britain: Usborne Publishing Ltd.

Student's Encyclopedia. (2008). *Dams.* [Online]. Retrieved March 27, 2008, from http://student.britannica.com/comptons/article-231142/dam.

Studdy S. (August 2004). *Geo3Fe3 Field Camp. Forestry Suppliers Field Book.* August 30, 2004.

Sydney Steel Corporation. (November 2004). Slag gallery. *SYSCO.* Retrieved August 26, 2009, from http://www.sysco.ns.ca/slag/slagpict2.htm.

Syque. (2009). Inductive reasoning. *Walden University*: Changing minds.org. Retrieved March 1, 2009, from http://changingminds.org/disciplines/argument/types_reasoning/induction.htm.

Tarbuck, E.J., and Lutgens, F.K. (1999). *Earth an Introduction to Physical Geology* (6th ed.). New Jersey: Prentice-Hall, Inc.

Turcotte, D.L. & Schubert, G. (2002). *Geodynamics* (2nd ed.). New York: Cambridge University Press.

Umpherson, D., Bennett, D., and Webb, J.R., (1991). *Bush Safety in Mineral Exploration; Ontario Ministry of Northern Development and Mines,* Education Series No. 2,73p. Ontario: Queen's Printer for Ontario.

Vanderkooy, C. (August 2004). *Geo3Fe3 Field Camp. Forestry Suppliers Field Book.* August 30, 2004.

van Hengstum,P. (August 2004). *Geo3Fe3 Field Camp. Forestry Suppliers Field Book.* August 30, 2004.

Walker, R.G & James, N.P. (July 2002). *Facies Models Response to Sea Level Change.* St. John's Newfoundland: Geological Association of Canada.

Walker, S. (August 2004). *Geo3Fe3 Field Camp. Forestry Suppliers Field Book.* August 30, 2004.

Weis, N.A. (2008). *Introductory Statistics* (8th ed.). New York: Addison Wesley.

Weller, A. (August 2004). *Geo3Fe3 Field Camp. Forestry Suppliers Field Book.* August 30, 2004.

Wells, J. (2000). *Icons of Evolution Science or Myth?* Washington: Regnery Publishing Inc.

Williams, Brian and Brenda. (1993). *The Random House Book of 1001 Questions and Answers about Planet Earth.* New York: Random House Inc.

Wilmsen, M., and Wood,Ch.J. (2004). The Cenomanian of Hoppenstedt, northern Germany. *Newsletter on Stratigraphy.* Gebruder Borntrager Berlin: Stuttgart. Vol 40.

Wilson, D. (2003). Sudbury Area Mining Railways: Recent aerial view of the massive INCO facility. *Bob Chambers/Kalmbach.* [Online]. Retrieved Oct 15, 2004 and June 16, 2007, from http://www.trainweb.org/oldtimetrains/Sudbury/Falconridge.htm.

Windus, Z. (August 2004). *Geo3Fe3 Field Camp. Forestry Suppliers Field Book.* August 30, 2004.

Xiujuan, M. (August 2004). *Geo3Fe3 Field Camp. Forestry Suppliers Field Book.* August 30, 2004.

Zelek, L. R. (August 2004). *Geo3Fe3 Field Camp. Forestry Suppliers Field Book.* August 30, 2004.

Zim, H.S. (1961). *Rocks and How They were Formed.* New York: Golden Press, Inc.

Maps

Converter. (2009). Lon/Lat UTM Converter Datum: NAD83. Retrieved April 22, 2009, from http://gis.wvdep.org/convert/llutm_conus.php

Google Earth. (2009). Version 5.0.11337.1968 (Beta). Build Date: January 29, 2009. Retrieved April 22, 2009, from Server: KH.google.com.

Map, Kidd Copper Propery Mine Tailings: 4 air photos, 2 hyperspectral image, 1970-2002.

Natural Resources of Canada. (2009). The Atlas of Canada: Perch Lake, Sudbury, Ontario. National Topographic Database. Topographic Map Sheet Number 041/06. Primary Map Data. 1:40 000. 46° 23' 45" N - 81° 25' 7" W. Retrieved April 11, 2009, from http://atlas.nrcan.gc.ca.

Editor's Note: the use of personal names or brand names in a peer-reviewed paper is for the identification purposes only and does not constitute endorsement by the named subjects or McMaster University on the subject matter presented by the author's views on the age of the earth represented by this report.

APPENDICES

APPENDIX A: Data Dictionary

A data dictionary is "a catalog of all the attributes for a data set, along with all the constraints placed on the attribute values during the data definition phase. It can include the range and type of values, category lists, legal and missing values, and the legal width of the field" (Clarke, 2003, p. 308).

A-1 Adjacency

Adjacency is "the topological property of sharing a common boundary of being in immediate proximity" (Clarke, 2003, p. 304).

A-2 Analysis

Analysis – "The determination of the components in a chemical sample" (Isaacs, *et. al.*, 2003, p. 40).

A-3 Architectural Elements

"The traditional descriptive stratigraphic scheme for subdividing ancient sedimentary rocks involves the definition of lithologically homogeneous units (formations) that can be further subdivided (members) or combined (groups). Formal methods and rules are given in the North American Stratigraphic Code (NACSN), (1983). Alternative stratigraphic schemes of increasing importance emphasize the *bounding discontinuities* rather than the internal lithological homogeneity. This type of subdivision has been formalized as *allostratigraphy*. Allostratigraphy is purely descriptive; *sequence stratigraphy* (Van Wagoner *et al.*, 1990) also recognizes units defined by discontinuities and unconformities, but relates them to cycles of sea level fluctuation. *Systems tracts* are linkages of contemporaneous depositional systems. However, different depositional systems occur at different positions of relative sea level, and allow the definition of three main systems tracts; *lowstand*, *transgressive* and *highstand*. Certain facies successions and facies geometries in the geological record appear to be characteristic of certain depositional environments. Similar successions and geometries of various ages can be compared with the facies, facies successions and facies geometries observed in modern sediments. In this way, the general characteristics of various depositional environments can be defined, and used to interpret other parts of the geological record. The comparison of modern and ancient depositional environments, and the search for the processes that control their facies successions and geometries, is now termed *facies modelling*" (Walker & James, 2002, p.1-3).

A-4 Attribute

"Attribute: A characteristic of a feature that contains a measurement or value for the feature. Attributes can be labels, categories, or numbers; they can be dates, standardized values, or field or other measurements. An item for which data are collected and organized. A column in a table or data file" (Clarke, 2003, p. 305).

A-5 Bedding Plane

"Bedding plane [GEOL] any of the division planes which separate the individual strata or beds in sedimentary or stratified rock" (Licker, 2003, p. 40).

A-6 Biofacies

Biofacies – "description is one in which the observations are concerned with the fauna and the flora present" (Nichols, 2003, p. 63).

A-7 Chronostratic Unit Interpretations

Chronohorizon or chronostratigraphic horizon – "Examples include coal beds … and many *biohorizons*" (MacDonald, *et al.*, 2003, p. 92).

Chronostratigraphy or time-stratigraphy "that area of *stratigraphy* dealing with the age and time relations of strata" (MacDonald, *et al.*, 2003, p. 92).

Chronotaxis is "a similarity of time sequence, such as the correlation of stratigraphical or fossil sequences on the basis of age and chronostratigraphical position" (MacDonald, *et al.*, 2003, p. 92).

A-8 Clarity

"Clarity: The property of visual representation using the absolute minimum amount of symbolism necessary for the map user to understand map content without error" (Clarke, 2003, p. 306).

A-9 Compaction

Compaction is "the decrease in pore space of a sediment and consequent reduction in volume or thickness. Compaction results from the increasing weight of younger sediment material that is continually being deposited or pressures caused by earth movements" (MacDonald, *et al.*, 2003, p. 101). "*Compaction* during burial affects fine-grained sediments much more than coarse-grained sediments. Grains are pushed closer together and bent or broken, and *pressure-solution* and redeposition occur at grain contacts, forming sutured contacts. The pore fluids expelled during compaction have an important role in the migration of ions and organic molecules within sediments. Mineral cements that are deposited in the pore spaces may be introduced by the fluid circulating through the sediment or redistributed from materials within the deposit. Reactions that occur between the fluid and minerals with which it is in contact, and the stability of diagenetic minerals, are affected by the E_H and pH of the circulating water. Common cementing minerals include *quartz*, calcite, *limonite*, *hematite* and *dolomite*" (MacDonald, *et al.*, 2003, p. 138, 139).

A-10 Competence

Competence is "the ability of a stream or wind current to carry *detritus*, as determined by particle size rather than amount. The diameter of the largest particle transported is the *competence value*" (MacDonald, *et al.*, 2003, p. 101).

"Competent – (of a sedimentary formation) strong and able to transmit compressive force much farther than a weak *incompetent* formation, i.e. the bed or stratum is able to withstand the pressure of folding without flowage or change in original thickness. Some of the factors that determine whether or not a stratum is competent are the *crushing strength* (resistance to crushing), the massiveness of the formation and the ability to 'heal' fractures" (MacDonald, *et al.*, 2003, p. 101, 102).

A-11 Compression

Compression is "a system of external forces that tends to shorten a body or decrease its volume" (MacDonad, *et al.*, 2003, p. 102).

A-12 Conformable

"When we observe layers of rock that have been deposited essentially without interruption we call them *conformable*" (Tarbuck & Lutgens, 1999, p. 193). Conformable sediment deposits are identified as having the same mechanism for deposition and occurring in the same unit of time. They are designated as a uniform layer of strata and become a set of similar sediments. "Set [GEOL] a group of essentially conformable strata or cross-strata, separated from other sedimentary units by surfaces of erosion, nondeposition, or abrupt change in character" (Licker, 2003, p. 333).

A-13 Connectivity

"Connectivity: the topological property of sharing a common link, such as a line connecting two points in a network" (Clarke, 2003, p. 307).

A-14 Consolidation

"Consolidation – Any process whereby soft or loose earth materials become firm, e.g. the cementation of sand or the *compaction* of mud" (MacDonald, *et al.*, 2003, p. 106).

A-15 Data Format

"Data format: A specification of a physical data structure for a feature or record" (Clarke, 2003, p. 308).

A-16 Datum

Datum –"a fixed or assumed point, line, or surface, in relation to which others are determined; any quantity or value that serves as a base or reference for other quantities or values. The top or bottom of a bed of rock, or other surface, on which structure contours are drawn" (Bates & Jackson, 1984, p. 127). "Datum [GEOL] the top or bottom of the bed of rock on which structure contours are drawn" (Licker, 2003, p. 91). Datum is "a fixed or assumed point, line or surface used as a base or reference for the measurement of other values" (MacDonald, *et al.*, 2003, p. 131). "Datum: A base reference level for the third dimension of elevation for the earth's surface. A datum can depend on the ellipsoid, the earth model, and the definition of sea level" (Clarke, 2003, p. 309).

A-17 Deposition

In lakes which vary in size, shape, salinity and depth affect deposition. Waves and storm currents are important in shallow water. Biochemical and chemical precipitation are common. Strong climatic control on lake sedimentation.

A-18 Diffusion

"Bulk flow of water with its contained solutes is the chief medium of material transfer in the upper crust. Differential movement of water and solutes by ionic and molecular diffusion is of secondary importance in moving large quantities of materials long distances but is one of the dominant factors in controlling reactions between waters and rocks. For example, if an aqueous solution is moving through a fractured rock and the solution is not in equilibrium with the rock, the rock usually tends to be altered at the solution-rock interface. The alteration process involves transport of ions from the solution through the altered material to react with fresh rock and back-transport of the soluble reaction products in to the external solution" (Garrels & Mackenzie, 1971, p. 54).

A-19 Erosion

"Erosion [GEOL] (1) the loosening and transportation of rock debris at the earth's surface. (2) The wearing away of the land, chiefly by rain and running water" (Licker, 2003, p. 111). Erosional or surface-run off processes causing these layers might be reasonable in a small area of the site but, highly unlikely to affect the entire land-site. Erosion would also affect the surface layers and leave the underlying layers intact. Erosion would remove the top-layers without changing the structure of the underlying formations. Aeolian winds could also cause surface changes, but highly unlikely to change the entire column.

A-20 Expected Error

"Expected error: One standard deviation in the units of measure" (Clarke, 2003, p. 311).

A-21 Face Method

"*Face method* – A mapping method used in quarries and *openpit mines* in which geological data are plotted in plan view or at a reference datum measured accurately near the *toe*, and data from higher levels are estimated" (Kearey, 2001, p. 93).

A-22 Facies

"*Facies* – All lithological and paleontological features of a particular *sedimentary* rock, from which depositional environment may be inferred" (Kearey, 2001, p. 93).

"Facies [GEOL] any observable attribute or attributes of a rock or stratigraphic unit, such as overall appearance or composition, of one part of the rock or unit as contrasted with other parts of the same rock or unit" (Licker, 2003, p. 117).

"In many studies, facies have been defined on a small scale – the units are only a few meters in thickness, and differences between facies are subtle, involving (for example) minor changes in proportion of silt and mud, the relative abundance and diversity of fossils, and minor differences in the style of lamination (Walker, 1983)" (Walker & James, 2002, p. 5).

"If one is to make inferences about events recorded in rock units, it is useful to employ the term *facies*. A sedimentary facies refers to the characteristic aspects of a rock from which its environment of deposition can be inferred. For example, a body of rock might consist of a bioclastic limestone along one of its lateral margins and micritic limestone elsewhere. Geologists might then delineate a 'bioclastic limestone facies' and interpret it as a near shore part of the rock body, whereas they might interpret the micritic limestone facies as a former offshore deposit. In this case, the distinguishing characteristic are lithologic (rather than biologic); therefore, the facies can be further designated as a *lithofacies*. In other cases, the rock unit may be lithologically uniform, but the fossil assemblages differ and permit recognition of different *biofacies* that reflect differences in the environment. A limestone unit, for example, might contain abundant fossils of shallow-water reef corals along its thinning edge and elsewhere be characterized by remains of deep-water sea urchins and snails. There would be thus two biofacies, one reflecting deeper water than the other. These could be designated the 'coral' and the 'echinoid-gastropod biofacies'" (Levin, 2003, p. 84, 85).

A-23 Facies Associations

"A group of sedimentary facies used to define a particular sedimentary environment. For example, all of the facies found in a fluviatile environment may be grouped together to define a fluvial facies association" (Allaby & Allaby, 2003, p. 199). Facies associations [are] generally deposited in the same broad environment. "Diamict facies are particularly difficult to give unambiguous environmental interpretations. Indeed, many facies defined descriptively in the field may at first suggest no particular interpretation at all. The key to interpretation is to analyze all of the facies communally, in context. Thus the *succession* in which they occur contributes important information that the facies, considered individually, cannot contribute" (Walker & James, 2002, p. 5, 6).

A-24 Facies Classification

Table 26 - Appendix A shows Miall's facies classification with a breakdown of the facies code, facies, sedimentary structures and textbook interpretations (Walker & James, 2002, p. 122).

Facies Identification. "Within a sedimentary sequence there may be many different facies present, but usually the number is not great. A particular facies is often repeated several or many times in a sequence. A facies may also change vertically or laterally into another facies by a change in one or several of its characteristic features" (Reineck & Singh, 1973, p. 98).

A-25 Facies Maps

"*Facies map* – A map illustrating lateral changes in the lithology of a *formation*, *group* or *system* within a sedimentary *basin*, which allows complex *stratigraphic* data to be presented and which may be used to construct *palaeogeographic* or palaeoenvironmental conditions" (Kearey, 2001, p. 93).

Table 26 - Appendix A: Miall Facies Classification

Miall Facies Classification			
Facies Code	**Facies**	**Sedimentary Structure**	**Interpretation**
Gms	Massive, matrix supported gravel	Grading	Debris flow deposits
Gm	Massive or crudely bedded gravel	Horizontal bedding, imbrication	Longitudinal bars, lag deposits, sieve deposits
Gt	Gravel, stratified	Trough cross beds	Minor channel fills
Gp	Gravel, stratified	Planer cross beds	Longitudinal bars, deltaic growths from older bar remnants
St	Sand, medium to very coarse, may be pebbly	Solitary or grouped trough cross beds	Dunes (lower flow regime)
Sp	Sand, medium to very coarse, may be pebbly	Solitary or grouped planer cross beds	Linguoid, transverse bars, sand waves (lower flow regime)
Sr	Sand, very fine to coarse	Ripple cross lamination	Ripples (lower flow regime)
Sh	Sand, very fine to very coarse, may be pebbly	Horizontal lamination parting or streaming lineation	Planer bed flow (upper flow regime)
Sl	Sand, very fine to coarse may be pebbly	Low angle (<10 °) cross beds	Scour fills, washed-out dunes, antidunes
Se	Erosional scours with interclasts	Crude cross bedding	Scour fills
Ss	Sand, very fine to very coarse, may be pebbly	Broad, shallow scours	Scour fills
Fl	Sand, silt, mud deposits	Fine lamination, very small ripples	Overbank or waning flood
Fsc	Silt, mud	Laminated to massive	Backswamp deposit
Fcf	Mud	Massive, with fresh water molluscs	Backswamp pond deposits
Fm	Mud, silt	Massive, desiccation cracks	Overbank or drape deposits
C	Coal, carbonaceous mud	Plant, mud films	Swamp deposits
P	Carbonate	Pedogenic features	Paleosol
Source: Facies Classification from Miall (1978b) (Walker & James, 2002, p. 122).			

A-26 Facies Models

Four main uses of facies models… assume four major functions: *norm*, *framework* and *guide*, *predictor*, and "*interpretation* for the system that it represents" (Walker & James, 2002, p. 7).

A-27 Facing

"*Facing* – The direction in which *beds* become younger" (Kearey, 2001, p. 93). "*Way-up criteria (way-up indicators)* – Phenomena used to determine the *younging* direction of *strata*, such as *sole marks*, *cross-stratification* and *graded bedding*" (Kearey, 2001, p. 292).

A-28 Formation

"Formation [GEOL] any assemblage of rocks which have some common character and are mappable as a unit" (Licker, 2003, p. 125).

A-29 Geologic Log

"Geologic log [GEOL] a graphic presentation of the lithologic or stratigraphic units or both traversed by a borehole; used in petroleum and mining engineering as well as geological surveys" (Licker, 2003, p. 134).

A-30 Geologic Province

"Geologic province [GEOL] an area in which geologic history has been the same" (Licker, 2003, p. 134).

A-31 Geologic Section

"Geologic section [GEOL] any succession of rock units found at the surface or below ground in an area" (Licker, 2003, p. 134).

A-32 Geologic Structure

"Geologic structure [GEOL] the total structural features in an area" (Licker, 2003, p. 134).

A-33 Horizons

A soil profile is "a vertical arrangement of the various discrete horizontal layers, called *horizons*, that make up any soil from the surface downwards to the unaltered parent material or *bedrock*" (MacDonald, *et al.*, 2003, p. 404). Horizons are commonly found in soils.

A-34 Humic Substances

"Humic substances can be defined as "a general category of naturally occurring, biogenic, heterogeneous organic substances that can generally be characterized as being yellow to black in color, of high molecular weight, and refractory" (Aiken *et al.*, 1985). Humic substances are found in soils, brown coal, fresh and marine waters, and marine and lacustrine sediments" (Eby, 2004, p. 141).

"There are three types of humic substances: humin, humic acid, and fulvic acid. *Humin* is that fraction of the humic material that is insoluble in water at all pH values. *Humic acid* is soluble in water at pH values greater than 2, and *fulvic acid* is soluble in water at all pH values. Humic and fulvic acids are the major components of both freshwater and marine DOC, although there are significant differences in the types of humic substances that occur in these two environments" (Eby, 2004, p. 141).

A-35 Ichnofacies

Ichnofacies – "description focuses on the trace fossils in the rock" (Nichols, 2003, p.63).

A-36 Informal Unit

"Informal unit – A body of rock or other stratigraphical unit that is referred to in casually descriptive but non-definitive terms using ordinary nouns, e.g. 'sandy formation', 'muddy formation'" (MacDonald, *et al*, 2003, p. 249).

A-37 Inset

"Inset: A map within a map, either at a smaller scale to show relative location, or a larger scale to show detail. An inset may have its own set of cartographic elements such as a scale and graticule" (Clarke, 2003, p. 315).

A-38 Interstitial

Interstitial "pertaining to the small spaces between the grains or crystals of a rock" (MacDonald, *et al.*, 2003, p. 253).

A-39 Isolith

Isolith is "an imaginary line connecting location points of similar *lithology* and excluding rocks of dissimilar features such as texture, colour or composition" (MacDonald, *et al.*, 2003, p. 259).

Table 27 - Appendix A: Thickness Scale

Thickness Scale for Beds (layers) and Laminae (in cm)	
Beds and Layers	**Laminae**
>300: extremely thick	3–10: very thick
100–300: very thick	1–3: very thick
30–100: thick	0.3–1: thick
10–30: medium	0.1–0.3: medium
3–10: thin	<0.1: thin
<3: very thin	
Source: Thickness scale for beds (Lucchi, 1995, p.11).	

A-40 Key Bed

"There are no formal upper or lower limits to the thickness and extent of rock units defined as a member, formation or a group" (Nichols, 2003, p. 236). "Key bed – A bed with sufficiently distinctive characteristics to make it easily identifiable in correlation" (Bates & Jackson, 1984, p. 281). Refer to Table 27 - Appendix A for the thichness scale for beds and laminae.

"Source bed [GEOL] the original stratigraphic horizon from which secondary sulfide minerals were derived" (Licker, 2003, p. 348).

A-41 Key Horizon

"Key Horizon – The top or bottom of an easily recognized, extensive bed or formation that is so distinctive as to be of great help in stratigraphy and structural geology" (Bates & Jackson, 1984, p. 281).

A-42 Laminations

Lamination is "the formation of laminae or the state of being laminated, specifically the finest stratification, such as is found in shale" (MacDonald, *et al.*, 2003, p. 269). Refer to Figure 54 - Appendix A. "The term cross-lamination is used for cross-sratification in which the set height is less than 6 cm and the individual cross-layers are less than 1 cm thick" (MacDonald, *et al.*, 2003, p. 118).

"Laminations – thin strata" (Albin, 2004, p. 19).

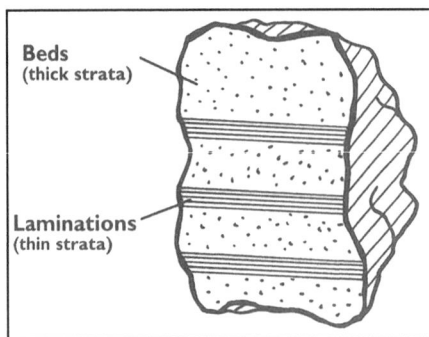

Figure 54 - Appendix A: Laminations -

Sedimentary Rock, (Albin, 2004, p. 19).

"Beds are commonly defined as layers greater than 1 cm in thickness and layers less than 1 cm in thickness are called laminae" (Blatt, Middleton & Murray, 1972, p. 111). Refer to Figure 54 – Appendix A, Figure 20 - Section 7.2.

A-43 Liquid Flow

"Liquid flow – Movement of a liquid, generally one of low viscosity, involving laminar and/or turbulent flow" (Bates & Jackson, 1984, p. 297).

A-44 Lithofacies

Lithofacies is "(1) a mappable subdivision of a specified *stratigraphical unit* differentiated from adjacent subdivisions by *lithology*. (2) The rock record of any sedimentary environment, including both physical and organic peculiarities" (MacDonald, *et al.*, 2003, p. 278). Lithofacies – "one confined to the characteristics of a rock which are the product of only physical and chemical processes" (Nichols, 2003, p. 62).

A-45 Lithology

"Lithology [GEOL] the description of the physical character of a rock as determined by eye or with a low-power magnifier, and based on color, structures, mineralogical components, and grain size (Licker, 2003, p. 186). Lithology is "the description of the characteristics of rocks, as seen in hand specimens and outcrops, on the basis of colour, grain size, texture and composition" (MacDonald, *et al.*, 2003, p. 278).

A-46 Lithofacies Codes

Lithofacies codes is a geologic nomenclature includes lithofacies, sedimentary structures and interpretation, refer to Figure 55- Appendix A.

A-47 Lithostratigraphy

Lithostratigraphy is "*stratigraphy* concerned with the organisation of strata into units based on lithological character and the correlation of these units" (MacDonald, *et al.*, 2003, p. 278).

"A lithostratigraphical unit has a three-part designation consisting of a locality name with the lithology and unit term, e.g. Appin Quartzite Formation, Snowdon Volcanic Group. Lithostratigraphical units, in descending order, are: *supergroup, group, formation, member, bed*" (MacDonald, *et al.*, 2003, p. 414).

	Lithofacies and Interpretation for Six Principal Facies Assemblages of Braided Stream Deposits		

Trolheim type: proximal rivers; predominantly alluvial fans subject to debris flows. Main facies: Gms, Gm. Also: St, Sp, Fl, Fm.

Scott type: proximal rivers; predominantly stream flows and including alluvial fans. Main facies: Gm. Also: Gp, Gt, Sp, St, Sr, Fl, Fm.

Donjek type: distal gravelly rivers; cyclic deposits. Main facies: Gm, Gt, St. Also: Gp, Sh, Sr, Sp, Fl, Fm.

South Saskatchewan type: sandy braided rivers; cyclic deposits. Main facies: St. Also: Sp, Se, Sr, Sh, Ss, Sl, Gm, Fl, Fm.

Platte type: sandy braided rivers; virtually noncyclic. Main facies: St, Sp. Also: Sh, Sr, Ss, Gm, Fl, Fm.

Bijou Creek type: ephemeral or perennial rivers subject to flash floods. Main facies: Sh, Sl. Also: Sp. Sr.

Facies Code	Lithofacies	Sedimentary Structures	Interpretation
Gms	massive, matrix-supported gravel	none	debris flow deposits
Gm	massive or crudely bedded gravel	horizontal bedding, imbrication	longitudinal bars, lag deposits, sieve deposits
Gt	gravel, stratified	trough crossbeds	minor channel fills
Gp	gravel, stratified	planar crossbeds	linguoid bars or deltaic growths from older bar remnants
St	sand, medium to very coarse, may be pebbly	solitary (theta) or grouped (pi) trough crossbeds	dunes (lower flow regime)
Sp	sand, medium to very coarse, may be pebbly	solitary (alpha) or grouped (omikron) planar crossbeds	linguoid, transverse bars, sand waves (lower flow regime)
Sr	sand, very fine to coarse	ripple marks of all types	ripples (lower flow regime)
Sh	sand, very fine to very coarse, may be pebbly	horizontal lamination, parting or streaming lineation	planar bed flow (lower and upper flow regime)
Sl	sand, fine	low angle ($<10°$) crossbeds	scour fills, crevasse splays, antidunes
Se	erosional scours with intraclasts	crude crossbedding	scour fills
Ss	sand, fine to coarse, may be pebbly	broad, shallow scours including eta cross-stratification	scour fills
Sse, She, Spe	sand	analogous to Ss, Sh, Sp	eolian deposits
Fl	sand, silt, mud	fine lamination, very small ripples	overbank or waning flood deposits
Fsc	silt, mud	laminated to massive	backswamp deposits
Fcf	mud	massive, with freshwater molluscs	backswamp pond deposit
Fm	mud, silt	massive, desiccation cracks	overbank or drape deposits
Fr	silt, mud	rootlets	seatearth
C	coal, carbonaceous mud	plants, mud films	swamp deposits
P	carbonate	pedogenic features	soil

Source: after Miall, 1977

Figure 55 - Appendix A: Lithofacies Codes: (Lewis, 1984, p.13).

A-48 Low-Angle Parallel Laminated Sand

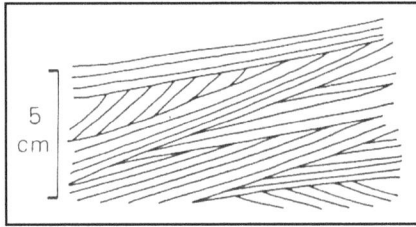

Figure 56 - Appendix A: Parallel-Laminated Sand

Illustrated by Figure 56 - Appendix A "Beach-foreshore bedding consisting of truncated sets of low-angle parallel-laminated sand (with parting lineation). Units of cross-lamination may occur, together with some burrows and heavy mineral concentrations" (Reineck & Singh, 1973, p. 61).

A-49 Map Types: Base Map

Base Map: The aggregate map is a collection of the point data used for reference compiled to form a base map.

A-50 Map Types: Dot Map

"Dot map: A map type that uses a dot symbol to show the presence of a feature, relying on a visual scatter to show spatial pattern" (Clarke, 2003, p. 310).

A-51 Map Types: Geologic Map

"Geologic map [GEOL] a representation of the geologic surface or subsurface features by means of signs and symbols and with an indicated means of orientation; includes nature and distribution of rock units, and the occurrence of structural features, mineral deposits, and fossil localities" (Licker, 2003, p. 134). "Geologic map. A map which shows the surface distribution of rock types, including their ages and relationships, and also structural features" (Allaby & Allaby, 2003, p. 228). "A Geological map on which is shown the surface distribution of rock units, their boundaries, their type and age relationships, and the occurrence of structural features. Geological maps are also used to indicate the presence of ores, petroleum, water, coal and other valuable subterranean materials" (MacDonald, *et al.*, 2003, p. 205).

A-52 Map Types: Geotechnical Map

"Geotechnical map. Map recording existing geology, and estimating likely conditions in terms helpful to the selection of construction techniques and ground treatment, and to the prediction of the reaction between ground and structure. Included in geotechnical maps are analytical maps, comprehensive maps, interpretive geologic maps, and slope category maps" (Allaby & Allaby, 2003, p. 230).

A-53 Map Types: Lithologic Map

"Lithologic map [GEOL] a kind of geologic map showing the rock types of a particular area" (Licker, 2003, p. 186). "Residual map [GEOL] a stratigraphic map that displays the small-scale variations (such as local features in the sedimentary environment) of a given stratigraphic unit" (Licker, 2003, p. 299).

A-54 Map Types: Real Map

"Real map: A map that has been designed and plotted onto a permanent medium such as paper or film with tangible form, and is a result of all the design and compilation decisions made in constructing the map, such as choosing the scale, setting the legend, selecting the colors, and so on" (Clarke, 2003, p. 323). "Base layer or map: A GIS data layer of reference information, such as topography, road network, or streams, to which all other layers are referenced geometrically" (Clarke, 2003, p. 305).

A-55 Mask

"Mask: A map layer intended to eliminate or exclude areas not needed for mapping and analysis" (Clarke, 2003, p. 318).

A-56 Matrix

"Matrix. Fine-grained material found in the pore space between larger sediment grains" (Plummer, McGery & Carlson, 2003, p. 560). "Rock matrix compressibility [GEOL] one of three types of rock compressibility (matrix, bulk, and pore); the fractional change in volume of the solid rock material (grains) with a unit change in pressure" (Licker, 2003, p. 309).

A-57 Offset

"Offset, also called horizontal normal separation (1) The horizontal component of *displacement* along a *fault*, measured perpendicular to the interrupted horizon" (2) In reflection shooting, the *horizontal displacement* of the reflection point from the vertical plane of the profile. The displacement is used as a correction for converting slant time to vertical time" (MacDonald, *et al.*, 2003, p. 318). This is beyond the scope of this report.

A-58 Page Co-ordinates

"Page coordinates: The set of coordinate reference values used to place the map elements on the map and within the map's own geometry rather than the geometry of the ground that the map represents. Often, page coordinates are in inches or millimeters from the lower left corner of a standard-size sheet of paper, such as A4 or 8-1/2 X 11 inches" (Clarke, 2003, p. 321).

A-59 Plane Bed

"Plane bed [GEOLOGY] a sedimentary bed without elevations or depressions larger than the maximum size of the bed material" (Parker, 1997, p. 280).

A-60 Permeability

"Permeable (of a rock or sediment) allowing a gas or fluid to move through it at an appreciable rate via large capillary openings" (MacDonald, *et al.*, 2003, p. 337).

"*Permeability*. The capacity of a rock to transmit a fluid such as water or petroleum" (Plummer, McGery & Carlson, 2003, p. 561). "Permeability – The capacity of a porous rock, sediment, or soil for transmitting a fluid; it is a measure of the relative ease of fluid flow under unequal pressure. The customary unit of measurement is the *millidarcy*" (Bates & Jackson, 1984, p. 377). "Rock permeability [GEOL] the ability of a rock to receive, hold, or pass fluid materials (oil, water, and gas) by nature of the interconnections of its internal porosity" (Licker, 2003, p. 309).

"In many applications the flow through a porous medium is linearly proportional to the applied pressure gradient and inversely proportional to the viscosity of the fluid. This behavior is known as *Darcy's law*" (Turcotte & Schubert, 2002, p. 374). (μ is not the actual velocity of the fluid in the small channels. It is the average velocity per unit area) (Turcotte & Schubert, 2002, p. 374).

$$q = k / \mu * dp / dx$$

The coefficient of proportionality k is called the permeability" (Blatt, Middleton & Murray, 1972, p. 9).

Permeability is "the measure of the ability of earth materials to transmit a fluid. It depends largely on the size of pore spaces and their connectedness; it is less dependent on the actual porosity. For example, a clay or shale with a porosity of 30 to 50 per cent may be far less *permeable* to water than gravel with a porosity of 20 to 40 per cent, because the pore interstices of the clay are so small that flow is inhibited. *Intrinsic permeability* is the property of a rock that controls the permeability (*hydraulic conductivity*). It is a property of the rock itself and remains constant whatever fluid flows through the rock. The customary unit of measurement of intrinsic permeability is the millidarcy" (MacDonald, *et al.*, 2003, p. 337).

"A measure of the ease of transmission of water through a rock is called its permeability. Movement of water is driven by a pressure gradient, commonly gravitational, and is retarded by friction with the grains of the rocks. The permeability can be tested by determining the rate at which the water will flow through a given cross-sectional area of a rock sample of a given length. The equation is analogous to that for heat flow,

$$q / t = \{K (P_2 - P_1) A\} / d,$$

Where q is the quantity of water delivered in cubic centimeters, t is time in seconds, $P_2 - P_1$ is the pressure drop across the sample in atmospheres, A is its cross-sectional area in square centimeters, d is its length in centimeters, and K is the coefficient of permeability. A permeability of 1 darcy is possessed by a rock that transmits 1 cm^3 of water in 1 sec through an area of 1 cm^2 down a length of 1 cm under a pressure gradient of 1 atm. Few rocks have permeabilities as high as 1 darcy; the mdarcy (darcy/1000) is in common use… The differences in permeability of several orders of magnitude between sandstones and shales is the major reason sandstones are the dominant carriers of underground water" (Garrels & Mackenzie, 1971, p. 52).

Table 28 - Appendix A: Average Permeability

Average Permeabilities of Some Sedimentary Rocks	
Sedimentary Unit and Age	**Average Permeability (mdarcies)**
Woodbine sandstone (Cretaceous)	1500
Leduc limestone (Devonian)	800
Smackover limestone (Jurassic)	737
Stevens sandstone (Miocene)	140
Bradford sandstone (Devonian)	50
Spraberry silty shale (Devonian)	0.5
Asmari limestone (Tertiary)	<0.5
Shale (Cretaceous)	0.004
Slate (Precambrian)	0.0013
Chert (Precambrian)	0.00019
Source: (Garrels & Mackenzie, 1971, p. 53).	

"Rain infiltrates the sandstone in its catchment area, and the underground waters (ground water) move downward through the sandstone beneath an impervious layer. Where a drill hole pierces the capping rock, the pressure in the sandstone is released and the water flows upward. If the permeability of the sandstone were 10 mdarcies, the head of water 500 m, and the distance from catchment area to discharge 200 km, the flow per year per square centimeter would be (100 m H$_2$O =10 atm):

$$q / 3.15 \times 10^8 = \{(0.01)*(50)*(1)\} / 2 \times 10^7$$
$$q = 7.9 \text{ cm}^3 \text{ H}_2\text{O}$$

(Garrels & Mackenzie, 1971, p. 52, 53).

"In other words, the water would be expected to move only 8 cm during a year. Such a rate is not atypical; ground-water flow rates of several centimeters to a few tens of meters per year are usual, and rates as high as 1000 m/year or more are attained in unconsolidated gravely sediments. Even a slow rate of movement of 8 cm/year becomes significant on a geologic time scale. A liter of water would be transmitted through each square centimeter of rock in 125 years and 8000 liters in a million years. Ground waters in rocks are highly variable in their chemical composition, but about 0.1 g/liter of dissolved solids in sandstone waters is not uncommon, and therefore about 800 g of dissolved material could be transported through each square centimeter in a million years" (Garrels & Mackenzie, 1971, p. 53, 54). Refer to Table 28 - Appendix A.

"The older a rock, the more chance it has had to be flushed out, with accompanying effects of solution and alteration of mineral grains, as well as cementation and pore filling (Garrels & Mackenzie, 1971, p. 54).

"…water flow down the rivers… All this water has at least a short period of contact with surface materials, and perhaps 25–35 percent percolates through these materials before draining into streams" (Garrels & Mackenzie, 1971, p. 54).

A-61 Pore space

"Porous – Containing voids, pores, or interstices, which may or may not interconnect. The term usually refers to smaller openings than those of *cellular* rock" (Bates & Jackson, 1984, p.398). "Pore [GEOL] an opening or channelway in rock or soil" (Licker, 2003, p. 270).

"*Pore space*. The total amount of space taken up by openings between sediment grains" Plummer, McGery & Carlson, 2003, p. 562).

Packing of the grains. "Some sediments (mainly sands of fine grain size) are deposited with a loose, unstable packing (Blatt, Middleton & Murray, 1972, p. 174).

A-62 Point Feature

"Point feature: a geographic feature recorded on a map as a location. Example: a single house" (Clarke, 2003, p. 322).

A-63 Pore water

Pore water – is between grains in sediment. "*Pore fluid (pore water)* – A solution occupying the pore spaces in soil or rock, whose composition reflects its origin, subsequent mixing and *diagenetic* interaction with the host sediment. Plays an important role in *diagenesis*" (Kearey, 2001, p. 209).

A-64 Porosity

"*Porosity*, the percentage of rock or sediment that consists of voids or openings, is a measurement of a rock's ability to hold water. Most rocks can hold some water. Some sedimentary rocks, such as sandstone, conglomerate, and many limestones, tend to have a high porosity and therefore can hold a considerable amount of water. A deposit of loose sand may have a porosity of 30% to 50%, but this may be reduced to 10% to 20% by compaction and cementation as the sand lithifies. A sandstone in which pores are nearly filled with cement and fine-grained matrix material may have a porosity of 5% or less. Crystalline rocks, such as granite, schist, and some limestones, do not have pores but may hold some water in joints and other openings" (Plummer, McGery & Carlson, 2003, p. 260).

"Although most rocks can hold some water, they vary a great deal in their ability to allow water to pass through them. Permeability refers to the capacity of a rock to transmit a fluid such as water or petroleum through pores and fractures. In other words, permeability measures the relative ease of water flow and indicates the degree to which opening in a rock interconnect. That distinction between porosity and permeability is important. A rock that holds much water is called porous; a rock that allows water to flow easily through it is described as permeable. Most sandstones and conglomerates are both porous and permeable. And impermeable rock is one that does not allow water to flow through it easily. Unjointed granite and schist are impermeable. Shale can have a substantial porosity, but it has low permeability because its pores are too small to permit easy passage of water" (Plummer, McGery & Carlson, 2003, p. 227).

"Porosity – the percentage of a rock's volume taken up by openings" (Plummer, McGery & Carlson, 2003, p. 562). "Total porosity [GEOL] the ratio of total void space in porous oil-reservoir rock to the bulk volume of the rock itself" (Licker, 2003, p. 377). "*Porosity* – Two common uses of this term are: (1) the property of containing pores, which are minute channels for open spaces in a solid; (2) the proportion of the total volume occupied by such pores" (DePree & Axelrod, 2003, p. 596). "When all soil pores are filled with water from rainfall for irrigation, the soil is said to be *saturated* with respect to water and at its *maximum retentive capacity*" (Brady & Weil, 2002, p. 206).

Archie's Formula (Archie's Law)

$$\rho = a\,\rho_w\,\phi^{-m}$$

"An empirical law relating the *electrical resistivity* of a porous rock ρ to its porosity ϕ and the resistivity of the *pore fluid* ρ_w where a and m are constants. (Kearey, 2001, p.15). "*Phi (ϕ) unit* – A unit used to express *clast* sizes in *clastic rocks* according to $\phi = -\log_2 x$, where x is the grain size in mm." (Kearey, 2001, p. 201, 202).

Jeremy Gabriel records the finding on Kidd Copper property as "sand – highly porous" (2004, p. 48).

A-65 Profile

"A profile may be dominated by a single structural pattern. More often, a number of types of aggregation are encountered as progress is made from horizon to horizon. It is at once apparent that soil conditions and characteristics—such as water movement, heat transfer, aeration, bulk density, and porosity – will be much influenced by structure" (Buckman & Brady, 1968, p. 56).

A-66 Profile Section

"Profile section [GEOL] a diagram or drawing that shows along a given line the configuration or slope of the surface of the ground as it would appear if it were intercepted by a vertical plane" (Licker, 2003, p. 276).

A-67 Quantitative Analysis

"Quantitative analysis involves measuring the proportions of known components in a mixture" (Isaacs, *et. al.*, 2003, p. 40).

A-68 Saturation

"Saturation – The degree to which the pores in a rock hold water, oil or gas; it is generally expressed as a percentage of the total pore volume" (MacDonald, *et al.*, 2003, p. 381).

A-69 Sheet Deposit

"Sheet deposit [GEOL] a stratiform mineral deposit that is more or less horizontal and extensive relative to its thickness" (Licker, 2003, p. 335).

A-70 Silt Composition

"Silt [GEOL] (1) a rock fragment or a mineral or detrital particle in the soil having a diameter of 0.002-0.05 millimeter that is, smaller than fine sand and larger than coarse clay. (2) Sediment carried or deposited by water. (3) Soil containing at least 80% silt and less than 12% clay" (Licker, 2003, p. 341). Silt is "(1) *detrital* particles, finer than very fine *sand* and coarser than *clay*, in the range of 0.004 to 0.062 millimeters. (2) an unconsolidated *clastic* sediment of silt-sized rock or mineral particles. (3) Fine earth material *in suspension* in water" (MacDonald, *et al.*, 2003, p. 400).

A-71 Silting

"Silting [GEOL] the deposition or accumulation of stream-deposited silt that is suspended in a body of standing water" (Licker, 2003, p. 341).

A-72 Silt loam

"Silt loam [GEOL] a soil containing 50-88% silt, 0-27% clay, and 0-50% sand" (Licker, 2003, p. 341).

A-73 Silt shale

"Silt shale [PETR] a consolidated sediment consisting of no more than 10% sand and having a silt/clay ratio greater than 2:1" (Licker, 2003, p. 341).

A-74 Silt soil

"Silt soil [GEOL] a soil containing 80% or more of silt, and not more than 12% percent of clay and 20% of sand" (Licker, 2003, p. 341).

A-75 Siltstone

"Siltstone [GEOL] indurated silt having a shalelike texture and composition. Also known as siltite" (Licker, 2003, p. 341). Siltstone is "a fine-grained *sedimentary rock* principally composed of *silt*-grade material. Siltstones are intermediate between sandstones and clays but less common than either" (MacDonald, *et al.*, 2003, p. 400).

A-76 Slag

Slag – "Material produced during the smelting or refining of metals by reaction of the flux with impurities (e.g. calcium silicate formed by reaction of calcium oxide flux with silicon dioxide impurities). The liquid slag can be separated from the liquid metal because it floats on the surface" (Isaacs, *et. al.*, 2003, p. 724). Slag is the "fused waste product from the smelting of ores" (MacDonald, *et al.*, 2003, p. 401).

"Basic slag—Slag formed from a basic flux (e.g.calcium oxide) in a blast furnace. The basic flux is used to remove acid impurities in the ore and contains calcium silicate, phosphate, and sulfide" (Isaacs, *et. al.*, 2003, p. 75). "$CaO + SiO_2$ $CaSiO_3$" (Machines Plc., 2003, p. 70). See Figure 56 - Appendix A.

Figure 57 - Appendix A, the image of iron being extracted from a blast furnace was taken from Stockley, Oxlade & Wertheim, 2001, p.174. "The slag sinks to the bottom of the furnace where it floats on the molten iron" (McGonagle, 1999, Blast furnace).

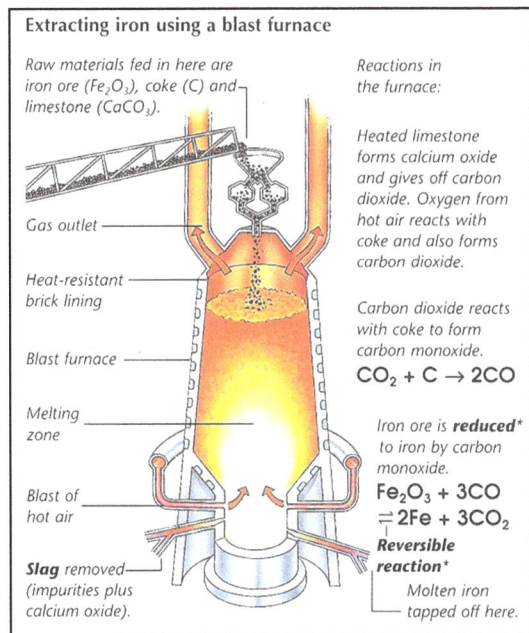

Figure 57 - Appendix A: Blast Furnace

Figure 57 - Appendix A is a representation of a blast furnace which allows the continuous production of molten iron. Slag is removed as fused waste product from the blast furnace. The furnace is charged with a mixture of coke, iron ore and limestone. At the bottom of the furnace coke and air (oxygen). Blast furnace is a "smelting furnace used to extract metals from their ores, chiefly pig iron from iron ore. The temperature is raised by the injection of an air blast. In the extraction of iron the ingredients of the furnace are iron ore (iron (III) oxide), coke (carbon), and limestone (calcium carbonate). The coke is the fuel and provides the agent (carbon monoxide) for the reduction of the iron ore"

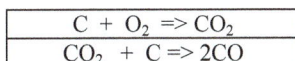

$C + O_2 => CO_2$
$CO_2 + C => 2CO$

(Research Machines Plc., 2003, p. 70).

"At the high temperature of the furnace the iron oxide is reduced to the metal.

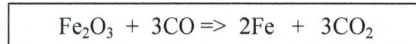

$Fe_2O_3 + 3CO => 2Fe + 3CO_2$

The limestone is decomposed to quicklime (calcium oxide), which combines with the acidic impurities of the ore to form a molten mass known as slag (calcium silicate)"

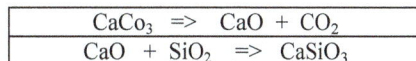

$CaCo_3 => CaO + CO_2$
$CaO + SiO_2 => CaSiO_3$

(Research Machines Plc., 2003, p. 70). Slag as defined from the Collins Gem Science Basic Facts: "At the bottom of the [blast] furnace coke and air (oxygen) react to produce carbon dioxide. This is an exothermic reaction and the temperature rises to $1800°$ C.

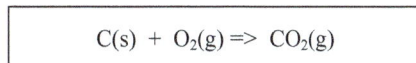

$C(s) + O_2(g) => CO_2(g)$

More coke then reacts with the carbon dioxide to produce carbon monoxide.

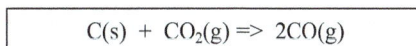

$C(s) + CO_2(g) => 2CO(g)$

The carbon monoxide reduces the iron oxides to iron. For example,

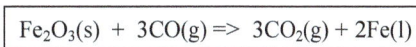

$Fe_2O_3(s) + 3CO(g) => 3CO_2(g) + 2Fe(l)$

The molten iron flows to the bottom of the furnace where it is tapped off. The limestone in the charge is decomposed by the heat, producing calcium oxide and carbon dioxide. The calcium oxide reacts with the impurities from the iron ore (mainly silica SiO_2 and forms a molten slag.

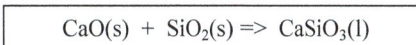

$CaO(s) + SiO_2(s) => CaSiO_3(l)$

The slag also sinks to the bottom of the furnace where it floats on the molten iron. (McGonagle, 1999, Blast furnace).

A-77 Slickens

"Slickens [GEOL] a layer of fine silt deposited by a flooding stream" (Licker, 2003, p. 343).

A-78 Soil Profile

The Soil Profile – "cut a section downward through the soil, horizontal layers … would be found. Such a section is called a *profile* and the individual layers are regarded as *horizons*" (Buckman & Brady, 1968, p. 4).

"The lower boundary of a bed is also called the *sole*" (Blatt, Middleton & Murray, 1972, p. 112).

"Soil Profile a vertical arrangement of the various discrete horizontal layers, called *horizons*, that make up any soil from the surface downwards to the unaltered parent material or *bedrock*" (MacDonald, *et al.*, 2003, p. 404). "Bedrock – "the solid rock that underlies gravel, soil, are other superficial material" (Bates & Jackson, 1984, p. 50). "As commonly defined, the ideal soil profile is divided into A, B and C horizons and their subdivisions. (a) A horizon or *topsoil*: the layer from which soluble substances have been removed by leaching or eluviation and carried downwards to the B *horizon*; (b) B *horizon* or *subsoil*: a zone of accumulation enriched by the clay minerals leached from the A horizon; (c) C *horizon*: the layer of weathered bedrock at the profile base, which has undergone little alteration by organisms" (MacDonald, *et al.*, 2003, p. 404).

A-79 Soil Properties and Behaviors

of Soil Separates

A soil profile has distinct layers from bottom up: (a) parent material; (b) substratum – usually more or less weathered.; (c) carbonate may gradually concentrate; (d) lower zone of accumulation iron and oxides, clay and even calcium; (f) and an upper zone of transition until it reaches the top layer – an upper zone of transition, ground/surface level. These horizons above the parent material are collectively referred to as the solum.

Table 29 - Appendix A shows the association betweenhgeneralized influences of soil separates.

A-80 Sole

"Sole [GEOL] (1) the bottom of a sedimentary stratum. (2) The middle and lower portion of the shear surface of a landslide. (3) The underlying fault plane of a thrust nappe. Also known as sole plane" (Licker, 2003, p. 347)

Table 29 - Appendix A: Soil Properties and Behavior of Soil (Brady &Weil, 2002, p. 127).

Rating Associated with Soil Properties			
Property / Behavior	Sand	Silt	Clay
Water-holding capacity	Low	Medium to high	High
Aeration	Good	Medium	Poor
Drainage rate	High	Slow to medium	Very Slow
Soil organic matter level	Low	Medium to high	High to medium
Decomposition of organic matter	Rapid	Medium	Slow
Warm-up in Spring	Rapid	Moderate	Slow
Compactability	Low	Medium	High
Susceptibility to wind erosion	Moderate (high if fine sand)	High	Low
Susceptibility to water erosion	Low (unless fine sand)	High	Low if aggregated, high if not
Shrink-swell potential	Very low	Low	Moderate to very high
Sealing of ponds, dams and landfills	Poor	Poor	Good
Suitability for tillage after rain	Good	Medium	Poor
Pollutant leaching potential	High	Medium	Low (unless cracked)
Ability to store plant nutrients	Poor	Medium to high	High
Resistance to pH change	Low	Medium	High
Exceptions to these generalizations do occur, especially as a result of soil structure and clay mineralogy.			
Source: Brady & Weil, 2002.			

A-81 Stratigraphic Column

"Stratigraphic column. A succession of rocks laid down during a specified interval of geologic time. The phrase 'the stratigraphic column' often refers to the whole sequence of strata deposited throughout geologic time" (Allaby & Allaby, 2003, p. 522).

A-82 Stratotype

"Stratotype [GEOL] a specifically bounded type section of rock strata to which a time-stratigraphic unit is ascribed, ideally consisting of a complete and continuously exposed and deposited sequence of correlatable strata, and extending from a readily identifiable basal boundary to a readily identifiable top boundary" (Licker, 2003, p. 355).

A-83 Stratum

Stratum is "a defined layer of *sedimentary rock* that is usually separable from other layers above and below; a *bed*" (MacDonald, *et al.*, 2003, p. 414, 415). "Stratum [GEOL] a mass of homogeneous or gradational sedimentary material, either consolidated rock or at unconsolidated soil, occurring in a distinct layer and visually separable from other layers above and below" (Licker, 2003, p. 355). "Stratum is a general term referring to all thicknesses and types of layers" (Blatt, Middleton & Murray, 1972, p. 111).

A-84 Thick Beds

"Thick-bedded [GEOL] pertaining to a sedimentary bed that ranges in thickness from 60 to 120 centimeters (2-4 feet)" (Licker, 2003, p. 373). According to the thickness scale for beds in Figure 20 - Section 7.2, the layers studied on the Kidd Copper property are classified as 3-10 cm [very thick laminae] or <3: very thin beds.

A-85 Thin Beds

"Thin-bedded [GEOL] pertaining to a sedimentary bed that ranges in thickness from 2 inches to 2 feet (5 to 60 centimeters)" (Licker, 2003, p. 373). Refer to Table 27 - Appendix A. "Thin-bedded (0.1-0.6 m) graded fine- and very fine-grained sandstones are commonly current –

and climbing-ripple laminated, with subordinate occurrences of parallel lamination and wavy lamination (forming T_{bc} sequences), with local convolutions. Typically, beds are tabular (sharp bases and tops), with local truncation by erosion surfaces. Commonly, thin-bedded sandstones form thickening-upward packages at the base, and thinning-upward packages at the top, of fan successions" (Hodgson, *et al*, March 2006, p. 25). "Laminations – thin strata" (Albin, 2004, p. 19). Refer to Figure 57 - Appendix A.86, thickness classes of strata.

A-86 Topology

"Topology: (1) The property that describes adjacency and connectivity of features. A topological data structure encodes topology with the geocoded features. (2) The numerical description of the relationships among geographic features, as encoded by adjacency, linkage, inclusion, or proximity. Thus, a point can be inside a region, a line can connect to others, and a region can have neighbors" (Clarke, 2003, p. 326).

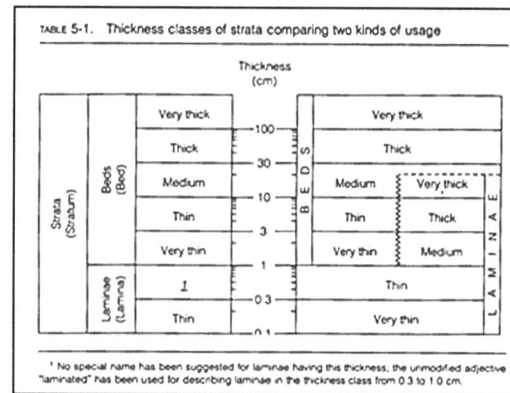

Figure 58 – Appendix A: Classes

Figure 58 - Appendix A.86 shows the thickness classes of strata. (Friedman-Sanders-Kopaska-Merkel, 1992, p. 162)

A-87 Turbulent Flow

"Many useful results for turbulent flow that have been developed up to now are semi-empirical: by judicious choice of what seem to be the most significant physical quantities and the most important physical effects, the investigator builds a framework he hopes will provide an adequate mathematical form for the relationships being sought, and then uses empirical data to make those relationships concrete" (Harms, *et al.*, 2004, p. 7).

"Mutual interaction between bed and flow… great variety of boundary geometries that we see in nature" (Harms, *et al.*, 2004, p.8)

"Simons and Richardson noted that for sands finer than about 0.5mm the sequence of bed phases with increasing velocity is no movement => ripples => dunes => flat bed => antidunes" (Harms, *et al.*, 2004, p. 10).

A-88 Variables to Reconstruct Dam and Flow

A model to reconstruct a spigot dam needs to define variables to characterize the system. Factors like the nature of sediment transport; structure of turbulence; distribution of velocities and shear stresses could be uniquely determined.

Flow is "the movement or rate of movement of water, or the moving water itself" (MacDonald, *et al.*, 2003, p. 185).

"Flow variables must involve the vertical scale of the flow; mean depth of flow… the other must be a measure of flow strength… or force exerted by the flow on the boundary. Flow depth is certainly a less important variable than flow strength" (Harms, *et al.*, 2004, p. 11).

"The velocity variable in most common use is mean velocity, U. The other kind of flow-trength variable, the boundary shear stress, usually denoted by t_0, is the average force per unit area exerted by the flowing fluid on the collection of surface grains that constitute the boundary at any given instant. This force is the aggregate of all of the viscous-shear forces, acting tangentially to the grain surfaces, and all of the pressure-distribution forces, acting normal to the grain surfaces… In flume experiments they are relatively simple to measure, but in natural flows this is often difficult to the point of being impossible, and the observer usually has to settle for approximate estimates" (Harms, *et al.*, 2004, p. 11).

Friction Factor: (Harms, *et al.*, 2004, p. 12).

"Velocity is sedimentologically more significant and more natural and conveys more potentially valuable information to sedimentologists than bed shear stress" (Harms, *et al.*, 2004, p. 12).

"…flow depths that are at a minimum at least twice the height of the waves" (Harms, *et al.*, 2004, p. 49).

Flow power is "stream power" (Harms, *et al.*, 2004, p. 12). "Steady flow in open, straight and very wide channels over full beds of loose sediment" (Harms, *et al.*, 2004, p. 14).

Seven Variables: (1) Flow depth d; (2) Flow strength U or t_o; (3) Flow viscosity μ; (4) Fluid density ρ; (5) Sediment size D; (6) Sediment density ρ_s; and (7) Acceleration of gravity g (Harms, *et al.*, 2004, p. 14).

Reynolds number $\rho U d/\mu$

Froude number $U/\sqrt{gd}$, size ratio d/D, and density ratio ρ_s/ρ (Harms, *et al.*, 2004, p. 15).

A-89 Vertical Accretion

Vertical accretion is "the vertical accumulation of a sedimentary deposit, such as the settling of sediment suspended in a stream that is prone to overflow" (MacDonald, *et al.*, 2003, p. 445).

A-90 Viscosity

"*Viscosity.* Resistance to flow" (Plummer, McGery & Carlson, 2003, p. 566).

A-91 Younging

"*Younging* – The property of *strata* by which they become *stratigraphically* younger in a certain direction, which can be determined from *way-up criteria*" (Kearey, 2001, p. 297).

APPENDIX B: Particles

mm	phi	Name	
256	-8	Boulders	
128	-7		
64	-6	Cobbles	Gravel Conglomerate
32	-5		
16	-4		
8	-3	Pebbles	
4	-2		
2	-1	Granules	
1	0	Very coarse sand	
0.5	1	Coarse sand	
0.25	2	Medium sand	Sand Sandstone
0.125	3	Fine sand	
0.063	4	Very fine sand	
0.031	5	Coarse silt	
0.0156	6	Medium silt	
0.0078	7	Fine silt	Mud Mudrock
0.0039	8	Very fine silt	
		Clay	

Figure 59 - Appendix B: Scale

In Figure 59 - Appendix B Udden-Wentworth scale of grain size classification where "four basic divisions [of grain size] are recognized:

1) Clay (<4μm);
2) Silt (4μm to 63 μm);
3) Sand (63μm to 2.0 mm);
4) Gravel/aggregates (>2.0 mm)" (Nichols, 2003, p. 12).

"Particles – The general dimensions, such as average diameter or volume, of particles or grains that make up a sediment or rock. The gauge of particle size is based on the premise that particles are spheres. This size range of particles in a sedimentary deposit can be determined by measurement of pebbles and larger fragments, sieving (for sands and gravels) and *elutriation* of silt and finer material.

Sedimentologists commonly use the Udden-Wentworth grade scale (commonly called the Wentworth scale) or the *PHI grade scale* to classify particle size" (MacDonald, *et al.*, 2003, p. 332).

B-1 Udden-Wentworth Grain Size Scale

"Wentworth classification [GEOL] a logarithmic grade for size classification of sediment particles starting at 1 millimeter and using a ratio of ½ in one direction (and two in the other), providing diameter limits to the size classes of 1, ½, ¼, etc. and 1,2,4, etc." (Licker, 2003, p. 395). Refer to Figure 59 - Appendix B.

"The four basic types of clastic sediment, based on the sizes of particles they contain, are *gravel* (pebbles, cobbles, and boulders); *sand* (particles from 0.06 to 2 mm [0.0025 to 0.08 in] in diameter); *silt* (particles from 0.004 to 0.06 mm [0.00016 to 0.0025 in] in diameter); and clay [particles less than 0.004 mm [0.00016 in] in diameter)" (Murck, 2001, p. 146).

Size classes (mm)	Components	Aggregate
>256	cobbles and boulders	rubble
256–4	pebbles	grave
14–2	granules	—
2–1/16 (0.063)	sand grains	sand
subdivisions		
2–1	sand grains	very coarse sand
1–1/2	sand grains	coarse sand
1/2–1/4	sand grains	medium sand
1/4–1/8	sand grains	very fine sand
1/16–1/256 (0.004)	silt grains	silt
<1/256	clay particles (minerals)	clay

Figure 60 - Appendix B: Grain Size

Figure 60 - Appendix B shows the grain size of sedimentary particles (Lucchi, 1995, p. 11).

B-2 Very Fine Sand

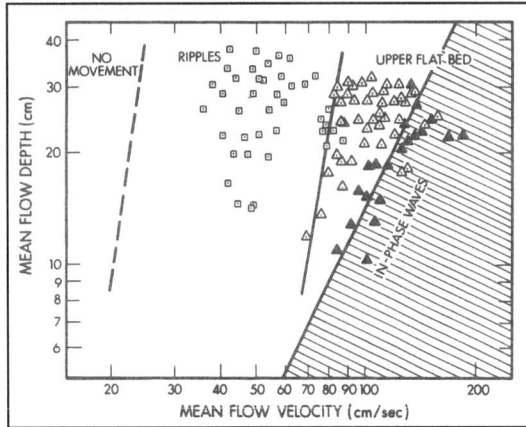

Figure 61 - Appendix B: Mean Flow Velocity

Very Fine Sand – "Depth-velocity diagram for sand size of 0.10 mm from flume data of Willis and others (1972). Squares, ripples; open triangles, upper flat bed. Hatched region lies to the right of line for Froude number = 0.84, which according to Kennedy (1963) is the minimum for development of in-phase waves; Froude number decreases toward upper left, increases toward lower right. Solid triangles represent in-phase waves" (Harms, *et al.*, 2004, p. 18). The mean flow velocity, see Figure 61 - Appendix B (Harms, *et. al.*, 2004, p. 18).

"If the dominant grains are sand-sized, then the rock is termed *sandstone*. The mineralogy determines the classification. Sandstone composed mainly of quartz grains is simply called quartz sandstone. When quartz is mixed with feldspar mineral grains, the rock is called an arkose" (Albin, 2004, p. 19).

"Should quartz be pulverized into very fine silt and then cemented together, the rock is called siltstone. If it remains as loose material, then it is still considered to be sediment" (Albin, 2004, p. 19).

B-3 Ripple-Drift

Suspension – "Fine particles of a solid (groups of atoms, molecules or irons) suspended in a liquid in which the solid does not dissolve" (Stockley, Oxlade &Wertheim, 2001, p. 145).

"The type of ripple-drift found in graded beds is commonly of the type that indicates very rapid 'fallout' of sediment from suspension. Many beds also contain a division of very fine sand showing highly contorted ripples ('convolute lamination'), which suggests rapid deposition from suspension and plastic deformation of partially liquefied sediment produced by shear stress on the bed" (Blatt, Middleton & Murray, 1972, p. 169).

B-4 Proximal or Distal turbidites

Proximal is "close to the source" (MacDonald, *et al.*, 2003, p. 355). "It has been suggested that the deposits closest to the point of initial deposition will be thick coarse beds, generally beginning with a massive or plane-laminated division ('proximal turbidites'). Those farthest from the point of initial deposition ('distal turbidites') will be thin, fine grained beds, beginning with a ripple cross-laminated division" (Blatt, Middleton & Murray, 1972, p. 170).

B-5 Mechanisms that cause sediment

deformation

"There are four main mechanisms that may cause sediment deformation: (a) gravity forces acting on a succession of strata showing a reverse density gradient, i.e., with denser sediment layers overlying less dense layers; (b) liquefaction of sediment; (c) gravity movement of sediment deposited on a slope (slumping); (d) shear stress exerted on recently deposited sediment by a flow moving above it" (Blatt, Middleton & Murray, 1972, p. 170).

B-6 Mechanical Sorting

"Mechanical sorting results from the fact that a river, or a marine current, has a range of speeds which could erode or transport detrital particles of certain size. Well-sorted sands have thus sand grains which almost all belong to the same phi. Such good sorting is evidence that the velocity of the current passing through the site of deposition varied within a narrow range. Coarser sediments had all been left behind before the current reached the site of deposition; finer particles remained in suspension and were carried away. Only sediments of a given size fraction were too coarse to be transported farther" (Hsü, 2004, p. 20).

"Sorting by chemical processes leaves an imprint on the mineralogical composition of a sediment, sorting by mechanical processes determines grain size" (Hsü, 2004, p. 22).

"Shear Sorting [GEOL] sorting of sediments in which the smaller grains tend to move toward the zone of greater shear strain, and the larger grains toward the zone of least shear" (Licker, 2003, p. 335).

B-7 Bedding Definitions

Bedding –"the arrangement of a sedimentary rock in layers; stratification. Also, the general character or pattern of the beds and their contacts within a rock mass, as *cross-bedding* and *graded bedding"* (Bates & Jackson, 1984, p.49).

"*Crossbedding* – Oblique lamination of certain beds in aeolian or water laid sediments, caused by current action" (DePree & Axelrod, 2003, p. 201).

"Bedding – A layered feature of sedimentary rocks characterized by differences in composition, texture or structure. The usual planar top and bottom surfaces that define the units (bedding planes or bedding surfaces) are parallel to the surface of deposition. Layers more than 1 cm thick are called beds (laminae are less than 1 cm thick). Within an individual beds o there may be internal layering approximately parallel to the bedding surfaces (planar bedding or lamination) or inclined at an angle (see cross-bedding). Some beds may show graded bedding; beds without any internal layering or other features are described as massive or structureless" (MacDonald, *et al.*, 2003, p. 56).

"Bedding plane – The dividing plane that separates each layer in a sedimentary or stratified rock of from layers below and above it. It ordinarily delineates a noticeable change in color, texture or composition" (MacDonald, *et al*, 2003, p. 56).

Bedding plane –"In sedimentary or a stratified rocks, the division plane that separates each successive layer or bed from the one above or below. It commonly marks a visible change in lithology or color" (Bates & Jackson, 1984, p. 50).

"Bedded – (of a formation, especially sedimentary) arranged in layers (of any mineral deposit) aligned with the bedding in a sedimentary rock" (MacDonald, *et al*, 2003, p. 56).

"Bed – The smallest formal lithostratigraphical unit that is lithologically distinguishable from beds beneath and above it. The term is usually applied to sedimentary strata but may be used for other types" (MacDonald, *et al*, 2003, p. 56).

Beds – "an informal term for strata that are incompletely known, constitute a lithologically similar succession, or are of local economic significance, e.g. 'beds of Permian age', 'key beds', 'cold beds'" (Bates & Jackson, 1984, p. 50).

Bed form is "any structure produced on the surface of a sediment by the action of a flowing medium (wind or water). The nature of the bed form depends on the flow velocity, for example *ripple marks* form at lower velocities than dunes" (MacDonald, *et al.*, 2003, p. 56, 57).

APPENDIX C: Clay Mineralogy

[Information included here is primarily beyond the scope of this report].

"A mineral is usually defined as a homogeneous, naturally occurring, inorganic solid" (Knopf, 1979, p. 13). Mineralogy is "the study of minerals, including their formation, composition, properties and classification (MacDonald, *et al.*, 2003, p. 300). This is outside the scope of this report. "The mineralogy and texture of preexisting rocks are changed when they are subjected to extreme pressures and temperatures. The objects that result from this process are called *metamorphic rocks*" (Albin, 2004, p. 14).

"Mineralogy 'mudstone, shale and clay' – too fine-grained for minerals to be distinguished with the naked eye, or usually even with the microscope. Clays consist of a mixture of clay minerals together with detrital quartz, feldspar, and mica. Iron oxides are usually abundant and contribute to the red and yellow colors. Black shales are rich in carbonaceous matter, and pyrite and gypsum commonly occur in them, sometimes as well shaped crystals" (Bishop, Woolley & Hamilton, 2001, p. 201).

C-1 Chlorine Content of Terrestrial Rocks

Table 30 - Appendix C: Chlorine Content of Terrestrial Rocks

Chlorine Content Terrestrial Rocks	
Rock Type	Cl (ppm)
Acidic rocks	200
Sedimentary Rocks	105
Shales and clays	100
Sandstones	20

- NaCl (Halite)
- Gypsum- $(CaSO_4 . 2H_2O)$
- Anhydrite $(CaSO_4)$

Refer to Table 30 - Appendix C for chlorine content on terrestrial rocks. An interesting point is what is considered a terrestrial rock – acidic rocks, sedimentary rocks, shales and clays as well as sandstones.

C-2 Normative Composition

"Determination of the normative composition of the average sedimentary rock is a difficult task. We chose to recast the chemical analysis into the minerals calcite, dolomite, quartz, illite, chlorite, montmorillonite, orthoclase, and albite. To make the calculation it was necessary to assign particular compositions to the minerals ilite, chlorite, and montmorillonite, each of which has a wide range of compositions in nature. Furthermore, a unique distribution of mineral percentages could be obtained only by arbitrarily setting the K-feldspar percentage" (Garrels & Mackenzie, 1971, p. 244).

C-3 Mineral Composition of a Rock

Norm is "the theoretical mineral composition of a rock, given in terms of anhydrous *normative mineral* molecules and calculated from the bulk chemical composition" (MacDonald, *et al.*, 2003, p. 314). "Normative mineral or standard mineral – *A mineral* the presence of which in an *igneous rock* is theoretically possible according to the bases of the *CIPW normative* calculation" (MacDonald, *et al.*, 2003, p. 315). See Table 31 - Appendix C for the chlorine content of terrestrial rocks.

C-4 Mineral Groups

"A mineral is a naturally occurring, inorganic, solid substance with a specific composition and arrangement of atoms" (Albin, 2004, p. 5). "A *crystal* is a form of mineral that has grown in an orderly and symmetrical manner" (Albin, 2004, p. 7). "Seven mineral groups can be found in nature: native elements, silicates, carbonates, sulfates, sulfides, halides, and oxides" (Albin, 2004, p. 13).

Table 31 - Appendix C: Diagnostic Properties

Mineral Groups	Clay Minerals (*Kaolinite* is a common example of this large group)
Chemical Composition	$XSi_4O_{10}(OH)_8$ (X is Al, Mg, Fe, Ca, Na, K)
Chemical Group	Sheet silicate
Diagnostic Properties	Generally microscopic crystals. Masses of clay minerals are softer than fingernail. Earthy luster. Clay-like smell when damp.
Other Properties	Seen as a chemical weathering product of feldspars and most other minerals. A constituent of most soils.
Source: Plummer, *et al.*, 2003, p. 543.	

Refer to Table 31 - Appendix C where the diagnostic properties of the common rock forming minerals are listed.

C-5 Clay Defined

"*Clay* – A sediment with particles < 4 μm in size" (Kearey, 2001, p. 49). Clay – "In geology the size of the constituent particles is usually taken to be less than 1/256 mm." (Isaacs, *et. al.*, 2003, p. 162). "Clay [GEOL] (1) a natural, earthy, fine-grained material which develops plasticity when mixed with a limited amount of water; composed primarily of silica, alumina, and water, often with iron, alkalies, and alkaline earths. (2) The fraction of an earthy material containing the smallest particles, that is, finer than 3 micrometers" (Licker, 2003, p. 71, 72). Clay is "a detrital mineral particle of any composition, having a diameter less than 0.004 millimeters" (MacDonald, *et al.*, 2003, p. 94). "Clay – A smooth, earthy sediment or soft rock composed chiefly of clay-sized or collodial particles and a significant content of *clay minerals*. Clays may be classified by colour, composition, origin or use" (MacDonald, *et al*, 2003, p. 94).

C-6 Clay Ironstone

"*Clay ironstone* – a sediment composed mainly of *siderite* occurring in thin *nodules* in some *argillites*" (Kearey, 2001, p. 50). Argillaceous "pertaining to rocks or sediments composed of

clay minerals or having a significant amount of clay in their composition" (MacDonald, *et al*, 2003, p. 38). "Clay ironstone [PETR] (1) a clayey rock containing large quantities of iron oxide, usually limonite. (2) A clayey-looking stone occurring among carboniferous and other rocks; contains 20-30% iron" (Licker, 2003, p. 72). Clay Ironstone is "a fine-grained sedimentary rock, grey or brown, consisting of clay and iron carbonate (*siderite*; it occurs in concretions or thin beds. Clay ironstone is usually associated with carbonaceous strata, in particular overlying coal seams in *coal measures* of the United States and Great Britain" (MacDonald, *et al*, 2003, p. 94).

C-7 Clay Minerals

"Clay minerals – Silicates of aluminum, iron and magnesium, belonging to the Phyllosilicate group. Their crystal structure consists of sheeted layers structures, with strong inter-and intra-sheet bonding but weak interlayer bonding. Clay minerals can be classified into six groups; the kaolinite-serpentinite group, the illite group, the smectite group, the vermiculite group, the chlorite group and the palygorskite-sepiolite group. The kaolinite-serpentinite group includes minerals close to $Al_2Si_2O_5(OH)4$, among which are anauxite, dickite, halloysite, kaolinite and nacrite. They form during pedogenesis. The smectic group minerals are close to $Al_2Si_4O_{10}OH_2 \cdot nH_2O$ with sodium, magnesium and iron variously substituted. The group includes Beidellite, Montmorillonite and Saponite. Smectites are swelling clays and expand in the presence of water. The vermiculite group result from the alteration of mica or chlorite, with a typical formula being $Mg_3(Si_3Al)O_{10}(OH)_2$. Vermiculites are swelling clays. The illite group are a group of mica-like clay minerals. The chlorite group consists of magnesium, aluminum, iron silicates and includes chlorite, dombassite and sudote. The palygorskite-sepiolite group are magnesium-rich, fibrous clays and include palygorskite (attapulgite) and sepiolite. Allophane and imogolite are non-crystalline hydrates silicates with properties similar to those of clays. In the formation of clays, the type of clay produced depends, during early stages of weathering, on the mineral composition of the parent rock, but in later stages it is climate-dependent" (MacDonald, *et al.*, 2003, p. 95).

"*Clay minerals – Phyllosilicate minerals* based on composite layers constructed from components with tetrahedrally and octahedrally co-ordinated cations. Mainly occurring as *microscopic,* platy particles in fine-grained *aggregates* which have varying degrees of *plasticity* when mixed with water. They are hydrous silicates, mainly of aluminum, magnesium, iron and potassium, which lose adsorbed and constitutional water at high temperature to yield refractory materials. The most important groups are *illites, kandites, smectites* and *vermiculites*" (Kearey, 2001, p. 50).

$Al_2Si_2O_5(OH)_4$	+	$2K^+$	+	$2HCO_3-$	+	$4SiO_2$
Kaolinite (residual clay)		potassium ion		bicarbonate ion		silica
				$\leftarrow$ in solution		$\rightarrow$|"

(Tarbuck & Lutgens, 1999, p.127).

"The clay minerals … concentrate in clay-size sedimentary deposits, which are generally higher in Al, Na, and K than sandstones" (Garrels & Mackenzie, 1971, p. 183). "Clay mineral [Mineral] One of a group of finely crystalline, hydrous silicates with a two-or three-layer crystal structure; the major components of clay materials; the most common minerals belong to the kaolinite, montmorillonite, attapulgite, and illite groups" (Licker, 2003, p. 72).

Clay Minerals are hydrated aluminophyllosilicates smaller than 0.004 mm (8Ø) in size (according to the Wentworth scale).

C-8 Claystone

"Claystone – Sedimentary rock indurated clay-sized silicate materials, having the texture and composition of shale but lacking its lamination and fissility" (MacDonald, *et al..*, 2003, p. 95).

C-9 Description of Mudstone, Shale and Clay

"Mud (1) a mixture of water with *clay* to *silt*-sized particles, ranging in consistency from semifluid to soft and plastic. (2) a fine-grained

red, blue or green marine sediment found mainly on the continental shelf; its colours are because of a predominance of various mineral substances. (3) the material of a *mudflow*" (MacDonald, *et al.*, 2003, p. 306, 307).

"Color – black, gray, white, brown, red, dark green or blue" (Bishop, Woolley & Hamilton, 2001, p. 200).

"Texture – grain size less than 0.0002 of an inch; individual grains are too small to be distinguished with the naked eye. Mudstone and shale feels smooth, and a pure clay is not gritty when smeared between the fingers. Clays are plastic and often sticky when wet" (Bishop, Woolley & Hamilton, 2001, p. 200, 201).

"Structure – when consolidated and relatively massive is known as mudstone (or claystone); if finely bedded so that it splits readily into thin layers it is called shale. When soft and uncompacted it is termed clay. Sun cracks, rain prints etc. Sometimes occur on bedding surfaces; and fossils and concretions are common" (Bishop, Woolley & Hamilton, 2001, p. 201).

C-10 Field Relations

"Clays tend only to occur in the younger geological formations, being consolidated into mudstones and shales with time. Being very fine-grained, clay is easily transported by water into the sea and lakes, where it accumulates with silt, sand, and calcareous organisms to form typical sequences of shales, siltstones, sandstones, and limestones. Some clays are *residual*, having formed *in situ* as soils; such are the bauxitic clays" (Bishop, Woolley & Hamilton, 2001, p. 201).

C-11 Geochemical Facies

"An area or region characterized by particular physic-chemical conditions that affect the production and accumulation of sediments …The respective E_H – pH diagrams of each type can indicate the depositional environment, chemical region of stability and conditions under which particular assemblages could have been formed" (MacDonald, *et al.*, 2003, p. 202).

Loreen Michele Sherman

C-12 Heavy and Light Minerals

"Heavy minerals – (1) dense accessory detrital minerals (e.g. tourmaline and zircon) from a sedimentary rock. These are separated from minerals with lower specific gravity, such as quartz, by means of high-density *heavy liquids*" … (2) samples of crushed igneous or metamorphic rock that sink in a heavy liquid. Light minerals are those that float" (MacDonald, *et al.*, 2003, p. 229). "Heavy minerals (HMs) are arbitrarily designated as minerals with a specific gravity greater than 2.85… Whereas such minerals generally constitute only roughly 1% of the sand-size fraction of sediments, they are important indicators of provenance (ultimate source lithology) and the geological history of the sediments (weathering, transportation, and deposition)" (Lewis, 1984, p.145). "Common heavy minerals include zircon, tourmaline, rutile, apatite, garnet" (Nichols, 2003, p.15). "Common minerals will float but the small proportion of dense minerals will sink" (Nichols, 2003, p.15).

"Mafic in general, it is synonomous with *dark minerals*" (MacDonald, *et al*, 2003, p. 283, 284).

C-13 Illite

Illite is "a group of *mica*-like *clay minerals* of three-sheeted structure, widely distributed in argillaceous shales" (MacDonald, *et al.,* 2003, p. 245).

C-14 Lutites

Lutite is "a general name for any fine-grained sedimentary rock composed of material that was once mud, such as *mudstone* or *shale*. More specifically, the proportion of silt contained should be between one-third and two-thirds of the total. The terms lutite, claystone, mudstone, siltstone and shale often overlap in usage" (MacDonald, *et al.*, 2003, p. 282).

C-15 Montmorillonite – Silicates

(Phyllosilicates, Clay Group)

System – monoclinic.

Appearance – never in crystals larger than 1µ in diameter. Usually in earthy, dusty and scaly amorphous-looking masses. White or gray.

Physical properties – very soft (1), very light, disintegrates easily, greasy feel. Opaque because of its extremely fine particles. Two interesting characteristics of montmorillonite are its capacity to expand by absorbing water or other liquids and its potential for ion exchange when, under certain conditions, the cations sodium, potassium and calcium present between layers in the structure are replaced by other cations present in a solution.

Environment – in a sedimentary environment in a tropical climate through alteration of feldspars in rocks poor in silica. In a hydrothermal environment of the expanse of volcanic glasses and tuffs.

Uses of montmorillonite "an important industrial mineral, used as an absorbent to purify and decolor liquids, as a filler in paper and rubber, as a base for cosmetics and medicines and as a mud for drilling and quarrying" (Mottana, Crespi & Lioborio, 1978, p. 234).

"Montmorillonite [Mineral] (1) a group name for all clay minerals with an expanding structure, except vermiculite. (2) The high-alumina end member of the montmorillonite group; it is grayish, pale red, or blue and has some replacement of aluminum ion by magnesium ion" (Licker, 2003, p. 209). "Montmorillonite – A dioctahedral clay mineral of the smectite group, $Na_{0.33} Al_{1.67} Mg_{0.33} Si_4O_{10} (OH)_2 \bullet nH_2O$. Hydrated sodium calcium aluminum silicate. It represents a high alumina end-member that has some slight replacement of Al^{+3} by Mg^{+2} and substantially no replacement of Si^{+4} by Al^{+3}" (Bates & Jackson, 1984, p.335). Montmorillonite is "a monoclinic clay mineral of the smectite group, $(Al, Mg)_8(Si_4O_{10})(OH)_8 \bullet 12H_2O$, the dominant clay mineral in *bentonite*. Two noteworthy characteristics are its capacity to expand by absorbing liquid and its potential for ion exchange when, under certain conditions, the cations sodium, potassium and calcium, present between the structural layers, are replaced by other cations present in solution" (MacDonald, *et al.*, 2003, p. 305).

"*Intermediate argillic alteration* – a type of low grade *wall rock alteration* in which *plagioclase* is altered to *kaolin* and *montmorillonite* group *minerals*" (Kearey, 2001, p. 136).

C-16 Monomineralic

"The amounts of other minerals tolerated under the definition vary with different authors" (Bates & Jackson, 1984, p. 335).

C-17 Pyrite

Pyrite is "a common yellow isometric sulphide mineral, FeS_2, dimorphous with *marcasite*. Pyrite forms under a wide range of pressure-temperature conditions, so is found in many geological environments. It occurs in masses, associated with chalcopyrite, among magmatic segregation deposits in *mafic* rocks. Pyritised concretions are formed by chemical deposition under water. It is the most abundant and widespread of the sulphide minerals and an important ore of sulphur. Its yellow-gold colour has often led to its being mistaken for gold" (MacDonald, *et al.*, 2003, p. 357, 358).

Alteration – pyrite oxidizes of either to iron sulfate or to the hydrated oxide, limonite.

Occurrence – "pyrite is one of the most widely distributed of sulfide minerals, occurring in a variety of environments… It is a common mineral in hydrothermal sulfide veins, in replacement deposits" (Bishop, Woolley & Hamilton, 2001, p. 32, 33).

Pyrite Sulfide FeS_2 (Iron Sulfide). Appearance – Striated, cubic, octahedral or pyritohedral crystals, sometimes occurring as "iron cross" twins. Compact, granular aggregates. Always fairly dark yellow, sometimes with an iridescent yellowish-brown film.

Physical properties – Dimorphic with marcasite. Hard (6 to 6.5), very heavy, very fragile, with poor cleavage.

Environment is "associated with sphalerite and galena in hydrothermal veins. On its own or associated with gold (auriferous pyrite) in hydrothermal, medium to low-temperature quartz veins … Pyritized concretions formed by chemical deposition under the water. As a diagenetic mineral replaces fossils or forms nodules of various sizes. Stable in metamorphic environments up to the granulite facies, where it

changes into pyrrhotite" (Mottana, Crespi & Lioborio, 1978, p. 31).

C-18 Residual Clay

"Residual clay [GEOL] very finely divided clay material formed in place by weathering of rock" (Licker, 2003, p. 299). Residual clay is "clay material formed in place by the weathering of rock; it is derived either from chemical decomposition of *feldspar* or by the removal of non-clay mineral constituents from clay-bearing rock" (MacDonald, *et al.*, 2003, p. 368).

"The more restricted chemistry of Paleozoic shales permits only two important clay minerals, illite and chlorite, with minor kaolinite and expanded clays. In younger rocks, kaolinite and mixed-layer clays add up to about 50 percent of total clay mineralogy" (Garrels & Mackenzie, 1971, p. 235).

C-19 Secondary Minerals

"Secondary (of rocks and minerals) formed by the alteration of pre-existing minerals (*primary minerals)*. Secondary minerals may be found at the site as *paramorphs* or *pseudo-morphs* or may be deposited from solution in rock interstices through which the solution is percolating, as, for example, *amygdales*" (MacDonald, *et al.*, 2003, p. 384).

"Secondary mineral [Mineral] a mineral produced in an enclosing rock after the rock was formed as a result of weathering or metamorphic or solution activity, and usually at the expense of a primary material that came into existence earlier" (Licker, 2003, p. 327).

"Secondary consolidation [GEOL] Consolidation of sedimentary material, at essentially constant pressure, resulting from internal processes such as recrystallization" (Licker, 2003, p. 327).

"Secondary enlargement [Mineral] overgrowth by chemical deposition on a mineral grain of additional material of identical composition in optical and crystallographic continuity with the original grain; crystal faces characteristic of the original mineral often result. Also known as secondary growth" (Licker, 2003, p. 327).

"Secondary enrichment [GEOL] the addition to a vein or ore body of material that originated later in time from the oxidation of decomposed ore masses that overlie the vein" (Licker, 2003, p. 327).

"Secondary interstices [GEOL] openings in a rock that formed after the enclosing rock was formed" (Licker, 2003, p. 327).

"Secondary enrichment or supergene enrichment – a near-surface process of mineral deposition by which a primary ore body or vein is later enriched to a higher grade. The enrichment process often consist of two phases: (a) oxidation of overlying ore masses above the *water table* (*zone of oxidation*); oxidation may then produce acidic solutions that leach out metals and carry them downwards towards the water table; (b) below the zone of oxidation, reaction of the solutions with the primary material by coating or replacement with an ore of more valuable content. The water table thus marks the lower limit of oxidation, near which there may be a zone of secondary enrichment. This process has been significant in the formation of many copper deposits" (MacDonald, *et al.*, 2003, p. 384, 385).

"Secondary porosity [GEOL] the interstices that appear in a rock formation after it has formed, because of dissolution or stress distortion taking place naturally or artificially as a result of the effect of acid treatment or the injection of coarse sand" (Licker, 2003, p. 327).

C-20 Sensitive Clay

"Sensitivity [GEOL] the effect of remolding on the consistency of a clay or cohesive soil, regardless of the physical nature of the causes of the change" (Licker, 2003, p. 332).

"Sensitive clay [GEOL] a clay whose shear strength is reduced to a very small fraction of its former value on remolding at constant moisture content" (Licker, 2003, p. 332). "Sensitive clay or quick clay – a *clay* that loses most of its strength after being remoulded. When the clay is disturbed, repacking of the grains results in an increase in pore-water pressure and the clay becomes mobile, i.e. it liquefies. Many sensitive clays are marine clays that have lost their original flocculated structure because the ionic species responsible for the strength of the bond

have been removed by fresh water passing through" (MacDonald, *et al.*, 2003, p. 392).

C-21 Sulfides

"An important mineral group is the *sulfides*, which contain the sulfur (S^{2-}) anion combined with different metal cations. Sulfide minerals typically are dense and have a metallic appearance. The most common are the iron sulfides pyrite (FeS_2, 'fool's gold') and pyrrhotite (FeS). Many sulfides are ore minerals, which means that they are sought and processed for their valuable metal content" (Murck, 2001, p. 41).

"Mine Tailings: Silicates & Sulphides, Puritite => when exposed to O_2 & H_2O quick reaction which generates sulphuric acid => dissolves minerals (ex U) and creates acid leachate" (Jack, August 2004, p. 30).

C-22 Thixotopic

"Thixotropic clay [GEOL] a clay that weakens when disturbed and increases in strength upon standing" (Licker, 2003, p. 373).

"Because of the very high concentration of solid particles in mass flows, the behavior of the fluid-solid mixture differs substantially from that of water or of any Newtonian fluid. Concentrations of up to 30% by volume of inert particles increase the apparent viscosity of the dispersion but the behavior remains essentially Newtonian. Higher concentrations of inert particles, or much lower concentrations of particles, such as clays, that form structured dispersions, cause departures from Newtonian behavior. Several types of different departures are possible: (a) The dispersion may be *non-Newtonian fluid* that has no strength but that has a viscosity that varies with the rate of shear. Commonly the viscosity decreases at higher rates of shear as the structure of the dispersion, caused by interparticle forces, is broken down. (b) The dispersion may have strength so that no deformation takes place until a critical shear stress for yield has been exceeded. If, after the strength is exceeded, the dispersion behaves as a substance with constant viscosity, it is called an ideal plastic, or *Bingham plastic*. A more common case is that the dispersion behaves when it flows as a substance with variable viscosity, or *pseudoplastic*. (c) A class of substances, called *thixotropic*, have strength until they are sheared, when the property of strength is lost and the substance behaves like a fluid (commonly a non-Newtonian fluid) until it is allowed to rest. After a period of rest the property of strength is reestablished. Dispersions of montmorillonitic clays (bentonite) commonly display this property of thixotropy. It is this property that makes them valuable as drilling muds. When circulation of the mud stops, the fluid 'jells'and its strength prevents the settling of the cuttings suspended in the mud" (Blatt, Middleton & Murray, 1972, p. 160, 161).

"The high viscosity and pseudo-plastic or thixotropic behavior of a concentrated muddy dispersion strongly influences the nature of mass flows and their deposits. Subaerial *debris flows* have densities of 2.0 to 2.4, indicating a water content of as little as 40% by volume. Debris flows commonly advance in a series of surges, with maximum velocities of the order of 1 to 3 m/sec. The thickness of the flow is of the order of 1 m and the slope may be as low as 1 degree, though it is commonly of the order of 5 degrees. The viscosity of the flow is not known from direct measurement but, by assuming viscous flow and knowing the thickness, slope and surface velocity, Sharp and Nobles (1953) calculated a probable value of the order of 1000 poises (as compared with 0.01 poise for pure water). Thus it appears that the Reynolds number is quite low (of the order of 10 to 100) so that it is probable that the flow is indeed laminar in nature" (Blatt, Middleton & Murray, 1972, p. 161).

"Minerals can usually be identified by such physical properties as color, hardness, density, crystal form, and cleavage. Cleavage refers to the tendency of some minerals to break smoothly along certain planes of weakness" (Levin, 2003, p. 35).

APPENDIX D: Colors

"Many minerals can display a wide range of colors, and a large number of minerals actually share the same color" (Albin, 2004, p. 11).

D-1 Rock Colour Chart – Munsell System

"Color is best described by reference to the standard *Rock Colour Chart* (Goddard *et al.*, 1951). The chart is based on the Munsell system, which is a widely accepted standard for color-identification. Colors are expressed by a shorthand notation: *hue* (given by a number) of red, yellow, green, blue or purple (given by letter); value of lightness or darkness of the color; and *chroma* of the strength or saturation of the color. For example, 5R ¾ means hue of 5 red, value of 3, chroma of 4" (Lewis, 1984, p. 1). Refer to Figure 62 - Appendix D rock color organization based on chroma or saturation.

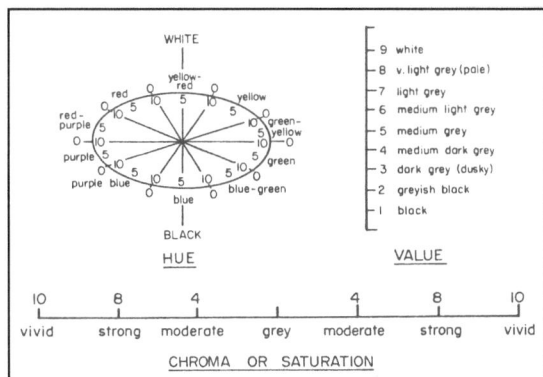

Figure 62 - Appendix D: Rock Color

"Whether the material was wet or dry should be stated; if practicable, both wet and dry colors should be given. If without a chart, keep to primary colors – shades like 'maroon', 'chocolate brown,' and 'lilac,'are visualized differently by different people. *Induration* is somewhat subjective, but the following terms provide a useful standard: *loose, friable* (grains break free with slight finger pressure); *indurated*; *well indurated* (grains not broken when rock is broken); *very well indurated* (grains break when rock is broken)" (Lewis, 1984, p. 2).

D-2 Clay

"Heavy clay – gray, dark

Sticky clay – white, yellow, dark

Stony clay – red" (Brady &Weil, 2002, p. 78).

D-3 Detritus

Detritus is "any loose rock or mineral material resulting from mechanical processes, such as disintegration or abrasion, and removed from its place of origin" (MacDonald, *et al.*, 2003, p. 136). "*Detritus* – General terms for unconsolidated sediments derived from preexisting rocks by natural agencies. Derived from the Latin word meaning worn" (DePree & Axelrod, 2003, p. 222).

D-4 Felsic Rocks

"*Felsic rocks* tend to be light in color and are composed chiefly of minerals enriched in the elements silicon and aluminum. They are associated with thick and slow-moving (*viscous*) magmas or lava flows. Felsic rocks occur primarily on continental landmasses" (Albin, 2004, p. 15).

D-5 Iron

"Iron is transported as hydroxides in colloidal suspension or it is attached to clay minerals and organic particles. Deposition occurs when the chemistry of the environments favours the precipitation of iron minerals … Further oxidization… give these sands the red colour seen" (Nichols, 2003, p. 33).

Iron – "A silvery malleable and ductile metallic transition element" (Isaacs, *et. al.*, 2003, p. 417).

Iron (ll) Chloride – a green-yellow deliquescent compound" (Isaacs, *et. al.*, 2003, p. 417).

Iron (ll) Oxide – "A black solid" (Isaacs, *et. al.*, 2003, p. 418).

Iron (ll) Sulphate – "An off-white solid" (Isaacs, *et. al.*, 2003, p. 418).

Iron (lll) Chloride – "A black-brown solid" (Isaacs, *et. al.*, 2003, p. 418). $FeCl_3$ "Their solutions are yellow or orange" (Stockley, Oxlade &Wertheim, 2001, p. 174).

Iron (lll) Sulphate – "A yellow hygroscopic compound" (Isaacs, *et. al.*, 2003, p. 418).

Ironstone – "A typical rusty color" (Isaacs, *et. al.*, 2003, p. 418).

D-6 Mafic

"*Mafic* igneous rocks are gray to black in color and are made up of mostly silicate minerals with magnesium and iron" (Albin, 2004, p. 16).

D-7 Ochre

Ochre is "a powdery red, brown or yellow iron oxide used as a pigment, e.g. brown or yellow ochre (*limonite*) and red ochre (*hematite*); also, any of various clays deeply coloured by iron oxides" (MacDonald, *et al.*, 2003, p. 317).

D-8 Pigmentation

"Most pigmentation in mudrocks results from the presence of either iron-bearing compounds (green, red, brown, yellow) or free carbon (gray, black)" (Blatt, Middleton & Murray, 1972, p. 398). "Hues of brown, red, and green often occur in sedimentary rocks as a result of their iron oxide content. Few, if any, sedimentary rocks are free of iron, and less than 0.1 percent of this metal can color a sediment a deep red" (Levin, 2003, p. 69).

D-9 Red Beds

"Red beds – Predominately red sedimentary strata composed primarily of sandstone, siltstone and shale; their color is the result of the presence of a coating of hematite (ferric oxide) on the sedimentary grains. They are generally considered to indicate sedimentation in an arid continental environment, with the hematite originating from the *in situ* oxidation of detrital iron minerals. Examples of red beds include the old red sandstone facies of the European Devonian and the new red sandstone of the Permian and Triassic" (MacDonald, *et al*, 2003, p. 365).

"*Red bed* – a sedimentary deposit whose *clasts* are coated with *haematite*, causing a red colour.

The presence of such beds in the *Proterozoic* is taken to indicate a major change from an oxygen-rich hydrosphere to an oxygen-bearing atmosphere" (Kearey, 2001, p. 222).

"*Red bed copper* – a cropper variety of the *sandstone-uranium-vanadium base metal deposit*, whose *country rocks* are often red" (Kearey, 2001, p. 222).

D-10 Red Clay

"Red clay [GEOL] a fine-grained, reddish-brown pelagic deposit consisting of relatively large proportions of windblown particles, meteoric and volcanic dust, pumice, shark teeth, manganese nodules, and debris transported by ice. Also known as brown clay" (Licker, 2003, p. 293). Red clay is "a fine-grained *pelagic deposit*, reddish-brown or chocolate-coloured, formed by the accumulation of material at depths generally greater than 3500 metres and at a far distance from the continents. Red clay contains rather large amounts of meteoric and volcanic dust, pumice, shark teeth, manganese concretions and ice-transported debris" (MacDonald, *et al.*, 2003, p. 365).

"*Red clay (brown clay)* – a red/brown *pelagite* primarily composed of *clay minerals* with minor *quartz*, *ash*, cosmic dust and *fish* teeth, the product of very slow deposition in the central parts of ocean *basins* at > 4 km depth and below the *calcite compensation depth*" (Kearey, 2001, p. 222).

D-11 Residual Ochre

"Residual ochre [GEOL] an earthy, red, yellow, or brownish iron oxide powder of iron oxide (usually the mineral limonite) produced during chemical weathering" (Licker, 2003, p. 299).

D-12 Sienna

"Sienna – Any brownish-yellow, earthy limonitic pigment for paint and oil stains. When burnt, it becomes dark orange-red to reddish brown" (MacDonald, *et al.*, 2003, p. 397).

D-13 Umber

Umber is "a brown earth consisting of manganese oxides, hydrated ferric oxide, alumina, silica and lime. It is widely used as a pigment, either in the greenish-brown natural state, *raw umber*, or as *burnt umber*, the dark brown-calcined state. It is darker than either *ochre* or *sienna*" (MacDonald, *et al.*, 2003, p. 440).

D-14 Wulfenite

"Wulfenite [Mineral] $PbMoO_4$ a yellow, orange, orange-yellow, or orange-red tetragonal mineral occurring in tabular crystals or granular masses; an ore of molybdenum. Also known as yellow lead ore" (Licker, 2003, p. 397).

D-15 Yellow Mud

"Yellow mud [GEOL] Mud containing sediment having a characteristic yellow color, resulting from certain iron compounds" (Licker, 2003, p. 401).

APPENDIX E: Ferric Oxide

"Oxidation is important in decomposing such ferromagnesian minerals as olivine, pyroxene, and hornblende. Oxygen readily combines with the iron in these minerals to form the reddish-brown iron oxide called hematite (Fe_2O_3) or in other cases a yellowish colored rust called limonite [FeO(OH)]." "However, oxidation can occur only after iron is freed from the silicate structure by another process called hydrolysis" (Tarbuck & Lutgens, 1999, p.127). "*Hydrolysis*, which basically is the reaction of any substance with water" (Tarbuck & Lutgens, 1999, p. 127).

E-1 Ferric Oxide Defined

"One of the most important reduction reactions for paleosols, because it controls their color, is the conversion of red, oxidized, ferric (Fe^{3+}) minerals, such as hematite, to reduced, gray, ferrous (Fe^{2+}) minerals such as siderite" (Retallack, 1990, p. 140, 141). "When metallic elements are combined with oxygen ions, *oxides* are produced" (Albin, 2004, p. 10).
Oxides include: Periclase MgO, Cuprite Cu_2O, Rutile TiO_2, Anatase TiO_2, Cassiterite SNO_2, Corundum Al_2O_3, Hematite Fe_2O_3, Ilmenite $FeTiO_3$, Perovskite $CaTiO_3$ (Nesse, 2004, p. 299-306).

E-2 Ferric Oxide

Ferric oxide - a red oxide of iron.

E-3 Ferriferous

"Ferriferous – (1) (of a mineral) containing iron. (2) (of a sedimentary rock) containing more iron than is usually the case" (MacDonald, *et al.*, 2003, p. 181).

E-4 Haemosiderin

Haemosiderin, hemosiderin - a granular brown substance composed of ferric oxide; left from the breakdown of hemoglobin; can be a sign of disturbed iron metabolism.

E-5 Iron

"Iron: in oxides, sulphides, (pyrite, marcasite, and pyrrhotite), chlorite, ironbearing clays, carbonate, tourmaline" (Moorhouse, 1959, p. 472).

"Iron – A heavy, chemically active metallic element, Fe. Native iron is rare in terrestrial rocks but occurs alloyed with nickel in crystalline form in iron meteorites. I ron occurs in a wide range of ores in combination with other elements. T he chief ores of iron consist mainly of the oxide of this element: Hematite (Fe_2O_3); Goethite (α-FeO(OH)); and Magnetite (Fe_3O_4)." (MacDonald, *et al*, 2003, p. 257).

Iron is a transition metal (Stockley, Oxlade &Wertheim, 2001, p. 172). Iron "is a fairly soft, white, magnetic metal which only occurs naturally in compounds. One of its main ores is haematite (Fe_2O_3), or iron (lll) oxide, from which it is extracted in a blast furnace. Iron reacts to form both ionic and covalent compounds and reacts with moist air to from rust" (Stockley, Oxlade &Wertheim, 2001, p. 174). "Iron made in the blast furnace is called pig iron. It contains about 5% carbon and 4 % other impurities, such as sulphur" (Stockley, Oxlade &Wertheim, 2001, p. 174).

E-6 Iron Formation

Iron Formation is "an iron-rich chemical *sedimentary rock*, thin-bedded or laminated and commonly containing layers of *chert*. Iron may be present as an oxide, silicate, carbonate or sulfide. Most iron formation is of Precambrian age" (MacDonald, *et al.*, 2003, p. 257).

E-7 Iron in Sedimentation

"Iron minerals are one of the major indicators of chemical conditions in sediments... and one of the main coloring agents" (Lewis, 1984, p. 187). "Many unresolved problems remain in accounting for their origin" (Lewis, 1984, p. 187). "Iron silicate minerals are formed under conditions not fully understood" (Lewis, 1984, p. 187). Iron bearing minerals include: Hematite, Goethite, Lepidocrocite, Ilmenite, Magnetite, Maghemite and Limonite (Lewis, 1984, p. 188). Siderite is considered a carbonate (Lewis, 1984, p. 188). The Silicates include: Glauconite, Chamosite, Greenalite, Thuringite, Minnesotaite, and Stilpnomelane (Lewis, 1984, p. 189). The Sulfides include: Pyrite, Marcasite, Pyrrhotite, Hydrotroilite and Melnikovite (Lewis, 1984, p. 190). Phosphates are iron bearing minerals.

E-8 Iron Ore Minerals

"The major iron ore minerals are hematite and magnetite" (Plummer, *et al.*, 2003, p. 535). "Most copper ores are sulfides. Chalcopyrite is the most important copper ore mineral" (Plummer, *et al.*, 2003, p. 535). "Hematite [Mineral] Fe_2O_3 An iron mineral crystallizing in the rhombohedral system; the most important ore of iron, it is dimorphous with maghemite, occurs in black metallic-looking crystals, in reniform masses or fibrous aggregates, or in reddish earthy forms. Also known as bloodstone; red hematite, red iron ore; red ocher; rhombohedral iron ore" (Licker, 2003, p. 150).

E-9 Ironstone

Ironstone is "a typically non-cherty, iron-rich Phanerozoic *sedimentary rock*, often oolitic. Ferriferous minerals found in ironstones include goethite, hematite, pyrite, siderite, limonite and chamosite" (MacDonald, *et al.*, 2003, p. 257). "There is a reciprocal relation between FeO and Fe_2O_3; young rocks are high in oxidized iron and old rocks are low, but total iron oxide remains almost constant" (Garrels & Mackenzie, 1971, p. 234).

E-10 Lithification

Lithification – "conversion of an unconsolidated (loose) sediment into solid sedimentary rock by compaction of mineral grains that make up the sediment, cementation by crystallization of new minerals from percolating water solutions, and new growth of the original mineral grains" (Research Machines Plc., 2003, p. 342).

E-11 Oxide

Oxide - any compound of oxygen with another element or a radical.

E-12 Oxidates

"Oxidates – Sediments composed of the oxides and hydroxides of iron and manganese that have crystallized from an aqueous solutio" (MacDonald, *et al*, 2003, p. 326).

E-13 Oxidised Zone

Oxidised zone is "the upper portion of an *ore* body that has been modified by a solution of surface waters bearing oxygen, soil acids and carbon dioxide. It extends from the surface to the *water table*. Primary ore minerals in this zone react with this solution to produce various secondary minerals containing oxygen. Chemical alteration of sulphide minerals within the zone of oxidation results in the formation of acids that then enable *meteoric water* to act as a solvent. For example, the weathering of *pyrite*, FeS_2, produces ferric sulfate, $Fe_2(SO_4)_3$, and sulfuric acid, H_2SO_4. The ferric sulfate reacts with sulfides of copper, lead and zinc to form soluble sulphates of these metals. The sulfuric acid dissolves carbonates such as the *gangue* minerals, *calcite* and *dolomite*. As a result of these processes, buried deposits of copper and copper sulphide will change to malachite, $Cu_2CO_3(OH)_2$, or azurite, $Cu_3(CO_3)_2(OH)_2$, under increasingly oxidizing conditions, i.e. towards the surface of the Earth" (MacDonald, *et al.*, 2003, p. 326). "If high-sulfide soils are drained, the sulfides and/or elemental sulfur quickly oxidize and form sulfuric acid" (Brady &Weil, 2002, p. 585).

E-14 Oxidization

"*Oxidation* – iron (and some other elements) can be chemically altered by oxidation, a process in which ferrous iron (Fe^{2+}) is changed to ferric iron (Fe^{3+}). Ordinary rust is a form of chemical weathering by oxidation of ferrous iron. Iron is present in many rock-forming minerals. When such minerals are chemically weathered the iron is released and oxidized. Oxidized iron can eventually form hematite (Fe_2O_3), a brick-red-colored mineral that gives most tropical soils of the world their distinctive red color" (Murck, 2001, p. 124).

E-15 Reacting Systems

"Sodium chloride, for instance, is one atom of sodium joined to one of chlorine. Compounds have different properties to the elements that make them up. Sodium, for instance, spits when put in water; chlorine is a thick green gas. Yet sodium chloride is ordinary table salt!" (Farndon, 2002, p. 10).

Soil formation "Minerals are continuously exposed to a new supply of CO_2-rich oxygenated water" (Garrels & Mackenzie, 1971, p. 371).

"Oxides crystallize relatively late in the crystallization history of the rock" (Moorhouse, 1959, p. 239).

E-16 Reacting Systems: Mineral Reactivity

"A final factor to consider is the rate at which buffering reactions will occur. Carbonate minerals are generally considered to be highly reactive, particularly when present as minute grains. Using the reactivity of carbonate minerals as a reference point … What is meant by relative reactivity is, under the given set of conditions, compared to calcite, how readily the mineral will react with the fluid. For minerals with low relative reactivity, there is little interaction between the fluid and the mineral. Even if these minerals were effective buffers, we would not expect them to have a significant effect on pH" (Eby, 2004, p. 87, 88).

E-17 Rust

Rust - a red or brown oxide coating on iron or steel caused by the action of oxygen and moisture.

E-18 Rusting

Rusting, rust - the formation of reddish-brown ferric oxides on iron by low-temperature oxidation in the presence of water.

Table 32 - Appendix E: Relative Reactivity of Common Minerals at pH = 5

Mineral Group	Typical Minerals	Relative Reactivity at pH = 5
Dissolving	Calcite, aragonite, dolomite, magnesite, brucite	1.00
Fast weathering	Anorthite, nepheline, forsterite, olivine, garnet, jadeite, leucite, spodumene, diopside, wollastonite	0.40
Slow weathering	Plagioclase feldspars (albite, oligoclase, labradorite), clays (vermiculite, montmorillonite)	0.01
Very slow weathering	K-feldspars (orthoclase, microline), muscovite	0.01
Inert	Quartz, rutile, zircon	0.004
Source: Eby, 2004, p. 88		

"*Sulfide oxidation* occurs under aerobic conditions. This oxidation is usually microbially mediated. The sulfide oxidation reactions are important in a number of environmental settings" (Eby, 2004, p. 113). Table 32 - Appendix E is a chart showing some typical minerals and some erosional factors including: dissolving, fast weathering, slow weathering, very slow weathering and inert in relation to relative reactivity at pH = 5.

Table 33 - Appendix E: Elemental Compositions for Soil Humic and Fulvic Acids

Elemental Compositions, Atomic Ratios, and Atomic Weights for Soil Humic and Fulvic Acids		
Element	Humic Acids	Fulvic Acids
Carbon (wt%)	53.6–58.7	40.7–50.6
Hydrogen (wt%)	3.2–6.2	3.8–7.0
Nitrogen (wt%)	0.8–5.5	0.9–3.3
Oxygen (wt%)	32.8–38.3	39.7–49.8
Sulfur (wt%)	0.1–1.5	0.1–3.6
Mol wt (daltons^2)	2000–5000	500–2000
H/C (atomic)	~0.8	~1.3
O/C (atomic)	~0.5	~0.8
Source: Eby, 2004, p. 141.		

Table 33 - Appendix E is an overview of the elemental compositions, atomic ratios, and atomic weights for soil humic and fulvic acids.

Loreen Michele Sherman

372

APPENDIX F: Experimental Future Reproducibility Projects

"The stage in science when measurements are sorted, tested, and examined visually for patterns and predictability" (Clarke, 2003, p. 305) is the part where reproducibility can validate the theorem or hypothesis. The following Appendix is a suggestion of future reproducibility projects that are beyond the scope of this report to do but with appropriate facilities and revenue streams these suggestions are doable.

F-1 Computer Models

Models can provide useful information and test theories by computing dirrerent attributes, factors and variables but are programmed; therefore models are limited within the constraints of inputs. "A *physical model* is a tangible constructed representation of a portion of the natural world" (Dingman, 2002, p. 25). Remote-sensing data could be further used to provide information for these models.

Computer models are available online at epa.gov/epahome/models.htm.

Geochemical Earth Reference Model earthref.org/GERM/main.htm.

The Geochemical society gs.wustl.edu.

To learn more about cracking dams refer to the following website http://simscience.org (Article [1], 2009, *p. nd*).

F-2 Slope-Area Measurements

In a similar environment slope and area measurements can be identified and examined as to the relationships they would have on flow and velocity rates. This would be useful in reconstructing how much "waste" sediment would be uplifted and mixed to verify the observations of this study. Other considerations would be determining the settlement lag, the settlement depth and the flood-plain characteristics. "It is often possible to estimate the discharge of a recent flood peak from observations of high-water marks in a reach. The procedure involves surveying the slope of the water surface as revealed by the high-water marks and the cross-sectional area beneath the

marks, and hence is known as the *slope-area method*" (Dingman, 2002, p. 619).

F-3 Choropleth Map

A future study. The suggested Choropleth map would consist of classifying the mineralogy associated at Kidd Copper. "Choropleth map: A map that shows numerical data (but not simply 'counts') or a group of regions by (1) classifying the data into classes, and (2) shading each class on the map" (Clarke, 2003, p. 306).

F-4 Delaunay Triangulation

The detection of patterns helps to formulate structural geologic patterns and the Delaunay triangulation could be a valuable tool. "An optimal partitioning of the space around a set of irregular points into nonoverlapping triangle and their edges" (Clarke, 2003, p. 309).

F-5 Dam Structures

In and Off-site Locations. The definition of dam is "a structure built across a river to stop the flow of water. A dam may be constructed … It may raise the water level to increase the depth for navigation purposes; divert water … flood control… Most dams are constructed of either concrete or of earth and rock. *Gravity dams* depend on the weight of their bulk for stability and usually have a flat vertical face upstream" (Isaacs, *et. al.*, 2003, p. 213). It would be expected that different dam structures would support the findings in this report, if the same sediment of waste-tailings was used for the input as to sediment composition and a turbulent flow was achieved.

F-6 Enbankment Dams

The dam on the Kidd Copper property waste site likely was an embankment dam with earth or rock fill since no evidence of masonry or concrete construction materials remained. "Enbankment dams are usually built of materials dug out of the ground, usually at or near the dam site" (Student's Encyclopedia, 2008, p. Enbankment Dams). Earthen embankments "near the site there must be clay to fill the trench

Copyright 2010 Star-Ting, Inc. All rights reserved. Worldwide.   1.877.896.7292   http://www.star-tingpublishing.com

and embanking material capable of standing safely, without slipping to hold up a clay core" (Dams, 2008, Enbankment Dams). However, these dams are not impermeable.

When the embankment dam broke the water would have swept the mine tailings up, there would have been an upwelling from the increased velocity of the water because of the inadequate design. An engineered dam would have had an adequate slope and been impermeable. A possible explanation is the embankment dam did not hold when it reached its water capacity and broke.

Understanding water capacity and the effect of deposition of sediment is useful for reclamation but is beyond the scope of this report.

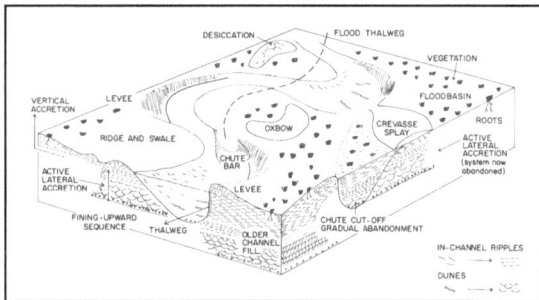

Figure 63 - Appendix F: Block Diagram

Figure 63 - Appendix F is a block diagram (Lewis, 1984, p. 8) used to represent data like: Depth Zone: penetration depths; Temperature; Turbulence; Textures; and /or Structures like: lenticular bdg.; variety-planar, cross and graded bdg, variable-sand, mud., or chaotic features.

F-7 Gravity Dams

Parts of a dam include: downstream face, heel, foundation and abutment. "Gravity dams use their weight to hold back the water in the reservoir. Gravity dams can be made of earth or rock fill or concrete. ...Generally, the base of a concrete gravity dam is equal to approximately 0.7 times the height of the dam:

$$\text{base} = 0.7 * \text{height}$$

The shape of the gravity dam resembles a triangle. This is because of the triangular distribution of the water pressure. The deeper the water, the more horizontal pressure it exerts on the dam. So at the surface of the reservoir, the water is exerting no pressure and at the bottom of the reservoir, the water is exerting maximum pressure" (Cracking Dams Advanced Level, 2008). Future studies of this type of dam would be useful to determine the pressure released and the effect upon the velocity of the dam break.

F-8 Wireline Logs

Running a wireline log could be a future test-study because "the recognition of characteristic log 'shapes' and distinctive vertical-profile character may be a useful tool for correlation purposes and an aid for interpreting fluvial style" (Miall, 1996, p. 273). Two interesting facts would be replicated: (1) Layers of different sediment chemistry, characters and descriptions would be found; and (2) The deposition event was known; therefore negating the interpretation for a fluvial style to be established. However, the information that is known could be used as a base-line to compare/contrast other wireline logs.

APPENDIX G: Satellite Site Data

G-1 Copper Cliff Satellite Site Data

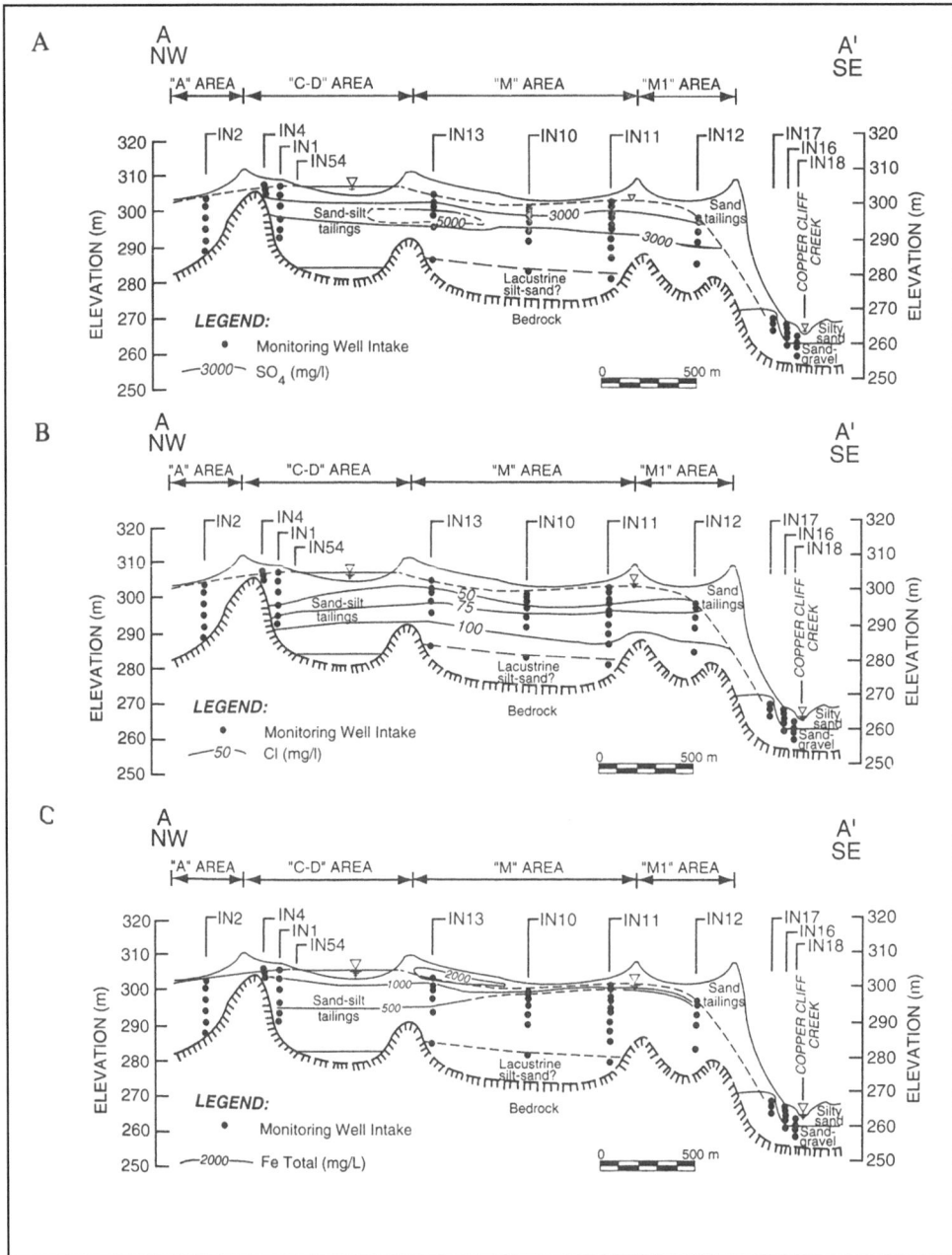

Figure 64 - Appendix G: Copper Cliff Satellite Site Data (Coggans, *et al.*, 1991, p. 453)

"The water table was raised along the cross section when water collection ponds were established on the surface of the older impoundments after they were filled with tailings. The moisture content of the unsaturated tailings increased as the water table rose, and reached the tailings surface at the pond locations" (Coggans, *et. al.*, 1991, p. 458). "The moisture contents in the vadose zone at IN12, is strongly influenced by a pond located 300 m to the north of cross section A-A', which was created after decommissioning of the impoundment" Refer to Figure 64 - Appendix G (Coggans, *et. al.*, 1991, p. 458).

G-2 Rainfalls for Various Durations

Figure 65 - Appendix G is a Depth–Duration – Frequency (DDF) Analysis chart to provide rainfalls (Dingman, 2002, p. 151).

Rainfall on the ground, various world locations:
At 20 mins – 206 mm = 8.10 in
At 42 mins – 305 mm = 12.00 in
At 2 hr 10 mins – 483 mm = 19.00 in

"The Water Resources Division (WRD) of the US Geological Survey (USGS) maintains over 5000 stream gages throughout the United States and publishes data on peak flows (floods), daily average flows, and water-quality parameters measured at these gages" (Dingman, 2002, p. 22).

Locations, Dates, and Amounts of Record Rainfalls for Various Durations. Data are Plotted in Figure 4-46.

Duration	Depth		Location	Date
	in.	mm		
1 min	1.50	38	Barot, Guadeloupe	Nov. 26, 1970
8 min	4.96	126	Füssen, Bavaria	May 25, 1920
15 min	7.80	198	Plumb Point, Jamaica	May 12, 1916
20 min	8.10	206	Curtea-de-Arges, Roumania	July 7, 1889
42 min	12.00	305	Holt, MO	June 22, 1947
2 hr 10 min	19.00	483	Rockport, WV	July 18, 1889
2 hr 45 min	22.00	559	D'Hanis, TX (17 mi NNW)	May 31, 1935
4 hr 30 min	30.8	782	Smethport, PA	July 18, 1942
9 hr	42.79	1,087	Belouve, Réunion	Feb. 28, 1964
12 hr	52.76	1,340	Belouve, Réunion	Feb. 28–29, 1964
18 hr 30 min	66.49	1,689	Belouve, Réunion	Feb. 28–29, 1964
24 hr	73.62	1,870	Cilaos, Réunion	Mar. 15–16, 1952
2 days	98.42	2,500	Cilaos, Réunion	Mar. 15–17, 1952
3 days	127.56	3,240	Cilaos, Réunion	Mar. 15–18, 1952
4 days	146.50	3,721	Cherrapunji, India	Sept. 12–15, 1974
5 days	151.73	3,854	Cilaos, Réunion	Mar. 13–18, 1952
6 days	159.65	4,055	Cilaos, Réunion	Mar. 13–19, 1952
7 days	161.81	4,110	Cilaos, Réunion	Mar. 12–19, 1952
8 days	162.59	4,130	Cilaos, Réunion	Mar. 11–19, 1952
15 days	188.88	4,798	Cherrapunji, India	June 24–July 8, 1931
31 days	366.14	9,300	Cherrapunji, India	July 1861
2 mo	502.63	12,767	Cherrapunji, India	June–July 1861
3 mo	644.44	16,369	Cherrapunji, India	May–July 1861
4 mo	737.70	18,738	Cherrapunji, India	Apr.–July 1861
5 mo	803.62	20,412	Cherrapunji, India	Apr.–Aug. 1861
6 mo	884.03	22,454	Cherrapunji, India	Apr.–Sept. 1861
11 mo	905.12	22,990	Cherrapunji, India	Jan.–Nov. 1861
1 year	1,041.78	26,461	Cherrapunji, India	Aug. 1860–July 1861
2 years	1,605.05	40,768	Cherrapunji, India	1860–1861

Data are plotted in Figure 4-46. From *Hydrology for Engineers* (3rd ed.), by R. K. Linsley, Jr., M. A. Kohler, and J. L. H. Paulhus. Copyright © 1982 by McGraw-Hill Book Company. Reprinted with permission of McGraw-Hill Book Company.

Figure 65 - Appendix G: Rainfalls for Various Durations

G-3 Crude Depositional Rate

"Deposit [GEOL] consolidated or unconsolidated material that has accumulated by a natural process or agent" (Licker, 2003, p. 94).

"Deposition [GEOL] the laying, placing, or throwing down of any material; specifically, the constructive process of accumulation into beds, veins, or irregular masses of any kind of loose, solid rock material by any kind of natural agent" (Licker, 2003, p. 94).

"Depositional fabric [PETR] arrangement of detrital particles settled from suspension or of crystals from a differentiating magma determined by the plane of the surface on which they come to rest" (Licker, 2003, p. 94).

G-4 Channel Shape on Velocity

"When a dam is built … the reservoir that forms behind it raises the base level of the stream. Upstream from the reservoir the stream gradient is reduced, lowering its velocity and, hence, its sediment-transporting ability" (Tarbuck & Lutgens, 1999, p. 240). The diagram below illustrates this point.

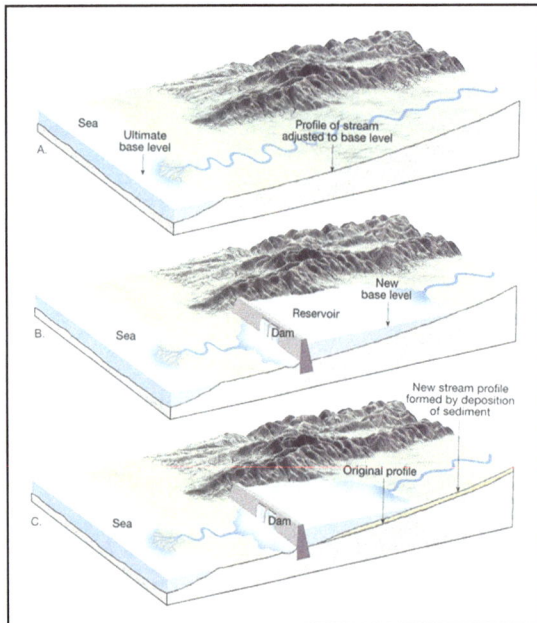

Figure 66 - Appendix G: Gradient Upstream

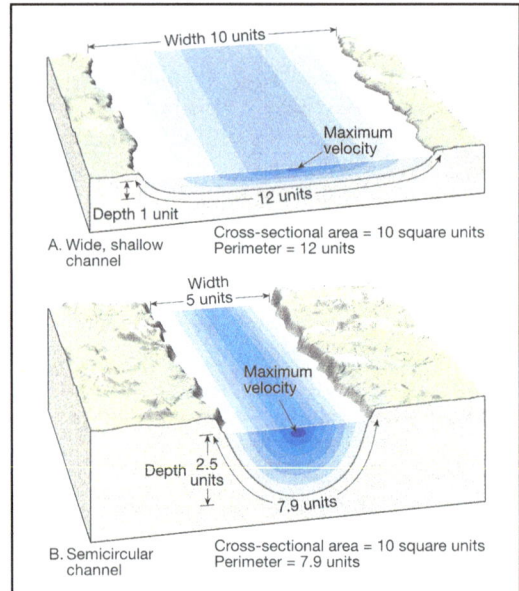

Figure 67 - Appendix G: Reservoir forms on a dam (Tarbuck & Lutgens, 1999, p. 235)

Influence of channel shape on velocity, Refer to Figure 66 - Appendix G and Figure 67 - Appendix G. "A: The stream in this wide, shallow channel moves more slowly than does water in the semicircular channel because of greater frictional drag. B: The cross-sectional area of this semicircular channel is the same as the one in part A., but it has less water in contact with its channel and therefore less frictional drag. Thus, water will flow more rapidly in channel B, all other factors being equal" (Tarbuck & Lutgens, 1999, p. 235).

G-5 Drainage

Drainage is "the removal of excess *meteoric water* by rivers and streams. A river network is related to the geological structure of a region. In the case of young *drainage systems*, it is determined by the existing rocks and superficial deposits. In mature systems, the courses may have been determined by stata that were subsequently removed" (MacDonald, *et al.*, 2003, p. 148).
Drainage basin is "a region drained by a particular stream or river system, i.e. an area that contributes water to a particular river or channel

system. The amount of water reaching the reservoir depends on the size of the basin, total precipitation, evaporation and losses caused by absorption. The term '*catchment area*' is often used instead by hydrologists" (MacDonald, *et al.*, 2003, p. 148).

Drainage density is "the relative spacing of a stream network, expressed as the ratio of the total length of all streams within a drainage basin to the area of the basin. Thus, it is a measure of the *dissection*, or of the *topographic texture*, of the area. The drainage density depends on the climate, geology and physical characteristics of the basin" (MacDonald, *et al.*, 2003, p. 148).

Drainage pattern is "the particular arrangement or configuration that is collectively formed by the individual stream courses in an area Drainage patterns reflect the influence of factors such as initial slopes, inequalities in rock hardness and structural controls, and the geological and the geomorphic history of the drainage basin. The most frequently encountered drainage patterns are … dendritic – pinnate – trellis – parallel – barbed – radial – centripetal – annular – deranged" (MacDonald, *et al.*, 2003, p. 148, 149). Refer to Figure 68 - Appendix G for the Manning Equation.

G-6 Manning Equation

		n	
Type of Channel and Description	Minimum	Normal	Maximum
Minor streams (top width at flood stage <100 ft)			
Streams on plain			
1. Clean, straight, full stage, no riffles or deep pools	0.025	0.030	0.033
2. Same as above, but more stones and weeds	0.030	0.035	0.040
3. Clean, winding, some pools and shoals	0.033	0.040	0.045
4. Same as above, but some weeds and stones	0.035	0.045	0.050
5. Same as above, but lower stages, more ineffective slopes and sections	0.040	0.048	0.055
6. Same as 4, but more stones	0.045	0.050	0.060
7. Sluggish reaches, weedy, deep pools	0.050	0.070	0.080
8. Very weedy reaches, deep pools, or floodways with heavy stand of timber and underbrush	0.075	0.100	0.150
Mountain streams, no vegetation in channel, banks usually steep, trees and brush along banks submerged at high stages			
1. Bottom: gravels, cobbles, and few boulders	0.030	0.040	0.050
2. Bottom: cobbles with large boulders	0.040	0.050	0.070
Floodplains			
Pasture, no brush			
1. Short grass	0.025	0.030	0.035
2. High grass	0.030	0.035	0.050
Cultivated areas			
1. No crop	0.020	0.030	0.040
2. Mature row crops	0.025	0.035	0.045
3. Mature field crops	0.030	0.040	0.050
Brush			
1. Scattered brush, heavy weeds	0.035	0.050	0.070
2. Light brush and trees, in winter	0.035	0.050	0.060
3. Light brush and trees, in summer	0.040	0.060	0.080
4. Medium to dense brush, in winter	0.045	0.070	0.110
5. Medium to dense brush, in summer	0.070	0.100	0.160
Trees			
1. Dense willows, summer, straight	0.110	0.150	0.200
2. Cleared land with tree stumps, no sprouts	0.030	0.040	0.050
3. Same as above, but with heavy growth of sprouts	0.050	0.060	0.080
4. Heavy stand of timber, a few down trees, little undergrowth, flood stage below branches	0.080	0.100	0.120
5. Same as above, but with flood stage reaching branches	0.100	0.120	0.160
Major streams (top width at flood stage >100 ft)[b]			
Regular section with no boulders or brush	0.025		0.060
Irregular and rough section	0.035		0.100

[a] The values of u_c and u_m for the common unit systems are

u_c = SI: 0.552;   cgs: 5.52;   English: 1.00;

u_m = SI: 1.00;   cgs: 4.64;   English: 1.49.

[b] The *n* value is less than that for minor streams of similar description, because banks offer less effective resistance.

Data from Chow (1959).

Figure 68 - Appendix G: Values of Manning Equation

Figure 68 - Appendix G shows values of Manning Equation for channels of various types (Dingman, 2002, p. 428).

Appendix H: Permission Statements

H-1 Permission Statements Collected

Collected from the McMaster University's school of Geography and Geology Field Camp course on August 30, 2004 at the Kidd Copper Mine site are written permission statements to use the primary data collected from 32 students' logs. Email records of the *Geo3Fe3* Geology students' written permission statements to collect their data from their Forestry Supplier's Field Books was given to Loreen Rudd. The data was stored for John Maclachlan to mark. The data assembled included: material, records, sketch maps, statements or log entries that were recorded within each student's entries. This information was collected blindly and without any prejudice. Loreen Rudd picked up the books in the GSB lab at 2:30 pm on Friday, October 15, 2004. Loreen photocopied the entries within sight of Susan Vajoczki (Burke Science Building 313); and Susan photocopied Loreen's entries so that no tampering with data could be accused.

H-2 Permission to Use 32 Students' Field Logs

(1)   Karl Belan – "Sure Thing Loreen" 02:49 PM 10/11/2004.

(2)   Andrew Benson – "That's fine Loreen You're welcome to use it" 09:43 AM 10/12/2004.

(3)   Ryan Brown – "Sure if you think my stuff will help, go nuts" 02:41 PM 10/12/2004.

(4)   Desmond Carroll – "You have my permission to use my notes" 12:55 PM 10/9/2004.

(5)   Erik Enders – "Sure Loreen, you can photocopy it" 08:05 AM 10/12/2004.

(6)   Cetina Farruggia – "Sure go right ahead" 03:18 PM 10/13/2004.

(7)   Jeremy (J.J.) Gabriel – "You have my permission to use my mine tailings assignment. I don't know how helpful it will be though" 05:27 PM 10/12/2004.

(8)   Justin Grenier – "Yes, if you like. I have no problem with you photocopying my field notes" 08:51 AM 10/8/2004.

(9)   Sarah Gyatt – "Hey… go right on ahead Loreen" 07:32 AM 10/12/2004.

(10)   Erica Hamilton – "Go ahead. As long as I get everything back, it's fine" 01:41 PM 10/12/2004.

(11)   Nicole Harper – "You are more than welcome to look at the information in my field book!" 10:39 PM 10/13/2004.

(12)   Eden Hynes – "I give permission to Loreen to photocopy my field notes Mine Tailings" October 18, 2004.

(13)   Toni Jack – "Feel free to use any of my notes from field camp, I handed my book in to Sue today" 11:27 PM 10/7/2004.

(14)   Daria Kilmenko – "I think that Van Phan photocopied it for you. But if you want mine it's okay. Let me know if yo need my map as well with all the processes" 06:40 M 10/12/2004.

(15)   Jen Lengert – "For my mine tailings assignment my final map had some discrepencies when compared to other students, but everything I wrote down in my logbook is what I saw… so sure, you can photocopy that section from my log book" 04:14 PM 10/12/2004.

(16)   Derrick Li – "I don't mind if you take my notebook" 02:35 PM 10/8/2004.

(17)   Karen De Medeiros – "Sure, I have no problem with that!" 11:25 PM 10/7/2004.

(18)   Lisa Melymuk – "Sure Loreen, no problem" 06:41 PM 10/12/2004.

(19) Heather Robertson – "Hey! It's no problem for you to use my field book" 11:22 AM 10/12/2004.

(20) Maddy Rosamond – "I handed in my original peer evaluation form with my field notebook, but here is a copy of the data … see attached file." "Sure" 08:28 AM 10/12/2004.

(21) Loreen Rudd (Sherman – author of this report) had Susan Vajoczki (Burke Science Building 313) witness the photocopying of the original untampered data.

(22) Tripti Saha – "Hi, you can have access to my Mine Tailings Assignment" 12:36 PM 10/12/2004.

(23) Saroj Singh – "U can photocopy my notes if u can access them from Sue no problem" 10:40 PM 10/7/2004.

(24) Stephanie Studdy – "Your welcome to" 02:11 AM 10/12/2004.

(25) Van Pham – "Sure, why not. But I got to hand it in today so I'll photocopy it for you and I can give it to you in Geo 3R03 class tomorrow" 10:48 AM 10/7/2004.

(26) Caitlin Vanderkooy – "Sure you can photocopy my mine tailing stuff" 10:20 AM 10/12/2004.

(27) Pete van Hengstum – "Yes you can photocopy my notes" 12:41 PM 10/9/2004.

(28) Sarah Walker – "Yes, you can photocopy my mine tailings notes… some info about the subsurface from the open area and a bit of a sketch map" 08:41 AM 10/12/2004.

(29) Andy Weller – "Loreen, I don't know what good they will be but you can use anything I have, Andy. ps. these might be interesting. whitefish falls 018.jpg; whitefish falls 013.jpg; and whitefish falls 017.jpg." 09:36 AM 10/12/2004.

(30) Zachary Windus – "Sure. Good luck" 08:53 AM 10/12/2004.

(31) May Xiujuan – "You are welcome to use the data" 11:59 PM 10/18/2004.

(32) Linda Zelek – "Yeah no problem" 12:05 AM 10/07/2004.

Editor's Note: the use of personal names or brand names in a peer-reviewed paper is for the identification purposes only and does not constitute endorsement by the named subjects or McMaster University on the subject matter presented by the author's views on the age of the Earth represented by this report.

APPENDIX I: Kidd Copper 'Dam' Field Entries

The following entries were attained from the field records that were submitted for marking (McMaster University, August 2004, p. *Geo3Fe3* Assignment #3: Field Notebooks). **Editor's Note:** the use of personal names or brand names in a peer-reviewed paper is for the identification purposes only and does not constitute endorsement by the named subjects or McMaster University on the subject matter presented by the author's views on the age of the Earth represented by this scientific report.

I-1 Belan

Karl Belan's entry from *Geo3Fe3* field camp on August 30, 2004 reads "Dam section: Man made dam broke at some point – water out of dam section reversed flow causing stream within existing gulley at the bottom dam section" (p. 41).

I-2 Carroll

Desmond Carroll records from the *Geo3Fe3* log entry on the same day, "Dam collapse, region flat valley plains reductions in flow and changes in base level" (2004, p.49). He notes "found a small stream: 15 cm/s Bearing 162°" (Carroll, 2004, p. 46).

I-3 De Medeiros

Karen De Medeiros was one of the thirty-two students present at Kidd Copper property on August 30, 2004 to complete a Mine Tailings Exercise assigned by Dr. Bill Morris from McMaster University. She records the dam's coordinates on her sketch map as "137200 546300" (p. 40). Karen's coordinates of the waste site area marked on her sketch map:

"NE corner: 5137600

SW corner: 467000

SE corner: 137200   546300"

I-4 Enders

Erik Enders' *Geo3Fe3* entry on the same day states "Dam broke ~ 1970 – want to keep tailings under water – pond created to keep tailings from being oxidized – miners dumped tailings in pond – algae blooms formed on the pond (dark mats) – dam broke and water flowed south out of the pond (evidence from direction of ravines)" (2004, p. 49).

I-5 Farruggia

Cetina Farruggia has a very descriptive insertion in her field book entry about the dam site on August 30, 2004 she writes, "Dam – core- more clay – white – clay rich ore – thixotropic (full of H_2O liquefies) – slumping of walls into the water plane – can tell since surface layer is still on top of walls. Basewall – the lowest point in the fluvial channel – as the dam breaks it clogs with sediment and flows more slowly" (p. 43).

I-6 Gabriel

From Jeremy Gabriel writings on August 30, 2004 "It is clear that water drained to SW corner – faint river beds can be seen in the Aeolian area – high concentrations of precipitated salt in shaded area" (p. 48).

I-7 Grenier

Justin Grenier recorded the following information in his *Geo3Fe3* field book on August 30, 2004: "UTM 0467209   5137455. Intersecting scallops topography driven by clay slumping of surface, underneath flowing out from under creep of lower layers. Dam collapsed … water poured out deposit just beyond dam lower water level, river now cuts deeper, wider cut through surface – erodes flat valley plane changing base level. Water level at mouth of river makes new generation of gulleys" (p. 81).

I-8 Gyatt

Sarah Gyatt writes "old dam" on the southeast corner of her sketch map of Kidd Copper site (2004, p. 57).

I-9 Harper

Nicole Harper marks the dam location on her sketch map (2004, p. 58).

I-10 Hynes

Eden Hynes' entry from *Geo3Fe3* field camp on August 30, 2004 reads: "Miners built dam to seal tailings pond. Tailings/waste dumped into area, area under water, algae blooms covered water surface. H_2O pumped in so minerals – sulphides don't oxidate. Dam broke →flow pattern" (2004, p. 45). Eden also writes "flow direction identified because small tributaries converge towards main perimeter channels and show us how the water found its way out" (2004, p. 47).

I-11 Jack

On August 30, 2004 Toni Jack records: "Slumping of walls into the water plane → can tell since surface layer is still on top of walls. Base level – the lowest point in the fluvial channel. As the dam breaks it clogs with sediment and flows more slowly→ a floodplain is created. A change occurs in the base level (gets lower) and creates steeper slumping channels within the other channels" (p. 38).

I-12 Kilmenko

Daria Kilmenko mentions the "damn collapsed" (2004, p.90). "damn collapsed (water poured out of the dam)" (2004, p. 93).

I-13 Lengert

Jen Lengert writes this record in her field book entry on August 30, 2004: "0467194 5137446. This is where dam may have collapsed" (p. 47).

I-14 Li

Derrick Li states "dam broken" on his entry on August 30, 2004 (p. 55).

I-15 Melymuk

Lisa Melymuk records "slumping along dam break area" (2004, p. 44).

I-16 Robertson

Heather Robertson another student provides these coordinates for the dam in her field book entry for August 30, 2004: "0467187 5137453" (p. 51, 53). She writes "dam broke and area exposed to air (30 years ago)" (p.50). Her notes continue "Dam broke here – area slumped in – ran out of space for lake – scalloped ledges on ridge – not usual topography – here intersecting scallops because clay (heavy) – water rich and can't support slope. Base level – lowest point H_2O can erode to where H_2O is at equilibrium. When dam broke – area became constricted water shed – with river w/ one exit therefore ended to base level. Center of valley levels out and walls collapsed in, eventually will erode to flat plain" (2004, p. 58). Her sketch map shows the dam positioned on the southeast corner of Kidd Copper property (2004, p. 59).

I-17 Rosamond

Maddy Rosamond writes her recollection of Bill Morris's story. The following excerpt is from her field book entry on August 30, 2004. It reads: "Bill's story – E valley – E side has rocks: an anthropogenic dam – valley is basin traps tailings – dam broke in corner (S) – clay liquefies easily (Xixotrophic) causes slumping on valley walls – causes clay zonation w/ altitude – rivers can only erode to base level, defined by dam level – when dam broke up river caused erosion, slowing down, getting wider – if base level changes, more erosion possible second channel found in larger channel … few ppl around here: not worth to fix dam – consider cost, risk, liability low" (p. 77).

Maddy goes on to record an excerpt from "Sue [Vajoczki's] lectures – mine tailings kept underwater to reduce oxidation, montmorillonite clay used in drilling, sent in with tailings – is swelling clay, will liquify" (2004, p.81). She adds" Dr. [Carolyn] Eyles' lecture – some bands not laminations, just due to oxidation (leisingangen)" (2004, p. 81).

I-18 Rudd

"Evidence of water being here: desiccation – sediment dried out shows things had water, algal blooms" (Rudd, 2004, p. 72). "Once was a mine-tailings pond waste-products" writes Loreen Rudd to describe the Kidd Copper property as she collected "data for the Mine Tailings Assignment #9" (2004, p. 73). On the southeast corner of her sketch map she marks "old dam" and labels a point where a "pipe" was visible on the mine-tailings map (2004, p. 78).

I-19 Singh

Saroj Singh supplies a reason for the dam being built. She writes in her field book on August 30, 2004. "The dam broke that was built to stop the river." She adds "Fluvial System: Base level, lowest point where water can erode to, all rivers try to erode to the dam base level, when the water dam gave away, it just became a small water shed and made its own fluvial system" (Singh, 2004, p. 85, 87).

I-20 Studdy

Stephanie Studdy recorded part of Bill Morris's story in her field book. On August 30, 2004 she writes: "Dam to trap tailings but the dam broke… not fixed because $ Company that built it no longer exists - Risk assessment – 20 people in all. Early 70's" (p. 86).

I-21 Van Pham

A similar report is recorded by Van Pham on August 30, 2004. "East side: Human-made dam to trap tailings so it doesn't come in contact with others – dam broke – very clay rich tailings comparative – liquifaction "thixotropic" – top surface creeping 'sliding' down into valley causes zonation – scallops is somehow being controlled – Base level: bottom of bed –'lowest level of erosion' – 'Tailing' river is controlled of local H_2O level – river fluvial system eroded down into lake and fluvial system will erode back into itself – river flow starts to meander – continue until flat plane is created – new base level developed" (p. 69).

I-22 van Hengstum

Pete van Hengstum on August 30, 2004 enters this description: "River Morphology – flood plains/oxbows/meandering – dam gave away – eroding back, new stable level – base level is changing through time – dam breaking" (p. 53). Pete places the dam on the southeast corner on his sketch map (2004, p. 55).

I-23 Walker

Sarah Walker describes the dam's location in her field book entry on August 30, 2004. She writes, "in South East corner there's a dam that broke a while back, water that entered area cause a lot of erosion and slumping → sands in area (subsurface) conduct water from the days therefore causing erosion underneath cap layer (vegetation/algae) and slumping and collapse occur later on → gullies form …→ grains fine towards dam coarse grains deposited 1st and fine ones deposited last before dam (clays)" (p. 27).

I-24 Weller

"Dam collapsed, depositing bedload as water continued river cuts down and gets wider" writes Andy Weller in his *Geo3Fe3* field book on August 30, 2004 (p. 43, 44).

I-25 Windus

Zachary Windus includes a "burst dam" on his sketch map (2004, p. 35).

I-26 Xiujuan

May Guan Xiujuan defines the word *thixotrophic* in her recorded field entry for August 30, 2004. She writes "Dam – SE (Talk by Morris) – to trap H_2O – tailing dam broke – a lot more clay in SE than other parts – clay's liquified (thixotrophic – full of H_2O)" (p.64). May adds "talk by Sue [Vajoczki] – tailing comes in through pipe under water" (2004, p. 66).

I-27 Zelek

Linda Zelek writes on August 30, 2004 "Mine tailings location 0467187 5137454. Site description: blocks of rock on right E side – use to build dam to make lake" (p. 35). She locates the dam on the southeast corner of her sketch map (2004, p. 36).

APPENDIX J: Geology and the Earth Sciences

"The body of knowledge called geology can be divided into *physical geology* and *historical geology*. The origin, classification, and composition of earth materials, as well as the varied processes that occur on the surface and in the deep interior of the Earth, are the usual subjects of physical geology. Historical geology addresses the Earth's origin, evolution, changes in the distribution of lands and seas, growth and destruction of mountains, succession of animals and plants through time, and developmental history of the solar system. The historical geologist examines planetary materials and structures to discover how they came into existence. He or she works with the tangible *results* of past events and must work backward in time to discover the cause of those events" (Levin, 2003, p. 2). "Geologist [GEOL] an individual who specializes in the geological sciences" (Licker, 2003, p. 135).

J-1 Biostratigraphy

"The science of biostratigraphy is applied essentially to rocks of the Phanerozoic Eon" (Gradstein, *et al.*, 2004, p. 140).

J-2 Geology

Geology is "the study of the Earth. The many branches of geology include the history of the Earth as a planet since its time of formation, its internal structures, the materials of which it is made, internal processes such as heat flow and the generation of magma, plate tectonics, the evolution of life forms, external processes such as weathering and erosion of the Earth's surface and the deposition of weathering products as sediments and their lithification to form rocks" (MacDonald, *et al.*, 2003, p. 206).
Geology – "science of the Earth, its origin, composition, structure, and history. It is divided into several branches, including:
Mineralogy – the minerals of the Earth;
Petrology – rocks;
Stratigraphy – the deposition of successive beds of sedimentary rock;

Paleontology – fossils;
Tectonics – the deformation and movement of the Earth's crust; Geophysics – using physics to study the Earth's surface, interior and atmosphere;

J-3 Geochemistry

Geochemistry – "the science of chemistry as it applies to geology" (Research Machines Plc., 2003, p. 260).
"Geomorphic – Pertaining or related to the form of the Earth or its surface features" (MacDonald, *et al*, 2003, p. 207).

J-4 Geomorphology

"Geomorphology – the science that treats the general configuration of the earth's surface; specifically the study of the classification, description, nature, origin, and development of landforms and their relationships to underlying structures, and the history of geologic changes as recorded by these surface features" (Bates & Jackson, 1984, p. 208).
Geomorphology is "the scientific discipline concerned with surface features of the Earth, including land forms and forms under the oceans, and the chemical, physical and biological factors that act on them, e.g. *weathering*, streams, *groundwater*, *glaciers*, waves, gravity and wind. As parts of geomorphology, *geomorphography* is the description of geomorphic features, and *geomorphogeny* deals with the origin and development of land features. *Geomorphic maps*, usually constructed with the aid of aerial photographs, indicate land forms of particular kinds or origins within a specific region" (MacDonald, *et al.*, 2003, p. 207).

J-5 Earth Sciences

Earth Science is "a broad term for all Earth-related sciences. Although it is often used as a synonym for *geology* or geological science, it includes, in its wider scope, fields such as the environmental aspects of geology, the physical geography of surface and near-surface processes, and physical *oceanography*. Earth system science studies the whole Earth as a system of interacting parts, the parts including the atmosphere, hydrosphere, the biosphere and the geosphere" (MacDonald, *et al.*, 2003, p. 157). Earth Sciences: "a group of sciences concerned with the study of the earth. The chief earth sciences are geology, geography, oceanography, meteorology, geophysics and geochemistry" (Isaacs, *et. al.*, 2003, p. 253).

J-6 Hydrology

Hydrology is "the science concerned with the occurrence, distribution, movement and properties of all waters of the Earth and its atmosphere. It includes surface water and rainfall, groundwater and the water cycle" (MacDonald, *et al.*, 2003, p. 237).

J-7 Petrography

"Petrography is the description and classification of rocks. Petrology embraces not only the systematic and descriptive aspect of rocks, but also the natural history and origin of rocks and rock masses" (Moorhouse, 1959, p. 1).

J-8 Lithology

"Lithology is a term sometimes applied to the identification of rocks in the field or megascopically" (Moorhouse, 1959, p. 1).

J-9 Mathematical Geology

"Mathematical geology [GEOL] the branch of geology concerned with the study of probability distributions of values of random variables involved in geologic processes" (Licker, 2003, p. 197).

J-10 Mining Geology

"Mining geology" The study of the geologic aspects of mineral deposits, with particular regard to problems associated with mining" (Bates & Jackson, 1984, p. 330).

J-11 Morphology

"Morphology – *Geomorphology*. The study of soil properties and distribution arrangements of *soil horizons*" (MacDonald, *et al.*, 2003, p. 306).

J-12 Physical Geology

"*Physical geology* is the division of geology concerned with Earth materials, changes in the surface and interior of the Earth, and the dynamic forces that cause those changes" (Plummer, McGery & Carlson, 2003, p. 6).

J-13 Sedimentology

"Sedimentology [GEOL] The science concerned with the description, classification, origin, and interpretation of sediments and sedimentary rock" (Licker, 2003, p. 329). "Sedimentology. The scientific study, interpretation, and classification of sediments, sedimentary processes, and sedimentary rocks" (Allaby & Allaby, 2003, p. 485).

"Sedimentology is the study of the processes of formation, transport and deposition of material which accumulates as sediment in continental and marine environments and eventually forms sedimentary rocks" (Nichols, 2003, p. 1).

J-14 Science Stratigraphy

"Sequence stratigraphy [GEOL] a branch of stratigraphy that subdivides the sedimentary record along continental margins and in interior basins into a succession of depositional sequences of as regional and interregional correlative units" (Licker, 2003, p. 332).

"Stratigrapher [GEOL] a geologist who deals with stratified rocks, for example, the classification, nomenclature, correlation, an interpretation of rocks" (Licker, 2003, p. 355).

"Stratigraphy [GEOL] a branch of geology concerned with the form, arrangement, geographic distribution, chronologic succession, classification, correlation, and mutual relationships of rocks strata, especially sedimentary. Also known as stratigraphic geology" (Licker, 2003, p. 355).

Stratigraphy "also called stratigraphical geology, the branch of *geology* concerned with all characteristics and attributes of rocks as they are in strata, and the interpretation of strata in terms of derivation and geological background" (MacDonald, *et al.*, 2003, p. 414).

"Stratigraphy is the study of rocks to determine the order and timing of events in Earth's history" (Nichols, 2003, p. 1).

"Stratigraphy provides the framework to unravel the events of the past few billion years" (Nichols, 2003, p. 229).

J-15 Strctural Geology

"Structural geology [GEOL] a branch of geology concerned with the form, arrangement, and internal structure of the rocks" (Licker, 2003, p. 358).

Structural geology is "the branch of *geology* concerned with the description, spatial representation and analysis of structural features, ranging from microscopic to the structure of regions. As such, it includes studies of the forces that produce rock deformation and of the origin and distribution of these forces. Structural features may be primary, i.e. those acquired in the genesis of a rock mass (e.g. horizontal layering), or secondary, i.e. resulting from later deformation of primary structures (e.g. folding or fracturing)" (MacDonald, *et al.*, 2003, p. 416, 417).

J-16 Structural Petrology

"Structural Petrology [PETR] the study of the internal structure of a rock to determine its deformational history. Also known as fabric analysis; microtectonics; petrofabric analysis; petrofabrics; petrogeometry; petromorphology; structural analysis" (Licker, 2003, p. 358).

APPENDIX K: Sedimentary Processes and Terminology

"The roots of sedimentology and stratigraphy extend back to the 16th century. Geologists continue to 'fine tune' sedimentologic and stratigraphic concepts through a variety of research avenues" (Boggs, 2006, p. Preface).

"Sedimentary – Pertaining to or containing sediment, or formed by its deposition" (Bates & Jackson, 1984, p. 453).

K-1 Sediment

"Sediment [GEOL] (1) a mass of organic or inorganic solid fragmented material, or the solid fragment itself, that comes from weathering of rock and is carried by, suspended in, or dropped by air, water, or ice; or a mass that is accumulated by any other natural agent and that forms in layers on the earth's surface such as sand, gravel, silt, mud, fill, or loess. (2) A solid material that is not in solution and either is distributed through the liquid or has settled out of the liquid" (Licker, 2003, p. 328). "Sediment – Solid material that has settled down from a state of suspension in a liquid. More generally, solid fragmental material transported and deposited by wind, water, or ice, chemically precipitated from solution, or secreted by organisms, and that forms in layers in loose unconsolidated form, e.g. sand, mud, till" (Bates & Jackson, 1984, p. 453). "The mechanism that produces most cross-stratification is encroachment by avalanching down the lee slope of dunes, ripples, bars, fans, or small deltas" (Blatt, Middleton & Murray, 1972, p. 117).

K-2 Sedimentation

"Sedimentation [GEOL] (1) the act or process of accumulating sediment in layers. (2) The process of deposition of sediment" (Licker, 2003, p. 329). Sedimentation is "the process of depositing sediments" (MacDonald, *et al.*, 2003, p. 387).

K-3 Science of Sedimentology

"Sedimentology [GEOL] The science concerned with the description, classification, origin, and interpretation of sediments and sedimentary rock" (Licker, 2003, p. 329). Sedimentology is "(1) the scientific study of *sedimentary rocks* and the processes responsible for their formation. (2) the origin, description and classification of *sediments*" (MacDonald, *et al.*, 2003, p. 387).

K-4 Sedimentation Delivery Ratio

"Sedimentary delivery-ratio [GEOL] The ratio of sediment yield of the drainage basin to the total amount of sediment moved by sheet erosion and channel erosion" (Licker, 2003, p. 329).

K-5 Sedimentation Diameter

"Sedimentation diameter [GEOL] The diameter of a sedimentary particle, determined from the measurements of a hypothetical sphere of the same gravity and settling velocity as those of a given sedimentary particle in the same fluid" (Licker, 2003, p. 329).

K-6 Sedimentary Differentiation

"Sedimentary differentiation [GEOL] the progressive separation (by erosion and transportation) of a well-defined rock mass into physically and chemically unlike products that are resorted and deposited as sediments in more or less separate areas" (Licker, 2003, p. 328).

K-7 Sedimentary Facies

"Sedimentary facies [GEOL] a stratigraphic facies differing from another part or parts of the same unit in both lithologic and paleontologic characters" (Licker, 2003, p. 328).

K-8 Sedimentary Insertions

"Sedimentary insertion [GEOL] the emplacement of sedimentary material among deposits or rocks already formed, such as by infilling, injection, or intrusion, or through localized subsidence due to solution of underlying rock" (Licker, 2003, p. 328).

K-9 Sedimentary Laccolith

'Sedimentary laccolith [GEOL] an intrusion of plastic sedimentary material (such as clayey salt breccia) forced up under high pressure and penetrating parallel or nearly parallel to the bedding planes of the invaded formation; characterized by a very irregular thickness" (Licker, 2003, p. 328).

K-10 Sedimentary Ore

"Sedimentary ore – A sedimentary rock of ore grade; an ore deposit formed by sedimentary processes, e.g. saline residues, phosphatic deposits, or iron ore of the *Clinton ore* type" (Bates & Jackson, 1984, p. 454).

K-11 Sedimentary Petrography

"Sedimentary petrography [PETR] the description and classification of sedimentary rocks. Also known as sedimentography" (Licker, 2003, p. 328).

K-12 Sedimentary Petrology

"Sedimentary petrology [PETR] the study of the composition, characteristics, and origin of sediments and sedimentary rocks" (Licker, 2003, p. 329).

K-13 Sediment-Production Rate

"Sediment-production rate [GEOL] Sediment yield per unit of drainage area, derived by dividing the annual sediment yield by the area of the drainage basin" (Licker, 2003, p. 329).

K-14 Sedimentation Radius

"Sedimentation radius [GEOL] one-half of the sedimentation diameter" (Licker, 2003, p. 329).

K-15 Sedimentation Rate

"Sedimentation rate or rate of sedimentation [GEOL] the amount of sediment accumulated in an aquatic environment over a given period of time" (Licker, 2003, p. 291). "The rate at which sediment is moved is called the *sediment discharge* and is defined as the sediment load moved past a given cross section of the stream in unit time" (Blatt, Middleton & Murray, 1972, p. 21). "The rates used in this compilation include those from direct observation of sediment accumulating from floodwaters, from measurement of survey points, from radiocarbon dating, and from other forms of isotopic dating" (Retallack, 1990, p. 283).

"In most studies of denudation rates, however, the bed load is ignored or is conventionally considered to comprise a certain fraction (commonly 10%) of the total load" (Blatt, Middleton & Murray, 1972, p. 21). "The range of channel pattern, sediment size, and discharge variation is so great in smaller rivers that it is difficult to make generalizations about dominant kinds of large-scale bed forms" (Harms, *et al.*, 2004, p. 33).

K-16 Sedimentary Rock

"Deposition – The laying-down of rock-forming material by any natural agent, e.g. the mechanical settling of sediment from suspension in water" (Bates & Jackson, 1984, p. 133).

Rock is "any aggregate of *minerals* or organic matter, whether consolidated or not. A rock may consist of only one type of mineral (a *monomineralic rock*) but more commonly contains a variety of minerals (a *polymineralic* rock). The geological definition includes materials such as *sands* and *gravels* as well as the consolidated *sedimentary*, *igneous* or *metamorphic rocks*, but in ordinary usage such materials are more commonly referred to as *alluvial* deposits, *drift,* etc." (MacDonald, *et al.*, 2003, p. 373).

"Sedimentary rock [PETR] a rock formed by consolidated sediment deposited in layers. Also known as derivative rock; neptunic rock, stratified rock" (Licker, 2003, p. 329).

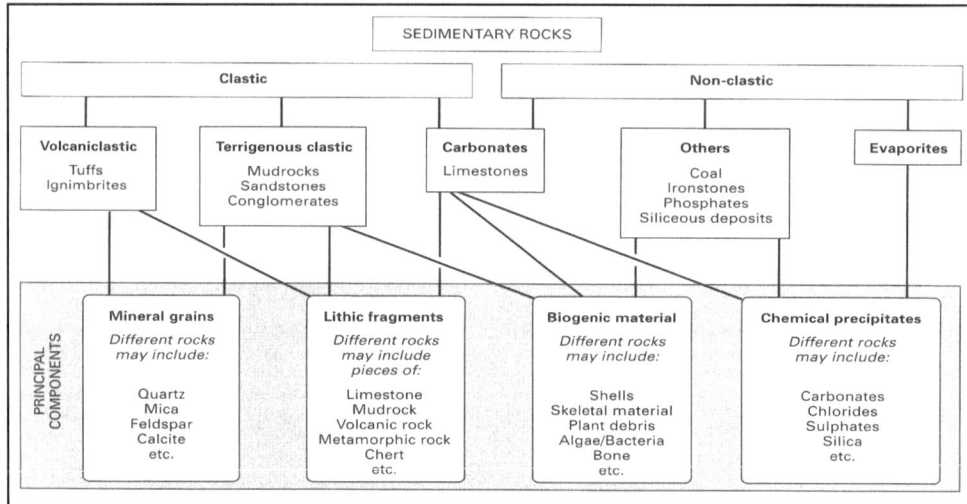

Figure 69 - Appendix K: Clastic and Non-Clastic Sedimentary Rocks (Nichols, 2003, p. 11)

"*Terrigeneous clastic sediments* are loose aggregates of clastic material which becomes a *terrigenous clastic sedimentary rock* when the material is lithified (*lithification* is the process of 'turning into rock'). Mud, silt and sand are loose *aggregates*; the addition of the suffix '-stone' (mudstone, siltstone, sandstone) indicates that the material has been lithified and is now a solid rock" (Nichols, 2003, p. 11). Refer to Figure 69 - Appendix K.

"Sedimentary rocks are either igneous or metamorphic rocks that are broken down by mechanical and / or chemical processes. The pieces of small rock, either detritus or sediment, are carried away by water and wind and later changed into a new form of rock by lithification" (Albin, 2004, p. 20).

K-17 Sedimentary Structure

"Sedimentary structure [GEOL] a structure in sedimentary rocks, such as cross-bedding, ripple marks, and sandstone dikes, produced either contemporaneously with deposition (primary sedimentary structures) or shortly after deposition (secondary sedimentary structures)" (Licker, 2003, p. 329).

K-18 Sediment Transport

Sediment transport is "the movement of *detrital* particles by air, water, ice or gravity. In the last case, sedimentary material moves downslope when the *shear stress* exerted by gravity exceeds the strength of the sediment. Mass gravity transport of material may occur sub-aerially or in the subaqueous environment, with processes ranging from rock falls and *slides* to *debris flows* and *turbidites*" (MacDonald, *et al.,* 2003, p. 387).

"Grains transported by air and water (*fluid transport)* move as *bed load* (by rolling and sliding and by *saltation*) or in *suspension*, when grains are kept up by turbulence. The different size/density populations of moving grains promote *sorting* of the material, and mechanical *abrasion* during transport removes corners and edges so the grains become more *rounded*. The lower density and viscosity of air means that air transports a smaller range of grain sizes than does water, and generally only the finest dusts move in suspension. As a result, *aeolian* deposits are usually much better *sorted* than water-laid sediments. Grain impacts are more vigorous during wind transport, so particles become rounded more quickly and commonly have a frosted or etched appearance" (MacDonald, *et al.,* 2003, p. 387).

K-19 Sedimentary Trap

"Sedimentary trap [GEOL] an area in which sedimentary material accumulates instead of being transported farther, as in an area between high-energy and low-energy environments" (Licker, 2003, p. 329).

K-20 Sedimentation Trend

"Sedimentation trend [GEOL] the direction in which sediments were laid down (Licker, 2003, p. 329).

K-21 Sedimentation Unit

"Sedimentation unit [GEOL] a sedimentary deposit formed during one distinct act of sedimentation (Licker, 2003, p. 329).

"Sedimentation unit – That thickness of sediment which was deposited under essentially constant physical conditions; a layer or deposit resulting from one distinct act of sedimentation" (Bates & Jackson, 1984, p. 454).

K-22 Sediment Yield

"Sediment yield [GEOL] the amount of material eroded from the land surface by runoff and delivered to a stream system" (Licker, 2003, p. 329).

K-23. Spatial Distribution Laminae: Close up of a bed: internal structures

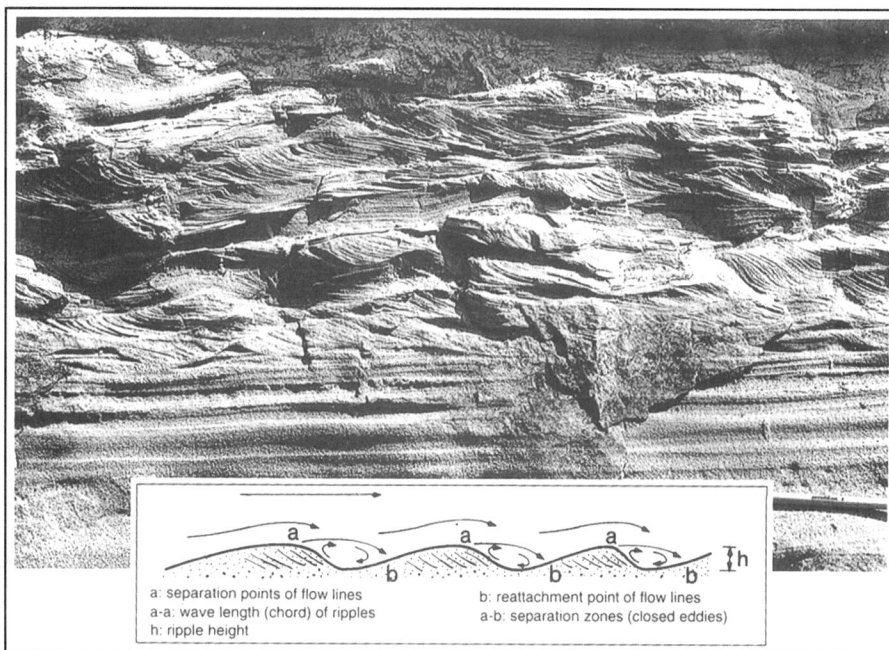

Figure 70 - Appendix K: Geometry of Bedding and Sedimentary Bodies (Lucchi, 1995, p. 28)

"Sedimentary structures can be easily discerned, up to the individual lamina, when looking at beds from a close distance. Laminae are depositional units and, at the same time, sedimentary strucures. Their spatial aggregation and arrangement creates, in fact, peculiar geometries and morphologies within a bed" (Lucchi, 1995, p. 28). "The bed shown in the photo [Figure 70 - Appendix K a close-up of the bed of internal structures] is again a turbidite" (Lucchi, 1995, p. 28). "The order of superposition of the two types of laminae is part of what is called a *Bouma sequence*" (Lucchi, 1995, p. 28).

Appendix L: The Big-Bang Theory

"The Sun and its planets were formed at about the same time. A whirling cloud of gas and dust collected in space. It grew denser and denser as gravity squeezed the gas and dust together. Most of the cloud formed the Sun. What was left over became the planets" (Williams, 1993, p. 5). "Cosmologists now believe that Newton's model was based on incorrect assumptions about the structure of space, time, and matter" (Isaacs, *et. al.,* 2003, p. 84). Hubble's law assumes "that the galaxies have always been moving apart at the speeds observed at present, the age of the universe (T) can be estimated, i.e.

$$T = 1/H_0.$$

On this basis the age of the universe would be 15-18 billion years. This implies that there must have been a time when space and time were minutely small, with all the matter and energy of the universe concentrated into this small volume" (Isaacs, *et al.*, 2003, p. 84-85).

Appendix M: The Geologic Timescale

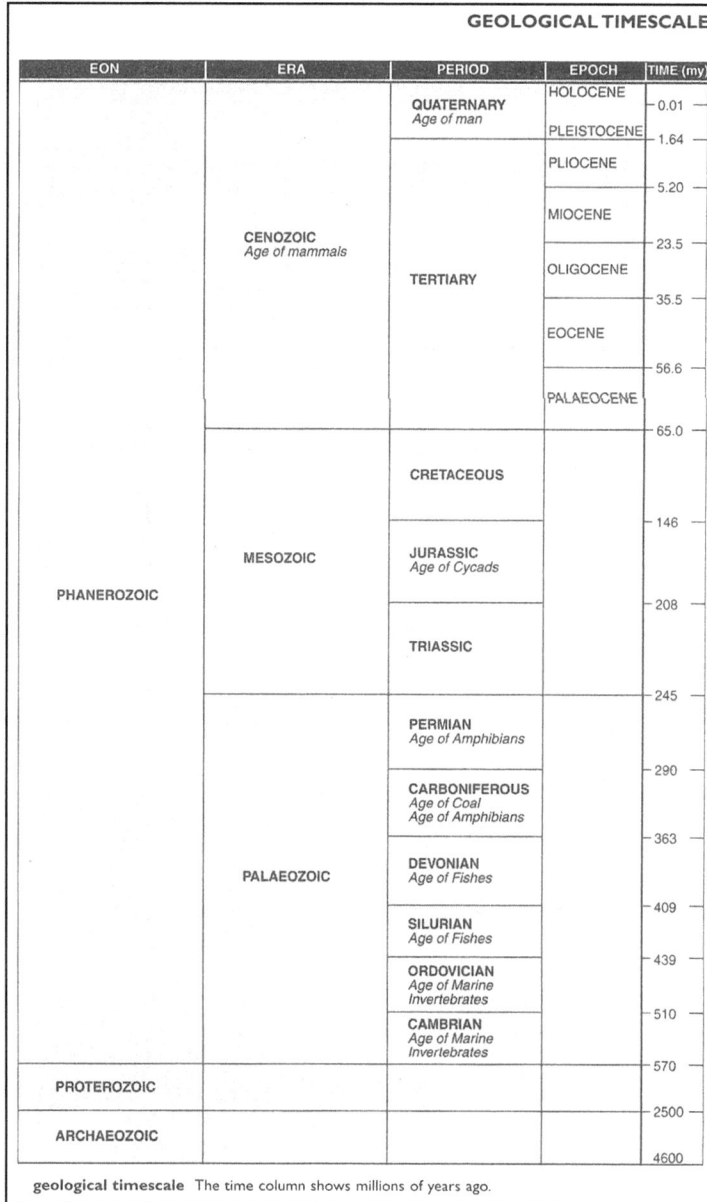

GEOLOGICAL TIMESCALE

EON	ERA	PERIOD	EPOCH	TIME (my)
PHANEROZOIC	CENOZOIC *Age of mammals*	QUATERNARY *Age of man*	HOLOCENE	0.01
			PLEISTOCENE	1.64
		TERTIARY	PLIOCENE	5.20
			MIOCENE	23.5
			OLIGOCENE	35.5
			EOCENE	56.6
			PALAEOCENE	65.0
	MESOZOIC	CRETACEOUS		146
		JURASSIC *Age of Cycads*		208
		TRIASSIC		245
	PALAEOZOIC	PERMIAN *Age of Amphibians*		290
		CARBONIFEROUS *Age of Coal* *Age of Amphibians*		363
		DEVONIAN *Age of Fishes*		409
		SILURIAN *Age of Fishes*		439
		ORDOVICIAN *Age of Marine Invertebrates*		510
		CAMBRIAN *Age of Marine Invertebrates*		570
PROTEROZOIC				2500
ARCHAEOZOIC				4600

geological timescale The time column shows millions of years ago.

Figure 71 - Appendix M: The Time Column Shows Millions of Years Ago

Figure 71 - Appendix M is the geological timescale with a "time column shows millions of years ago" (Research Machines Plc., 2003, p. 261).

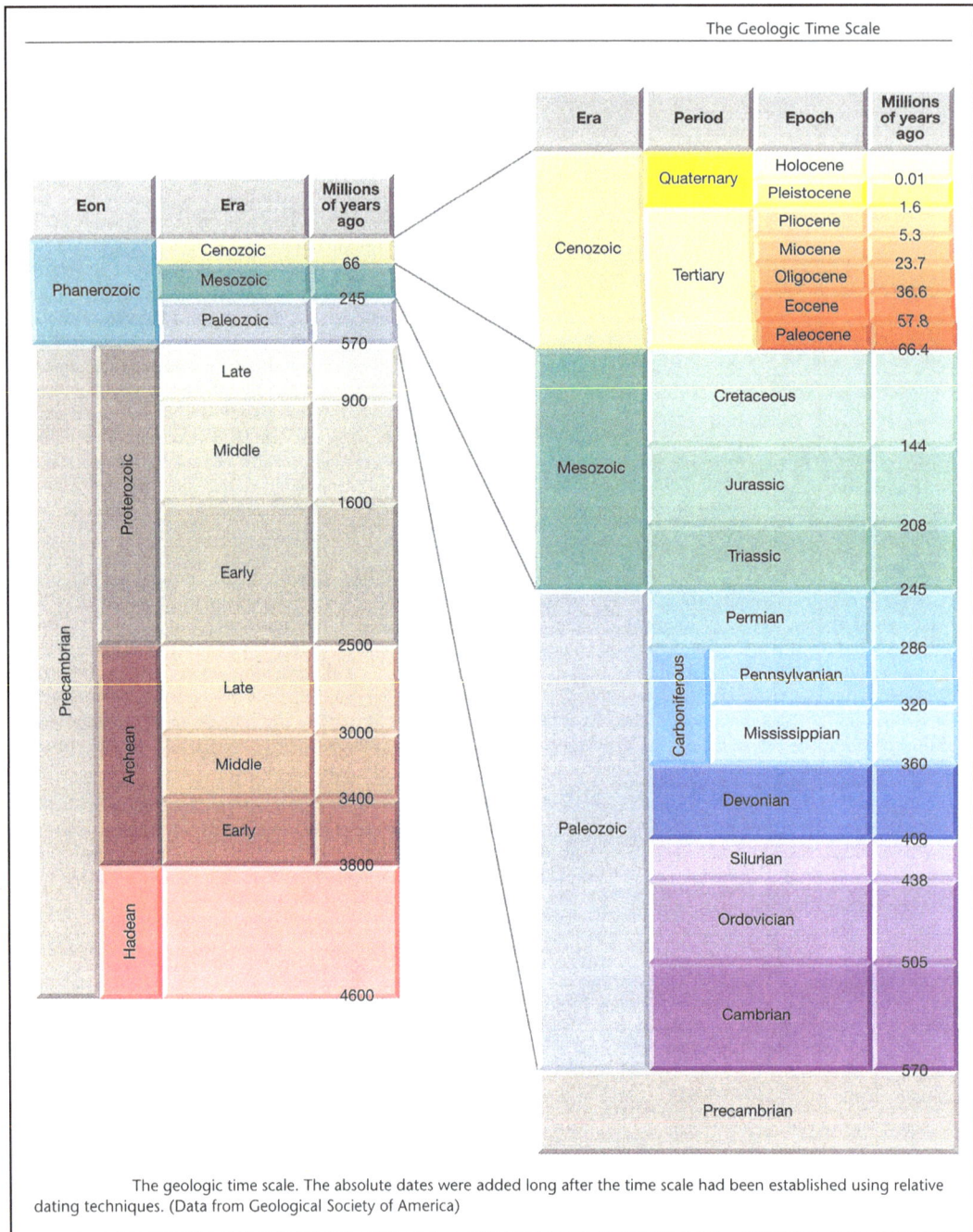

The geologic time scale. The absolute dates were added long after the time scale had been established using relative dating techniques. (Data from Geological Society of America)

Figure 72 - Appendix M: Time Units of Geologic Time Scale (Tarbuck & Lutgens, 1999, p. 205)

Missing information is apparent in Figure 72 – Appendix M that is used to build the geologic time scale, how reliable is this method?

INDEX

Actualism 7, 8, 181, 182, 243

Catastrophic 'events' 5, 28, 44, 45, 47, 49, 52, 86, 92, 198, 261, 274, 306, 314

Deposition 4, 5, 7, 25, 26, 28, 34, 35, 36, 41, 47, 49, 51, 53, 54, 65, 66, 69, 71, 76, 84, 85, 86, 87, 88, 90, 91, 93, 94, 98, 102, 104, 134, 137, 138, 139, 141, 142, 143, 144, 147, 148, 150, 153, 155, 156, 157, 158, 159, 160, 161, 162, 163, 165, 166, 167, 168, 169, 170, 171, 173, 174, 175, 176, 177, 178, 180, 181, 183, 184, 185, 186, 187, 192, 193, 194, 195, 197, 198, 199, 200, 201, 202, 203, 206, 207, 208, 209, 210, 211, 220, 231, 244, 246, 247, 249, 250, 252, 253, 256, 257, 258, 260, 261, 262, 263, 265, 266, 269, 270, 271, 272, 273, 274, 275, 276, 277, 278, 281, 282, 283, 284, 285, 286, 287, 288, 289, 290, 294, 299, 300, 301, 302, 304, 305, 306, 307, 309, 310, 311, 312, 31`3, 314, 315, 316, 317, 318, 319, 320, 321, 322, 323, 324, 326, 327, 328, 329, 330, 331, 338, 339, 340, 341, 350, 356, 357, 360, 361, 362, 363, 365, 366, 373, 376, 383, 384, 386, 387, 388, 389

Dam Breach 7, 25, 26, 34, 56, 61, 65, 67, 82, 134, 142, 262, 301, 313, 329, 330

Diagenesis 7, 27, 66, 98, 142, 197, 263, 285, 286, 307, 311, 314, 320, 329, 330, 349

Equal Declivities 7, 25, 93, 94, 210, 271, 272, 273, 277, 316, 317, 330

Geologic Laws 5, 7, 27, 34, 92, 263, 278, 288, 317, 329, 330

Geological Time Scale 7, 213, 216, 218, 220, 223, 227, 229, 230, 246, 293, 295, 296, 310, 321, 322, 330

Interpreting the Past 91, 165, 274, 282

Original Continuity 7, 138, 148, 208, 315

Original Horizontality 4, 7, 34, 100, 141, 156, 166, 167, 168, 169, 170, 174, 175, 199, 214, 254, 268, 271, 273, 278, 289, 290, 306, 307, 317, 318, 319, 330

Rapid Processes 5, 90, 163, 164, 186, 300

Rapid Stratification 1, 2, 5, 7, 25, 26, 27, 51, 73, 93, 94, 139, 142, 143, 144, 147, 159, 162, 163, 174, 186, 198, 206, 221, 254, 263, 267, 269, 270, 273, 274, 276, 278, 282, 286, 287, 288, 289, 290, 299, 300, 305, 306, 307, 308, 309, 310, 314, 324, 325, 328, 329, 330, 331

Relative Dating 7, 213, 303, 321, 322

Superposition 4, 6, 7, 8, 34, 91, 92, 100, 156, 159, 166, 170, 171, 172, 173, 174, 175, 176, 177, 199, 207, 209, 210, 213, 214, 215, 268, 271, 272, 296, 307, 317, 319, 320, 322, 323, 330, 389

Uniformitarianism 7, 48, 144, 180, 181, 182, 183, 214, 243, 251, 270, 278, 290, 291, 292, 307, 318, 323, 328, 330

Walther Facies Law 207, 317

Younger Earth 1, 7, 93, 94, 95, 144, 178, 179, 198, 221, 231, 266, 287, 288, 290, 300, 305, 308, 309, 323, 324, 325, 326, 328, 329, 331

OTHER BOOKS by LOREEN SHERMAN, MBA

The Book on Formulaic Communication and Its Action Guide

- Formulaic Communication is language skills for global communication in the Information Age.
- Formulaic Communication is an executive level communication skill for today's workforce.
- Formulaic Communication is higher level communication training to give your company a competitive advantage. By understanding the modern age of communication and its media forms, your company will communicate better with its internal and external resources.

Visit http://www.loreensherman.com/formulaic-communication .

The Book on Leadership Skills for Today

You Will Learn:

- What leadership skills leaders need in today's Information Age.
- Five steps for strategic leadership deployment.
- What makes an organization's culture?
- What influences today's business environment?
- How does workplace demographics impact organizational culture?
- How has the market created new complexities for leaders?
- How does a leader manage change prompted by social media? And much more …

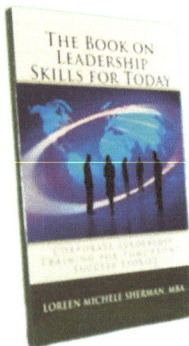

Visit http://leadershipmodules.com/ to get your PDF copy today.

The Book on Business Skills for Today

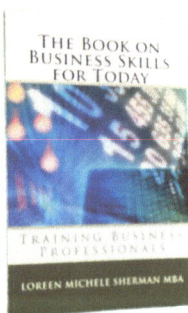

Changes in the environmental landscape led Loreen Sherman to note that new business skills are needed for today's workforce. Each chapter provides helpful information on how to develop strong business skills for today's business professionals. The new business environment is considered. Topics and strategies for the deployment of growth are considered. Communication, leadership and motivation are business skills that require a new approach for the new business landscape for companies to maintain a competitive advantage.

Visit http://www.businessskillsfortoday.com and leave a comment.

About the Author, Loreen Sherman, MBA

Born in Calgary, Alberta, Canada, Loreen attended the University of Calgary in 1979 but it wouldn't be until 2002 that she returned to start the journey towards being a Geologist. In 2005, Loreen graduated from McMaster University in the Faculty of Social Sciences, School of Geography and Geology, Department of Earth and Environmental Sciences, Programme Level III. Her course specialization was GeoSciences. She earned her BA in Geography and Geology. While a student at McMaster, Loreen gathered most of the primary research used in this scientific report.

Loreen completed another level of academic achievement by attaining her Master of Business Administration from the University of Phoenix in 2009. After graduation, Loreen continued her research project again. This time she worked on gathering secondary research on sedimentary processes and geologic subject material. Loreen's scientific library grew as she carefully compiled references. She has endeavored to provide current and relevant information to explain the depositional processes of sedimentation. This report was carefully written over a six-year period.

She says, "You know you have found truth when all the pieces fit into the puzzle perfectly; and the final picture results clearly for all to see as the same picture."

Loreen attained Faculty certification from Meritus University where she taught business courses to CEOs, directors, executives and managers. Her courses of specialization were Managerial Decision Making and Organizational Behavior. She is a business expert.

Loreen qualified as a candidate for the National Astronaut Recruitment campaign. She is an honored member of the Heritage Registry of Who's Who. She has volunteered as: Zone Captain for the Heart and Stroke Foundation; SIFE Judge for the ACE National and Regional Exposition; Ambassador for the Calgary, Chamber of Commerce; and is a strong supporter of Diabetes Association and VeteransOfCanada.

Loreen is currently CEO and Knowledge Manager for Star-Ting Incorporated. Loreen speaks on high-impact and influential leadership and welcomes an invite from you to speak at your next corporate or business function or special event.

www.ingramcontent.com/pod-product-compliance
Lightning Source LLC
Chambersburg PA
CBHW050800220326
41598CB00006B/79